'This remarkable book, informed by unusually wide-ranging and sophisticated scholarship, is highly creative theoretically. Demertzis makes cultural trauma central to a political sociology of emotions, demonstrates how *ressentiment* illuminates populism and nationalism, and explains how forgiveness can be a key psychological-cum-moral action for rebuilding the fractured solidarities that threaten contemporary society.'

— *Jeffrey C. Alexander, Lillian Chavenson Saden Professor of Sociology, Yale University, US*

'In this work of wide-ranging and deep scholarship, Nicolas Demertzis provides a foundational text for the emergent field of the political sociology of the emotions. At the same time, in his rich analyses of the concepts of cultural trauma and *ressentiment*, and in their application to Greece, he draws not only on history but on philosophy and psychoanalysis, and shows how interdisciplinary work is central to the understanding of today's major issues. Anyone studying the emotional bases of politics will have much to learn from this book.'

— *Barry Richards, Professor of Political Psychology, Bournemouth University, UK*

The Political Sociology of Emotions

The Political Sociology of Emotions articulates the political sociology of emotions as a sub-field of emotions sociology in relation to cognate disciplines and sub-disciplines.

Far from reducing politics to affectivity, the political sociology of emotions is coterminous with political sociology itself plus the emotive angle added in the investigation of its traditional and more recent areas of research. The worldwide predominance of affective anti-politics (e.g., the securitization of immigration policies, reactionism, terrorism, competitive authoritarianism, nationalism and populism, etc.) makes the political sociology of emotions increasingly necessary in making the prospects of democracy and republicanism in the twenty-first century more intelligible.

Through a weak constructionist theoretical perspective, the book shows the utility of this new sub-field by addressing two central themes: trauma and *ressentiment*. Trauma is considered as a key cultural-political phenomenon of our times, evoking both negative and positive emotions; *ressentiment* is a pertaining individual and collective political emotion allied to insecurities and moral injuries. In tandem, they constitute fundamental experiences of late modern times. The value of the political sociology of emotions is revealed in the analysis of civil wars, cultural traumas, the politics of pity, the suffering of distant others in the media, populism, and national identities on both sides of the Atlantic.

Nicolas Demertzis is Professor at the National and Kapodistrian University of Athens and Director of the National Centre for Social Research (EKKE). His research focuses on political sociology, cultural sociology, political communication, and emotions sociology. He edited *Emotions in Politics. The Affect Dimension in Political Tension* (2013).

Routledge Studies in the Sociology of Emotions

Mary Holmes *Professor at the University of Edinburgh, UK* and
Julie Brownlie *Senior Lecturer at the University of Edinburgh, UK*

The sociology of emotions has demonstrated the fundamental and pervasive relevance of emotions to all aspects of social life. It is not merely another specialized sub-discipline; rather it aims to reconfigure bases of mainstream sociology. This book series will not only be of interest for specialists in emotions but to sociology at large. It is a locus for developing enhanced understandings of core problems of sociology, such as power and politics, social interactions and everyday life, macro-micro binaries, social institutions, gender regimes, global social transformations, the state, inequality and social exclusion, identities, bodies and much more.

Emotions in Late Modernity
Edited by Roger Patulny, Alberto Bellochi, Sukhmani Khorana, Rebecca Olson, Jordan McKenzie and Michelle Peterie

Love as a Collective Action
Latin America, Emotions and Interstitial Practices
Adrian Scribano

Interactional Justice
The Role of Emotions in the Performance of Loyalty
Lisa Flower

Governing Affects
Neo-Liberalism, Neo-Bureaucracies, and Service Work
Otto Penz and Birgit Sauer

The Political Sociology of Emotions
Essays on Trauma and *Ressentiment*
Nicolas Demertzis

For more information about this series, please visit: www.routledge.com/sociology/series/RSSE

The Political Sociology of Emotions

Essays on Trauma and *Ressentiment*

Nicolas Demertzis

Taylor & Francis Group

LONDON AND NEW YORK

First published 2020
by Routledge
2 Park Square, Milton Park, Abingdon, Oxon OX14 4RN

and by Routledge
52 Vanderbilt Avenue, New York, NY 10017

Routledge is an imprint of the Taylor & Francis Group, an informa business

British Library Cataloguing-in-Publication Data
A catalogue record for this book is available from the British Library

Library of Congress Cataloging-in-Publication Data
Names: Demertzis, Nikos, author.
Title: The political sociology of emotions : essays on trauma and ressentiment / Nicolas Demertzis.
Description: Milton Park, Abingdon, Oxon ; New York, NY : Routledge, 2020. |
Series: Routledge studies in the sociology of emotions | Includes bibliographical references and index.
Identifiers: LCCN 2020008354 (print) | LCCN 2020008355 (ebook) | ISBN 9780815380733 (hardback) | ISBN 9781351212472 (ebook)
Subjects: LCSH: Political sociology. | Emotions–Political aspects.
Classification: LCC JA76 .D365 2020 (print) | LCC JA76 (ebook) | DDC 306.2–dc23
LC record available at https://lccn.loc.gov/2020008354
LC ebook record available at https://lccn.loc.gov/2020008355

ISBN: 978-0-8153-8073-3 (hbk)
ISBN: 978-1-3512-1247-2 (ebk)

Typeset in Times New Roman
by Wearset Ltd, Boldon, Tyne and Wear

To my family and friends.
The stepping-stones of a life cycle.

Contents

Preface

This book can be considered as a piece of work following the end of the 'emotions-proof' period in sociology. Its aim is two-fold: first, by putting together previously scattered initiatives and sociological research agendas related to the emotion-politics nexus, the book delineates the emerging political sociology of emotion as a sub-field of the sociology of emotions. Second, it focuses on some key political emotions and emotions-driven experiences such as trauma, national identity, forgiveness, populism, and the mediatization of emotionality.

The book consolidates my previous body of theoretical and empirical research over different themes and topics which have been re-elaborated and articulated in a systematic and more thorough manner. In addition, it includes new material drawn from secondary and primary quantitative and qualitative research on various drama trauma cases, the mediatization of emotions, resentment and *ressentiment*, and the traumatic experience of the Greek civil war. Examples from more than one national context will be referred to with Greece being of special interest because this country has recently gained undivided attention from the media and political analysts alike due to the repercussions of the sovereign debt crisis and the impact of neo-liberal fiscal policies on citizens' value priorities and processes of collective remembering. In addition, Greece's populist and nationalist politics offer the right playground for reviewing the politics-emotion nexus.

The book is organized around two central themes: trauma and *ressentiment*. They both constitute fundamental experiences of late modern times. Trauma is considered as a key cultural-political phenomenon of our times evoking both negative and positive emotions, and *ressentiment* as a key individual and collective political emotion allied to traumas and insecurities. More analytically, the discourse of progress and development has been more or less replaced by the discourse of crisis, and 'trauma' tends to develop into one of the master signifiers of late modern societies. It is not accidental that scholars talk about our times and culture as 'trauma time' and 'the culture of trauma'. Moreover, the sub-fields of memory studies, trauma studies and disaster studies have appeared in social science and humanities in tandem with the sociology of emotions. Issues related to trauma will be theoretically approached according to the tenets of the Yale cultural trauma middle-range theory.

Interwoven with identity formation and reformation and stocks of social memory, traumatic experiences elicit negative and positive emotions and sentiments. Traumas break and remake social bonds and, in that sense, they evoke, *inter alia*, hatred, anger, fear, sorrow, resentment or *ressentiment*, on the one hand, and solidarity, dignity, compassion, and sympathy, on the other. As such, trauma and resentment/*ressentiment* should be considered alongside each other. Notwithstanding the confusion between resentment and *ressentiment* in contemporary political analysis and International Relations theory, the latter has come to be seen as a global negative emotion that permeates neo-liberal societies in at least two aspects: first, it is a compensatory emotional stance vis-à-vis failures faced by members of lower classes; second, it is thought of as an emotional motivation for Islamic terrorism.

The book makes a strong case for the establishment of political sociology of emotions as a sub-field of the parenting field of the sociology of emotions. The need for a political sociology of emotions stems from two factors: first, although emotions and affectivity were banned from mainstream political analysis ever since the late 50s, a good deal of political sociological research on political culture and the 'social bases of politics' has engulfed emotions already in its conceptual repertoire. In the meantime, the emotionalization of social theory and the so-called 'affective turn' in social sciences and humanities cannot but affect political sociology itself which has also been constantly expanding its scope of interests. In this respect, the political sociology of emotions is not epistemologically different from late modern political sociology; it could not possibly break away from the general field of political sociology. On the contrary, the political sociology of emotions should be coterminous with political sociology plus the emotional 'filter' through which it attempts to causally explain and interpretatively understand its subject matters.

Second, the political sociology of emotions comes as a 'remedial' gesture to the excessive success of the sociology of emotions (or emotions sociology). Over the last ten years or so the divergence and the number of themes and topics the sociologists of emotions are coping with has become so great that this really dynamic field would be better off if 'formatted' in conjunction with particular – though interlinked – spheres of social action and socio-analytic interest like economy and finance, academia, culture, politics, globalization, social philosophy, etc. In this sense, the political sociology of emotions is expected to contribute to the improvement of the academic division of labor within the sociology of emotions, in its way to become a 'normal paradigm'. Of course, there are certain contiguities between the political sociology of emotions and other sectors within emotions sociology, on the one hand, and cognate fields and disciplines like – among others – political psychology, cultural studies, and ethnography, on the other. This poses the double challenge of inter-sectional thematizations and delicate disciplinary boundaries to be addressed by this book as well as by other, and perhaps younger, colleagues willing to come to terms with this emergent sub-field.

This book is a benchmark of a study program on emotion carried out over the last 15 years from a primarily sociological and, to a lesser degree, from a psychological and/or psychoanalytic and philosophical perspective being implicitly in a 'dialogue' with my much earlier – but never abandoned – preoccupations with political sociology and political communication. The central theoretical approach of this book is weak social constructionism as it attempts to bridge subjectivism and objectivism in understanding the ways social agents, being both rational and emotional, are carrying out their political projects in concrete historical conjunctures and space dimensions not fully acknowledging the prerequisites and consequences of their own action. In this sense, weak construction is kin to Giddens' structuration theory. Phenomenological method and historical-comparison are the main methods employed in the analyses of the subsequent chapters. A few words are needed herein: to really comprehend the situated experiences of resentment and *ressentiment*, requires the phenomenological attitude of *epoché*, namely, the suspending or bracketing of the researcher's prior knowledge of the phenomenon at hand and immersion into the subjects' experience as a disinterested sociological observer (Schutz 1978; Husserl 1978; Natanson 1978). In this way one transcends both the pure conceptual analysis of, let's say, resentment and *ressentiment*, and the ordinary givenness of their manifestations naively lived, as it were, by people under populist or nationalist political cultural settings. This is done through the employment of the reconstructive process of 'double hermeneutic', the second order interpretation of an already interpreted set of significations and emotions constituting everyday actors' life-world (Giddens 1984: 284, 374). Phenomenologically 'speaking reality' in terms of double hermeneutic ends up in a sort of co-consciousness, a fusion of hermeneutic horizons which by-passes solipsism and objectivism as methods of investigation (Gadamer 1979). This fusion takes place in time-space settings which are not to be seen as straightforward occurrences or as determining externalities, but as historical facts endowed with a deep inner structure of meaning and feeling. Hence the postulate that emotions are historically constructed and expressed while mediating culture and body. To grasp the historicity of emotions, it is necessary to go beyond analysis of cultural artefacts and compare different historico-political cases and phenomena. In this respect, comparisons were made in order to clarify, among others, experiences of trauma, civil war emotionality, and the affectivity of national habitus by using material from auto-biographies, novels, and poems as indicatively meaningful sources for the documentation of the emotional underpinning of the political cultural phenomena discussed in the chapters to follow. This material was selectively drawn from influential authors' work which symbolically condenses collective experiences and shared emotions. Among these authors are Dostoyevsky and Elytis, a great Greek Nobel Laureate. Apart from this, with regards to the Greek civil war and its traumatic repercussions, research material was collected through qualitative interviews with informants who took part on both sides of that war. Lastly, the analysis of populism and *ressentiment* is indirectly supported by quantitative work (e.g., Capelos and Demertzis 2018; Capelos, Katsanidou and Demertzis 2017).

Overall, half of the book chapters are more conceptually rather than research oriented.

The introductory first chapter delineates the theoretical and methodological tenets for launching the political sociology of emotions as a promising sub-field within emotions sociology. Part I is about the politics of trauma, leveraged by chapters dedicated to the theory of cultural trauma (Chapter 2), the trauma drama of the Greek civil war (Chapter 3), the moral implications incurred by the mediatization of traumatic emotional experiences (Chapter 4), and forgiveness as a possible moral-emotional response to victimization (Chapter 5). The second part refers to the politics of *ressentiment* and contains a long analysis of the twin emotional terms of resentment/*ressentiment* and their relevance to political action (Chapter 6), an examination of the emotional underpinnings of populism (Chapter 7), and an investigation of the emotional bases of the nation-state (Chapter 8). The emotional analysis of populism and nation-ness are principally premised on the elaboration of *ressentiment* made in Chapter 6. Finally, the Postscript wrap ups the book's argumentation and discusses the boundaries and the links between the political sociology of emotions and cognate disciplines and sub-disciplines.

Acknowledgments

Although the author is ultimately responsible for the published end product, I have to seek comfort in the multiple ways dear friends and colleagues have contributed in the creation of this book. Either during the meetings and conferences held by the Sociology of Emotions Research Network of the European Sociological Association, or in academic symposia, conferences, and seminars hosted in British, Irish, German and Greek Universities and Research Centers, as well as during private conversations, I have benefited from discussing with them my ideas and tentative theoretical propositions. Some of them have also read earlier versions of some of the chapters, offering valuable suggestions and criticism. In this respect, I would like to express an intellectual debt of gratitude to Alex Afouxenidis, Jack Barbalet, Tereza Capelos, Sean Homer, Alexandros-Andreas Kyrtsis, Gerasimos Moschonas, Mikko Salmela, and Yannis Stavrakakis. I am thankful to my theoretical interlocutors, first and foremost my *Doktorvater* Ron Eyerman for his constant support, as well as to Jeffrey Alexander, Bettina Davou, Jonathan Heaney, Thanos Lipowatz, Ilias Katsoulis, Thomas Koniavitis, and Nicos Mouzelis. I am also grateful to Nicolas Christakis, Petros Gougoulakis, Socratis Koniordos, Stamatis Poulakidakos, Christian von Scheve, Katarina Scherke, Eva Tomara, Charalambos Tsekeris, Panagiotis Vassiliou and Phaidon Vassiliou for their prompt bibliographical help. I also want to thank Mary Holmes and Julie Brownlie for encouraging me to go on with the book and hosting it in the Routledge Studies in the Sociology of Emotions.

In various places in the book, I have drawn on material from previously published work of mine: 'The Drama of the Greek Civil War Trauma', in R. Eyerman, J. C. Alexander, and E. Breese (Eds) *Narrating Trauma. On the Impact of Collective Suffering*. Colorado: Paradigm Publishers, 2013: 133–162; 'Mediatizing Traumas in the Risk Society. A sociology of emotions approach', in D. Hopkins, J. Kleres, H. Flam, and H. Kuzmics (Eds) *Theorizing Emotions. Sociological Explorations and Applications*. Frankfurt/New York: Campus Verlag, 2009: 143–168; 'Forgiveness and *Ressentiment* in the Age of Traumas'. In the *Oxford Research Encyclopedia on Politics* (Demertzis, 2017). This material has been re-worked to fit the rationale of the chapters in support of my attempt to carve out the political sociology of emotions as a new sociological sub-field.

Nicolas Demertzis
Athens, February 2020

Chapter 1

The political sociology of emotions

An outline

Political sociology in search of emotionality

The term 'political sociology of emotions' or 'political sociology of emotion' is only rarely mentioned in academic literature (Berezin 2002; Demertzis 2006, 2013; Heaney 2019). If not used as a figure of speech, it alludes to the fact that it is possible to carve out a sub-field within the sociology of emotions. The question, however, is whether such a sub-field is really needed for doing better sociological work and more adequately analyzing the emotions-politics nexus or is destined to end up as another epistemic fractal which does no more than contribute to the disciplinary chaos (Abbott 2001) and the crisis of international sociology. This book argues for a genuine need for the creation of a political sociology of emotions and it is my conviction that there are two converging vectors pointing in this direction. The first refers to the evolution of political sociology itself; the other vector is the status of the sociology of emotions.

In one way or another, the relationship between social institutions and politics, policy, and polities has been an issue of concern for the nineteenth century social theorists premised on the relative separation of state, society, and economy, signaled by the advent of modernity. In brief, Karl Marx's political writings, which stand at odds with the once prominent historical materialist credos, as well as Max Weber's types of power and the theory of politics as vocation, shaped the space for political sociology to appear. The same holds of course for Alexis de Tocqueville's analysis of American democracy, Lorenz von Stein's study of the French Revolution, Robert Michels and Mosei Ostrogorski's pioneering studies on political parties, movements and bureaucracy, Vilfredo Pareto's sociology of political elites, and Georg Simmel's theoretical propositions on conflict, group affiliation and mobilization premised on emotional arousal (Turner 1975). Just before World War I, André Siegfried's electoral geography in France, and soon after World War II the electoral behavior studies at Columbia University (Paul Lazarsfeld, Bernard Berelson and Hazel Gaudet), made headway in producing a fully-fledged political sociology. The 1950s and early 1960s were actually the era of the emergence of political sociology as a separate and institutionalized field of theory and research.

Next to the Columbia School of electoral studies, the influential group of the Michigan School (Angus Campbell, Gerald Gurin and Warren Miller) pushed the politico-psychological and socio-psychological study of voting preferences in the USA further. Due to the Cold-War political climate at the turn of the 1950s on the one hand and, on the other hand, the western concern for the cultural prerequisites of democracy in the Third World (as it was then called) and its legitimate consolidation in affluent countries in the 1960s, William Kornhauser's *The Politics of Mass Society*, Gabriel Almond and Sidney Verba's *The Civic Culture*, and Gabriel Almond and Bingham Powell's *Comparative Politics* edited in 1959, 1963, and 1966 respectively, were some strong forerunners for the establishment of political sociology.

Disciplinary-wise, however, the decisive milestones were Seymour Lipset's *The Social Bases of Politics* (1960) and Seymour Lipset and Stein Rokkan's *Party Systems and Voter Alignments: Cross-national Perspectives* (1967). As a product of the efforts made by the Committee of Political Sociology, founded at the 4th World Congress of the International Sociological Association in 1959, each book has been a landmark in the relevant literature. Since then, political sociology has ceased to be an underrated area of sociological investigation (Allardt 2001: 11704).

Inspired by Aristotle, Lipset's central concern was the conflict vs consensus dilemma permeating modern democratic politics. In this respect, acknowledging the intellectual debt to Marx, Tocqueville, Weber and Michels, and focusing mainly on the USA realities he established political sociology along four main tenets: the preconditions of democracy; voting as democratic class struggle; the sources of political behavior and the social bases of party support; the input of trade unions in political life. Even after the pathbreaking *Party Systems and Voter Alignments*, political sociology strove to defend itself as an autonomous and genuine field of its own, not to be seen as an intellectual transit station through which new(ish) issues and perspectives travel before being absorbed by the two 'parent' disciplines, i.e., sociology and political science (Nash and Scott 2001: 1).

There has been therefore a matter of disciplinary boundary which became all the more obvious in the early 1980s when political sociology engulfed post-structuralist, feminist and cultural studies perspectives. This happened because the main focus of political sociology ceased to be the social-class bases of the nation-state political power and its central institutions, namely, parties, elites and trade unions (Bottomore 1979). Sharply differentiated from the rational choice paradigm, the 'new political sociology' of the 1980s enlarged its scope far beyond formal politics to include a capillary re-conceptualization of power alongside the Foucauldian post-structuralist tenets. It was also influenced by the 'cultural turn' in sociology (Alexander and Smith 1998) placing emphasis thereafter on the meaning-making endeavors of political subjects as they are affected by post-modernization processes such as individualization, life-politics, post-materialism, informationalization, cognitive mobilization, etc. (Nash 2000: x–xiii). The 'cultural turn' in political sociology resulted in an increasing trend towards anti-essentialism and the rejection of grand narratives that emphasized the decentered nature of power and political mobilizations and the emotional repertories of political actors (Taylor 2010: 18–21, 85–90, 110, 196). In other words, in a global world of incremental complexity political sociology after the 1980s conceptualizes, and to a certain extent contributes to, the enlargement of the boundaries of the political (Gransow and Offe 1982; Offe 1985). That is why identity politics, contentious politics and social movements action, gentrification processes, the mediatization of politics, and citizenship rights center-staged the political sociology agendas in the USA and Europe.

No doubt, one of the most crucial factors that drove this re-direction comprises the intensified processes of globalization and/or glocalization ever since the 1990s; as a consequence, political sociology has been all the more disjoined from methodological nationalism which was anyway about to lose its grip by the '60s because of the rapid development of comparative politics. This has not only been inscribed in the widening of its thematic agendas, including thereafter transnational social and political movements, digital politics, and multinational corporate business next to more traditional topics such as political participation, power and authority, political parties and civil society (Orum and Dale 2009).[1] It is also expressed via the emergence of International Political Sociology, a combination of political sociology and international relations to be studied through con-current empirical and theoretical analysis and research of transversal issues and topics. Thus, global governance, international citizenship, political geographies, and big data surveillance are, among others, the prime object of attention studied from a more or less processual and relational perspective (Basaran et al. 2017), much owed to Emmanuel Wallerstein's world-systems theory and his conceptualizations of geopolitics and geo-culture, as well as of the anti-systemic social movements (Wallerstein 2000; Mateos and Laiz 2018).

Apparently, the proliferation of themes and perspectives conferred added value to political sociology which in the meantime had become a highly established section within sociology even if it lacks disciplinary rigidity (Hicks et al. 2005). Thematic, theoretical and methodological fluidity, disciplinary contiguity and interconnectivity have been called forth by the extant de-differentiation of the scientific serendipity employed in our post-modern era (Crook et al. 1992: 197–219). Notwithstanding cross-fertilization and multifaceted analyses, it seems that since the 1980s political sociologists have not let themselves open to the insights of, nor did they valorize the conceptualizations drawn from, the fast-growing sociology of emotions. Even if a good number of them have been affected by the cultural turn of the '80s in the first place, they have not systematically incorporated the affective dimension in the analysis of political phenomena. Indicatively, in the highly inclusive Handbook of Janoski et al. (2005) only two chapters (out of 32) adequately address emotionality with respect to the interpretation of the politics of culture, protest movements and revolutions. Yet, the authors, James Jasper and Jeff Goodwin, were already established sociologists of emotions and social movements (Jasper 2005; Goodwin 2005) who promulgated the need for political sociologists to embrace cultural and emotional concepts in their theoretical toolkit.

More recently, in *The SAGE Handbook of Political Sociology*, a two-volume and 67 chapter piece of comprehensive work (Outhwaite and Turner 2018), there is hardly any substantive reference to the sociology of emotions, or to emotion or affect, or to particular emotional terms such as fear, anger, resentment, shame, and hatred, especially in the analysis of hot topics like violence, movements, revolutions, and war where such a reference would reasonably be expected. It seems that when emotions meet politics in scientific investigation there is still suspicion or even negligence of their importance and they continue to be treated as something 'other' to mainstream analyses (Heaney 2019).

At any rate, treating emotionality as a side issue runs counter to the number of contributions in which emotions were a part of the analysis both before and after the formal disciplinary setting of political sociology where emotions were already brought into the analysis. Although Marx is considered among the forerunners of political sociology, his assertion while criticizing Hegel's philosophy of right that, for the identification of the particular interest of a class with the interest of society as a whole there has to be 'a moment of enthusiasm' (Marx 1844/1978: 27–28) usually passes unnoticed. In equal measure, although Antonio Gramsci's notion of hegemony is frequently referred to, scholars seem to overstress its moral-intellectual aspect while overlooking the necessity of "organized political passion" which prompts the overcoming of individual calculations in an "incandescent atmosphere" of emotions and desires (Gramsci 1971: 138–140). Against the rationalist understanding of politics, even more explicit is the inclusion of emotionality in the theory of political culture proposed by Almond and Verba (1963) who say that, alongside cognitive and evaluative orientations, a political culture is also composed of affective orientations towards such political objects as the structures of inputs and outputs, and the self as political agent. The same holds true for the Michigan School's pivotal notion of party identification as the main driver of voting preference,[2] as well as for the widespread academic interest in trust as supportive and cynicism as undermining elements of democracy (Cappella and Jamieson 1997; Warren 1999). However, political sociologists were not prepared to push the point further and systematically cope with the politics-emotion nexus. Accordingly, they paid, as it were, more attention to Giddens's 'Third Road' than to the possibility of a 'democracy of emotions' which can be cultivated reflexively in intimate relationships in order to reinforce at a later stage the 'life politics' of social groups struggling for alternative expression beyond the traditional public-private distinction (Giddens 1992: 184, 1994: 16, 119–121).

Fifteen years ago, commenting on its future, Hicks *et al.* (2005: 4) maintained that despite its success in focusing on the social bases of politics, political sociology 'needs to be more inclusive of recent developments' and undertake possible syntheses of new developments into existing theories. Today we may claim that this need has been partially met; if anything, and even if one sticks to its hard core, i.e., political power, one would expect that the emotional element could be given more chances by political sociologists, to the extent that it is also through emotions that in every society the mode of political domination is reproduced and ideologically justified. Or, as recently argued, to the extent politics is about the creation, maintenance and use of power, political actors understand the relevance of the creation of relatively stable affective dynamics to further political projects, both as a target for destabilization when it comes to their political opponents, and as a goal to achieve for themselves (Slaby and Bens 2019: 345–346). The point therefore is to complement the study of the social bases of politics with an analysis of its emotional bases which, nevertheless, have been initially explored ever since the late 1970s by sociologists of emotions like Theodore Kemper, Randall Collins, Arlie Hochschild, David Heise, Susan Shott, etc. As it is rightly put, 'a research agenda for a renewed and revitalized political sociology needs to focus on the ways in which people "feel", "experience" and "live" the complex contradictions of the social and political world' (Taylor 2010: 198).

The sociology of emotions at a crossroads

In both sides of the Atlantic, the sociology of emotions has now come of age and the fast-growing research on political neuroscience, affective political psychology, and affective science for that matter, have rendered the demarcation between emotion and reason in analysing politics a thing of the past. As a field, the sociology of emotions has been growing in a steadfast way and since the early '80s, it has been developing into a 'normal' but no less multifaceted, if not amorphous, scientific paradigm.[3] As a consequence, a very large number of social scientists worldwide have come to realize that any description, explanation or interpretation of social phenomena is incomplete, or even false, if it does not incorporate the feeling and embodied subject into their study of structures and social processes (Bericat 2016). This is an achievement of the sociology of emotions.

Notwithstanding pertinent definitional matters, i.e., how to conceptualize emotional terms like affect, sentiment, passion, feeling, emotion, mood etc. (Demertzis 2013: 4–6; Slaby and von Scheve 2019b), the sociology of emotions has made us social scientists become fully aware of at least two aspects of the human condition. First, emotions are culturally mediated; namely, any sort of emotions, be it moral, primary, secondary, programmatic, anticipating etc. is: (a) elicited and experienced relationally and situationally, (b) expressed according to social conventions (feeling rules) and structures of feeling which make for its valence, arousal, and potency, (c) discursively narrated within and through language games partaking thus in identity and will formation processes. Second, any socio-political phenomenon is permeated by emotionality which means that, in different degrees, social agents are always both rational actors and sentimental citizens. In this respect, the sociology of emotions contributes to the general sociological theorizations by systematically providing an emotions-centered lens on the world and on social order (Hochschild 2009; von Scheve 2013).

When the pioneers of the field (Hochschild, Scheff, Kemper, Collins) coined the term in the early and mid-1970s, they probably could not foresee the steadfast pace with which the sociology of emotions would move forward over the next few decades. Suffice it to mention that in 1975 Thomas Scheff in San Francisco organized within the American Sociological Association (ASA) the first session ever on emotions.[4] The very same year Randall Collins published his *Conflict Sociology* wherein emotions are given a central explanatory role for the emergence and the resolution of conflicts and the micro-dynamics of social stratification and, while reflecting on the relationship between emotion and gender, Arlie Hochschild was the first to use the term 'sociology of emotions' (Hochschild 1975). Four years later she published a highly promising article on the sociological analysis of emotions which finally led to her classic *The Managed Heart* in 1983. Those works resonated with the American and European *Zeitgeist* of the 1960s where expressivity and the self were center-staged (Kemper 1991: 303).

Given that strong interest, in 1986 the ASA formally established a standing section on the sociology of emotions and the following year in Chicago it organized two thematic sessions which resulted in the *Research Agendas in the Sociology of Emotions* edited by Kemper (1990). Prompted by the emotional turn in social sciences and humanities, in 1990 the British Sociological Association set up a study group for the sociological investigation of emotions and in the 1992 annual meeting of the Australian Sociological Association, a session on the sociology of emotions was organized for the first time. Also, in 2004 a Research Network for the sociology of emotions was formed in the European Sociological Association, and has expanded significantly in terms of membership and activities (Kleres 2009). Network members have been instrumental in launching a new book series on emotions (Routledge Studies in the Sociology of Emotions) and the new sociological journal *Emotions & Society*.

Apart from different theoretical perspectives (e.g., constructionism, structural theories, etc.), a glance at the conference programs of the European Sociological Association Research Network of the Sociology of Emotions[5] over the last ten years or so is indicative enough for someone to realize that this field has been developing in many directions: economics and finance, politics and civic action, law and justice, counseling, international relations, media of communication, gender relations and identities, organization, intimate relations and personal life, migration, globalization, post-conflict societies, language and literature, history, body and sexualities, social geography, religion, morality and philosophy. Such a variety is the unavoidable and, to a certain extent, welcome consequence of the fervent emotional/affective turn in sociology (Clough and Halley 2007; Hopkins *et al.* 2009). One wonders, however, if this 'let a hundred flowers bloom' sort of development will go on forever, thus compelling the sociology of emotions to become coterminous with sociology itself.

In view of such a rather bizarre prospect, the sociology of emotions needs to be somewhat formatted. This can be done by clustering emotionality around central spheres of social action such as economy, culture, politics, religion, and so on.

As a relatively separate field, then, the political sociology of emotions comes forth as a remedial response to the overexpansion of its disciplinary matrix.

Toward a political sociology of emotions

Given the growth in the sociology of emotions as a generic field of study, a robust political sociology of emotions has not yet been established although academics have come to realize both the importance of affect and emotions as a micro-foundation of political action and macropolitical institutionalization, and the necessity of studying the politics–emotion nexus in a rigorous way. In Bourdieusean terminology, the political sociology of emotions is needed for the comprehension of the emotionality ingrained into the political field and the ways politicians' emotional capital is converted into political capital, with 'emotional capital' understood herein as the 'embodied' part of cultural capital (Heaney 2019). At any rate, however, just as the field of sociology of emotions existed for a long time without the name, as rightly argued by Hochschild (2009: 29), the same is true for the political sociology of emotions. A good number of scholars have already carved out the space for its emergence as a distinct sub-field wherein their work may be embedded retroactively. Their work is found in influential edited volumes or published as articles in refereed journals. The space here does not permit exhaustive or detailed treatment. It suffices, however, to selectively brief on some areas of research which have made an impelling contribution toward a would-be political sociology of emotions.

After many years of under-evaluation, bringing emotions back in the analysis of social movements was characterized as the 'return of the repressed'. Movement analysis could not remain unaffected by the affective turn and, from the late 1990s on, a growing number of related works have reset the agenda in movement research. It has been postulated that if political opportunities delineate the condition of the possible for the genesis of a movement, people will not be able to participate in it without feelings of solidarity, loyalty, efficacy, anger, hope, frustration, vengeance, enthusiasm, or devotion (Benski and Langman 2013). The analysis of emotions has contributed to the systematic understanding of the emergence, duration, action, decline and effectiveness of social movements. This is the direction taken, among others, by Goodwin et al. (2001a, 2004), Flam and King (2005), Goodwin and Jasper (2006), Van Stekelenburg and Klandermans (2011) and Van Troost, Van Stekelenburg, and Klandermans (2013). In effect, these works have been focusing on two planes: (a) since 'every cognitive frame implies emotional framing' (Flam 2005: 24), scholars have been preoccupied with the re-framing of emotional reality as part of movements' strategic planning; (b) they scrutinized the emotional energy created in the internal dynamics of a movement.

Apart from the socio-political movements' analysis, the multi-layered emotions-politics nexus was discussed in an edited volume based on papers presented at a conference held at Bristol in 2004 (Clarke, Hoggett, and Thompson 2006).

Most of the chapters are influenced by psychoanalytic approaches to specific emotions like envy, *ressentiment*, hatred, pity, compassion, etc. These emotions were linked to political phenomena such as populism, justice, voting, and mobilizations while most of the chapter authors (Jack Barbalet, James Jasper, Nicolas Demertzis, Paul Hoggett) addressed the emotionality-rationality dichotomy. Two of the editors reappeared six years later with another collective volume which further expanded the scope of their discussion. Most of the chapters point to the crucial role of specific emotions like fear, forgiveness, love, despair and enthusiasm in national and international affairs (Thompson and Hoggett 2012).

Bearing in mind that the thematic hardcore of traditional political sociology is the causal explanation and interpretive understanding of the societal processes behind the distribution, legitimation, and the exercise of political power, the special issue on emotions and power edited by Jonathan Heaney in the *Journal of Political Power* (2013) comes as no surprise. Two years later, Heaney and Flam (2015) turned it into a book, the chapters of which (re)connect power and emotion as two fundamental constituents of the human condition which, as argued, until recently have been studied in parallel. The contributors accomplish this on a diverse array of topics such as education, organizations, social movements, politics, the media, rhetoric, and comparative intellectual history.

Precisely because global risks and insecurities are permeating contemporary societies, fear and anxiety have been scrutinized with respect to their origins, uses and consequences. Relevant pointers can be found in historic approaches to fear in different political-cultural settings (Burke 2005), socio-psychological research on objectless and free-floating emotional climates of fear (de Rivera 1992), and in the post-modern 'liquid fear' and anxiety as a kind of derivative or second-degree fear which is deemed a steady frame of mind that, even in the absence of an imminent and tangible threat, instils systematically a sense of insecurity and vulnerability (Bauman 2006: 3–4, 132; Glassner 1999). They can also be found in many approaches of 'political fear' and/or the 'politics of fear' and the securitization of migration flows (Robin 2004; Barbalet and Demertzis 2013; Kinnvall and Nesbitt-Larking 2011) where, among others, fear is seen as a political tool of the power elite rule as well as of subordinated groups who can occasionally jeopardize the elite's privileges.

In her insightful argumentation, Sara Ahmed (2004) foretells issues and topics for a future political sociology. Based on a relational approach, in what she calls 'the sociality of emotion', she combines psychoanalytic, post-colonial and critical feminist theoretical postures in an attempt to answer the question 'why are relations of power so intractable and enduring, even in the face of collective forms of resistance?' (Ahmed 2004: 12). In this attempt, among others she scrutinizes pain, hatred, fear, disgust, shame, and love in relation to violence, national identities, heteronormativity, and trauma.

Two other research areas of equal importance anticipating, as it were, the political sociology of emotions, are the media of communication and the public sphere. Including emotions in media studies is nothing new; ever since propaganda research and media violence research in the first quarter of the twentieth century emotional appeals and interpellations have been in many ways a central theme of analysis. Yet, the more recent emotional turn in social sciences ensued even more academic thrust; it has been understood, for example, that the media, old and new alike, contribute to the sharing of emotions, the forging of intimacy relations at a distance (Thompson 1995), and the construction of public emotional agendas (Döveling 2009). It has also been realized that media usage is not only a means but also a cause of affect and emotion regulation (Schramm and Wirth 2008). In addition, the place of emotion in journalism, either for gaining access to personalized authentic story-telling or when it comes to the hidden emotional labor performed by journalists themselves, has been highly estimated in our contemporary environment of mediatized politics, be it mainstream, populist, identity-centered, or post-identity politics (Wahl-Jorgensen 2019; Yates 2015). At any rate, a large block of evidence about the role of emotions in communications is found also in the new/social media studies (Papacharissi 2015; Benski and Fisher 2014) and, up to a certain point, it is systematically arranged in the *Handbook of Emotions and Mass Media* edited by Katrin Döveling, Christian von Scheve, and Elly A. Konijn (2011).

The more the publics become affective through Twitter and Facebook usage, the more emotionality is infused in the public sphere (Slaby and von Scheve 2019b: 21–22). Oscillating between being a forum of deliberation and serving as a space of cacophony, the national and international public sphere is multidimensional and ambivalent regarding political and social identity formation processes. This is not just the result of social media networking; it has deep roots in cognitive and emotional capitalism (Moulier-Boutang 2012) and the concomitant individualization coupled with the increased informalization of feeling rules in late modernity (Wouters 2007) and neo-liberal governmentality (Illouz 2007). It is by no means coincidental that the concept of the 'emotional public sphere', alongside the 'emotionalization of public life' and 'emotional governance' – and 'emotional resistance' for that matter – has come to the fore as a field intertwined with the traditional public sphere of rational debate and of formal democratic institutions and processes (Richards 2007). The emotional public sphere calls forth a distinction between democratic emotionality and demagogic emotional manipulation. Albeit not an easy distinction, it is of importance for the formation of not just responsive, but also responsible leaders and publics, who are to keep alive the tradition of *res publica*. For the point is not to efface or disavow emotionality from political culture and the public sphere. This would indeed be naïve and obsolete, an old-fashioned rationalistic regression; the point is to substantiate the so-called sentimental citizen and emotional reflexivity (Demertzis and Tsekeris 2018).

As already mentioned, not every research area deemed to foreshadow the political sociology of emotions can be described, let alone to be overhauled, in this section. In broad strokes, it can be said that a political sociology of emotions is premised on over-researched areas and that it is expected to gain saliency when less researched and researchable issues and topics will be dealt with in the oncoming years. That being said, two glaring questions are called forth, an epistemological and an epistemic one: (a) if it is high time to delineate a sociology of emotions, what is its epistemological status? and (b) what are its boundaries with other cognate fields and sub-disciplines?

First and foremost, we should take note of not conflating the political sociology of emotions with an emotive political sociology. The latter would reduce, in a monistic way, political phenomena to emotions and feelings, it would render them affective phenomena *in toto*. On the contrary, a political sociology of emotions would explicitly integrate the emotional/affective perspective in its conceptual toolkit and analytics. More to the point, any political sociology of emotions could not possibly break away from the 'parental' field of political sociology and, paradoxically, the political sociology of emotions should be in principle coterminous with political sociology which is, as we showed earlier, an ever-growing discipline in itself. It would ultimately be a mistake to construe the political sociology of emotions as emanating only from the dynamics of the 'social movements society' and the post-modern mediatization processes. Hitherto the emotionalization of science and society has been highly salient as the readdressing of core questions, themes and topics of political sociology such as the state, party competition, and political power becomes all the more necessary. As it is rightly put,

> if the political sociology of emotion is to fulfil its potential, it will need to move beyond its current focus on movement and protest – on what might be considered "exceptional" politics, important as these are – to (re)address these core components of political life that were so central to the concerns of the founding fathers, including, and especially, the relationship between emotions and 'the state', political parties, and embodied politicians.
>
> (Heaney 2019: 3)

In other words, the political sociology of emotions is none 'other' than its original scientific discipline (namely, political sociology). To illustrate the point further, epistemologically the political sociology of emotions is double-edged: affectively, it is a subfield of the sociology of emotions; politically, it is coterminous with political sociology in the making. It could be seen then as a theoretical hybrid; however, the political sociology of emotions is not to subvert or undermine the political sociology research agendas. Remaining analytically at the middle-range level of theorization, it just adds an emotional/affective filter through which political sociology(ies) approaches its objects. This is what I shall be doing in the following chapters.

The boundary-issue referred to above regarding political sociology is concomitantly, not to say inevitably, reappearing in the political sociology of emotions. It seems that the relationship/difference between political psychology(ies)[6] and the political sociology of emotions, as far as the analysis of political behavior and the illumination of the emotion-politics nexus are concerned, is the most striking case in point. Over the last twenty years or so a tremendous amount of theoretical and empirical work has been conducted by American and European political psychologists and political neuroscientists focusing, among others, on party identification and identity politics, electoral preferences, political participation, opinion making processes, decision making, intergroup relations, and conflict analysis (e.g., Sears *et al.* 2003; Nesbitt-Larking *et al.* 2014).[7] From an ideal type perspective, the political psychology of emotions points to a more individual-level analysis of opinionation, decision making, and electoral choice remaining for the most part at the micro-analytical level. In contrast, the political sociology of emotions perspective is expected to employ historical, cultural and socio-psychological conceptualizations and set its thematizations on a more or less macro level, paying attention to shared or collective vis-à-vis individual emotions, pondering over concepts like feeling rules, structures of feeling, emotional regimes, emotional cultures, emotional climates and emotional atmospheres.

Not infrequently, however, epistemic boundaries are transgressed and what actually counts is the degree of interdisciplinarity and crisscrossing thematizations in the analysis of the emotion-politics nexus. Insofar as emotions do not emanate *within* the individual as much as *between* the individuals and their social situations (Barbalet 1998: 67), or as long as they are not 'in' either the individual or the social, but are the interfaces for the very constitution of the psychic and the social as objects (Ahmed 2004: 10), the micro-macro dualism collapses as an epistemological, let alone ontological, postulate. It can remain only as a methodological distinction and as an analytic tool that helps us 'to view the same social processes or social practices both from the point of view of actors and from that of systems' (Mouzelis 2008: 226–227). Thus, the micro-macro distinction becomes one of scale and ratio (Ellis 1999: 34) and that is why social scientists have tried to connect them in multiple ways (Alexander and Giesen 1987: 14) or to bridge them by inserting the meso level of analysis (Girvin 1990: 34–36; Charalambis and Demertzis 1993; von Scheve and von Luede 2005; Turner and Stets 2005: 312–313). With regard to the analysis of emotions, the micro level concerns the intrapersonal dimensions of emotive life, and the meso level corresponds to social interaction in groups, institutions, everyday encounters and the emotional dynamic therein, whereas the macro level entails norms, rules, law, traditions and socio-economic structures which provide the path dependency for emotional cultures and social emotions to be formed. An attempt will be made to follow this line of methodological approach while analyzing topics and issues in the following chapters about populism, anti-politics, memorialization politics, and nationalism.

The political sociology of emotions' raw material: political emotions

In its attempt to dismantle the affective/emotional bases of politics, the political sociology of emotions is destined to always come across political emotions.

Probably due to the sentimentalism of the rising 'Trumpism' and right-wing populism, this term has only recently gained currency academically or otherwise; on 10 January 2013 a Google search gave 18,900 results for 'political emotions', and just 8,470 for 'political emotion' (Demertzis 2014). On 15 February 2020 the figures totalled 160,000,000 and 132,000,000 respectively. In one way or another this term is employed as a means to designate the crucial role affectivity plays in politics.

As a theoretical notion, before its current popularity, political emotion had been used more often by political philosophers, not however without some vagueness and ambiguity. For instance, Barbara Koziak defines them circularly as the kind of emotions enacted in political processes (Koziak 2000: 29). In an implicit dialogue with John Rawls regarding the role of positive emotions like love and compassion for the promotion of a secular, just, and liberal society where everyone should enjoy the same rights that are granted to others, Martha Nussbaum does make use of the term 'political emotions' but she hardly gives a systematic definition or conceptualization of them (Nussbaum 2006, 2013). Scrutinizing Aristotle's theses on emotions, Marlene Sokolon describes them as those emotions that are relevant to a political community, fulfilling at least one of the following conditions: a) habituating into ethical dispositions, b) connecting to the question of justice, c) altering an individual's judgment (Sokolon 2006: 17).

Its usage by social scientists, political sociologists, and communication scholars spans from total un-theorization (e.g., Berezin 2001; Weber 2018; Wahl-Jorgersen 2019: 34, 89, 129), ending up with conceptual refinement (e.g., Protevi 2014). At any rate, the concept stands unclear as it is frequently conflated with the cognate notion of 'political affect' (Protevi 2009; Gould 2010). Inspired by Spinoza's affection-affect interplay, as something which acts upon the body and as something which is acted upon by consciousness, the media philosopher Brian Massumi (2002), a founder of the 'cultural affect theory', refers to 'affect' as a pre-social, unnamed and never fully conscious nonlinear complexity, prior to passivity and activity. Seen in this light, Massumi's affect serves as a whole and overarching emotional climate which is linked metonymically to the particular emotions that may, as it were, spring from it. In his own words, emotion is 'a subjective content ... qualified intensity, the conventional, consensual point of insertion of intensity ... into narrativizable action-reaction circuits ... it is intensity owned and recognized' (Massumi 2002: 28).

In a similar vein, Deborah Gould, drawing on Massumi's work, makes a distinction between '*affect*, as bodily sensation that exceeds what is actualized through language or gesture, and an *emotion* or *emotions*' (Gould 2010: 27).[8] A cognate approach of political affect was adopted by John Protevi, a philosopher influenced by Gilles Deleuze; he conceptualizes political affect as a non-individualistic embodied affective cognition and as an 'imbrication' of the socio-political, the psychological, and the physiological (Protevi 2009). In concretising this highly abstract notion, he analyzes specific cases taken from American socio-political history (the Louisiana slave revolt, Hurricane Katrina, the Columbine High

School shootings, etc.) where fear, rage, and love sit at the centre of what he calls body politic. Five years later he appears to have abandoned the 'political affect' designation; in a chapter on 'political emotion', he adopts a situationist and interpersonal processual approach to emotions 'encompassing brain, body, and world'. Upon this premise he defines political emotion as a collective emotion elicited within a political context by an event or an issue which is necessarily the focus of the emotional person (Protevi 2014: 327). To the extent that it is the relational-processual property of an emergent collective subject and not just the aggregate of individual affective reactions, a political emotion, precisely because of its *political* nature, sets three interlocking thymotic situational axes: the transversal between opposing groups, the horizontal within each competing group, and the adjunctive relation to bystanders (Protevi 2014: 328). Recently, Slaby and Bens (2019) proposed a generic philosophical, but no less amorphous, notion of political affect derived from the web of Spinoza's passive and active affects which eventually comprises the ontology of the Political.

Wrapping up the entire argument of the book, in the Postscript I will come back to the discussion of affect and emotion. For the time being it suffices to say that, due to its conceptual vagueness and visceral nature, affect matches conceptually better to cultural studies, ethnography and psychoanalytic sociology; instead, a componential concept of emotion, and political emotion for that matter, is compatible with the middle-range theoretical status of the sociology of emotions. In this respect, 'emotion' is a boundary marker between the political sociology of emotions and cognate disciplines and sub-disciplines. Keeping thus a distance from the cultural affect theory, my own definition goes as follows: political emotions are lasting affective predispositions supported reciprocally by the political and social norms of a given society, playing a key role in the constitution of its political culture and the authoritative allocation of resources (Demertzis 2014: 227–228). Accordingly, political emotions are:

a   Inherently (but not exclusively) relational and social: they are elicited from asymmetrical figurations of power and are triggered by appraisals that make a reference to other people (Elster 1999: 139) or institutions engaging people: fear, hope, gratitude, anger, vengeance, disgust, awe, trust, distrust are only some of the extant affective reactions within a polity.
b   Programmatic, in the sense that they are long-term affective commitments and dispositions toward political figures (for example adoration and devotion for a charismatic leader), political symbols, ideas or institutions (Sears 2001; Barbalet 2006). Take for instance patriotism and solidarity when it comes to the nation-state, and social movement activity which is also fuelled by resentment, hatred, displaced shame, and *ressentiment*. As lasting affective states, political emotions include what Goodwin *et al.* (2001b: 10–11) call 'moods', as well as what de Rivera (1992) calls 'emotional climate'. Moods or emotional climates are equally lasting and stable experiences that differ from (political) emotions in not having a definite object.

c   Not necessarily consciously felt; one may be thoroughly proud of one's coun-
    try while at the same time hating unconsciously other nationalities. Or, as
    Scheff (1994b) has demonstrated, one's conscious anger against a political
    opponent may be symptomatic of suppressed shame. Although they speak
    of unconscious affective systems rather than distinct emotions, theorists of
    'affective intelligence' have convincingly shown that much of political judg-
    ment and decision making is taking place below the threshold of awareness
    (Marcus *et al.* 2000).

d   Self-targeted or introjected (e.g., shame, pride, fear, anxiety, sorrow) as much
    as they are other-targeted or extroverted (e.g., admiration, compassion, anger,
    hatred, generosity, loyalty).

e   Individual as well as collective or shared.

f   Oftentimes ambivalent, antithetical or incongruent; incongruent emotions
    are elicited when subjects experience simultaneously incompatible affective
    tendencies. Namely, people may experience intense fear, which normally
    leads to 'flight mode' behaviour, and at the same time anger, which activates
    aggressiveness. Class resentment may co-exist with degrees of admiration
    linked to emulating tendencies (Benski 2011; Benski and Langman 2013;
    Maldonado 2019: 20–21).

A strong case should be made at this point as to the fact that there are no prefixed
and inherent political emotions. Any emotion whatsoever may acquire political
character as long as it is involved in the political antagonism and insofar as political practices, public policy shaping and implementation, as well as polity institutional arrangements, are inexorably but not exclusively founded on emotional
grounds (Revault d' Allonnes 2008). To put it differently, there are no exclusively
political emotions but only 'scenarios' and modalities of political involvement
and emotionality (Koziak 2000: 29) occurring in various figurations wherein any
emotion may acquire political significance. Politicians and political parties valorise their emotional capital when they trigger and manage political emotions. Thus,
political emotions are not a generically different kind of affective experience; they
are emotions that play a crucial role in the formation of political reality.

If this is actually the case, disregarding short-term emotions altogether is erro-
neous. Apart from enduring programmatic (political) emotions or sentiments,
short term reflex emotions play a considerable role too, especially in periods of
political tension (demonstrations, riots, petitions, etc.) as well as in political information processing and impression management. The above-mentioned definition
encompasses 'higher order' emotions or sentiments linked to central functions of
a political system and the basic tenets of a political culture (or subculture).[9] They
may be designated as 'political emotions proper' or as 'salient political emotions'
(Sokolon 2006: 181), differentiated from what I would call 'politically relevant
emotions', i.e., urges, reflexes and highly transient affective experiences which
in general play a marginal role in the *longue durée* of the Political. The political
emotions proper are culturally-socially constructed and have a strong cognitive

and moral component (Goodwin *et al.* 2001b: 13). In all likelihood the political sociology of emotions, as attempted in the following chapters, is bound to valorise the former type of political emotions more than the latter. In order to illustrate the above, the next section of this chapter focuses on (political) cynicism.

Cynicism as political emotion

Contemporary political systems are facing pertinent legitimation and rationality crises. This is often manifested in decreasing electoral turnout and volatile voting behavior, negativism and dissatisfaction with politics and political personnel, political apathy, civil disobedience, decreasing civic engagement or even *incivisme*, and decreasing party membership and partisanship, as well as pessimism regarding the future of western democratic polities (Dalton 1988: 225–244; Gibbins 1990). All these issues go along with a subjective sense of political inefficacy, alienation and cynicism, three interrelated aspects of contemporary political malaise. Usually, political cynicism is understood as disbelief in the sincerity, honesty or goodness of political authorities, political groups, political institutions or even the political system itself as a whole (Milbrath and Goel 1977; Listhaug 1995; Barnes, Kaase *et al.* 1979: 575). Also, it is conceived as a lack of civic duty, i.e., the obligation and responsibility to engage in public affairs (Woshinsky 1995: 118). For Cappella and Jamieson (1997: 19, 141–2, 166) the center of the concept of political cynicism is the absence of trust independently of evidence pro or con. In political life, the cynic, sealed within her/his own self-reinforcing assumptions, begins with mistrust and must be persuaded to the opposite view. As a kind of judgment, they define political cynicism as mistrust generalized from particular leaders or political groups to the political process as a whole – a process perceived to corrupt the persons who participate in it, and that draws corrupt persons as participants.

Keeping a distance from this conceptualization, I argue that political cynicism is more than a kind of political judgment and is not coeval to skepticism and distrust, either obsessive or paranoid. It is a very complex concept which brings into play a variety of cognitive and affective orientations towards politics. It is a cluster political emotion, one that is not easily operationalized in empirical political analysis. For one thing, the current discussion on political cynicism is rooted in the post-modern condition which leaves contemporary western societies appearing as 'post-ideological' and 'post-deontic' (Lipovetsky 1992). As the subjects in the West are experiencing the loss of a steadfast rational and – what is more – moral canon guiding their judgment (Bauman 1993: 9–10), the binding power of ideologies is shrinking (Žižek 1989: 7, 28–30). Since 'the grand narrative has lost its credibility' bringing on de-legitimation and radical suspicion towards 'preestablished rules' (Lyotard 1984: 37, 81; Rosenau 1992: 133–137), individuals become all the more aporetic and distrustful of the offered ideological truths. Thus, in the gradual absence of ethical restrictions (Baudrillard 1983), the moral situation of post-modernity is comprised of the standpoint that 'yesterday's idealists have become pragmatic' (Bauman 1993: 2) and that 'liars call liars liars' (Sloterdijk (1988: xxvii).

Further on, the post-ideological character of contemporary western societies lies exactly in that, despite all the distrust towards ideologies and 'great narratives', people continue to act *as if* they believed them. It is a sort of 'enlightened false consciousness', a contemporary version of the discontent in civilization described by Freud. It is a diffuse and almost all-embracing cynicism in an era when the distance and difference between progress and regression, rationalism and irrationalism, Right and Left have been suppressed in social consciousness, therefore losing much of their past social and political validity (Žižek 1989: 29; Bauman 1993: 5). In the worst case, this ethical quandary leads to decisionism, to floating responsibility, and to a fortification of the blasé feeling; it is also conducive to negativism, narcissistic frustrations, narrow commitment and to a cool indifference regarding the distinction between illusion and reality (Bauman 1993: 21ff.; Lash 1978: 47–48, 91–92). At best, the post-modern other-directed subject is left with a *cynical disillusionment*; namely, one experiences a contradictory state of mind where one is aware of the illusionary nature of ideology and behaves in everyday life as if one had never gotten rid of it. According to Žižek (1989: 33), this renders cynicism the prevailing ideology of today's society: even if people do not take ideological propositions seriously, even if they keep an ironical distance, they are still putting them into effect. To put it otherwise: in the contemporary cynical attitude, ideology can afford to reveal the secret of its functioning without affecting its efficiency (Žižek 1996: 200–201). From an emotions sociology point of view it could be argued that the *as if* stance is a feeling rule or a structure of feeling regulating emotionality and manners not only in late but also in early modernity to say the least. Let me illustrate this by referring to Bazarov, the nihilist hero of Ivan Turgenev's *Fathers and Sons* (1861) and the romantic Anna Sergeevna who, in an attempt to endure their longed for but miscarried love affair, frankly cast their relationship as a bygone dream and thought of love itself as a spurious imaginary feeling. They both believed they were telling the truth, Turgenev tells us. *But 'was the truth, the whole truth, to be found in their words? They themselves did not know, much less could the author. But a conversation ensued between them, just as if they believed one another completely'.*[10] But why does all this happen? At the risk of simplification, in the following pages I will propose some interpretative comments.

THE EXISTENTIAL QUESTION OF POSTMODERN LIFE

In a way the postmodern condition can be interpreted as 'modernity emancipated from false consciousness' in the sense that individuals disenchant and dissolve the imaginary supports of their existence, and demystify ideological explanations by uncovering the 'truth of the truth' (Bauman 1992: ix, 188). This is a process of self-institution without illusions, as the subject learns how to live along with risks, uncertainty, contingency and negativity. If in the pre-modern and modern societies individuals used to misrecognize contingencies and negativities of any kind (risks, death, traumas, etc.), masking and mystifying them with imaginary ideological constructions, in late or post-modernity they tend to give meaning to their experiences through demystification and disenchantment, reconciled all the more, as it were, with the Impossible,. In view of backlash politics, the moral and existential question of the postmodern political condition is how can we transform our contingency into our destiny without giving up freedom, without relapsing into experiments of social engineering or of redemptive politics (Heller and Fehér 1988: 19).

Yet, there are certain existential boundaries to demystification; at the end of the day, one is compelled to retreat to fantasy once again, even if this is no more than a mystification of the demystification itself (Vattimo 1992: 42). To recall Nietzsche (1887/1994: 120), 'man still prefers to *will nothingness*, than *not* will'. In this respect, the cynical position offers such an imaginary cure: the cynic of today is not someone who has no illusions at all. Instead, her/his very cynical outlook, the radical skepticism, sarcasm and distrust towards all that appears as solid but melts into air, is an imaginary means to go on with his/her life in the absence of any firm and enduring moral foundation. One is compelled to think and feel that social reality exists, not in the name (or for the sake) of moral principles but simply because this is the way things are; and yet one cannot but behave as if these principles were really operative. I would say that this *as if* attitude is perhaps one of the last resorts for the search of meaning in the wrecked everyday life; the cynics 'know that, in their activity, they are following an illusion, but still, they are doing it' (Žižek 1989: 33). In other words, cynicism is a mode of dampening the emotional impact of anxieties through either a humorous or a world-weary response to them (Giddens 1990: 136).

LEGITIMATION THROUGH DISBELIEF

It would be too much to suggest that this sort of cynicism, which is actually an under-cover pragmatism, is the only dominant (post) ideology of our times. Both Žižek and Sloterdijk are probably one-sided, influenced by the experience of 'the really existed socialism'. Traditional hard and soft ideologies have not vanished. On the contrary, they return under new circumstances, interwoven with the electronic image, acquiring new and modified forms. Hence, I would claim that contemporary cynicism is an *emotional climate* that mediates all current ideological discourses. Furthermore, as a permeating emotional climate, cynicism is longterm and highly complex. As a way of defining it, I would describe cynicism, and political cynicism for that matter, as a diffused cluster emotion or sentiment which consists of a number of interwoven negative feelings and cognitions such as melancholy, pessimism, latent desperation, irony, self-pity, hopelessness, scepticism, sarcasm, guffawing, mocking, cheering, aloofness, distrust, mistrust, calculating behaviour, misanthropism, blasé relativism, bad faith, fatalism, nihilism, discontent, indignation, *ressentiment*, resignation, and powerlessness.

It is not accidental, therefore, that cynicism is described as 'a position of excessive sentimentality' that leads to a melancholic and self-pitying reaction to the apparent disintegration of political and social reality (Bewes 1997: 7, 171). The cynical reaction, though, is 'insufficiently contemptuous' as it accepts the account of the world conveyed by various ideologies while rejecting it (Bewes 1997: 166, 171, 199). In this respect, cynicism makes things worse than they are as it perpetuates the current condition (bureaucratisation, technologization, etc.), leaving the individuals with no hope (Stivers 1994: 13, 90, 180). The common feature of cynicism and idealism is that both accept reality as it is; the cynic masquerades as a realist and the idealist pretends to be hopeful. By the same token, cynicism becomes a form of legitimation through disbelief, according to Jeffrey Goldfarb's (1991: 1) ingenious suggestion. This is not a paradox; rather, it is a self-fulfilling prophesy. To the extent that no alternative is seen, the modern cynic reacts critically, sarcastically and skeptically. But the cynic's critical reaction does not focus on the distance between ideals and social and political practices; should s/he do that s/he would become resentful or *ressentiment*-ful.[11] Instead, s/he casts doubt on the existence of the ideals themselves leaving the existing order of things intact. In a fatalistic way, the cynics do not believe in the ideals and principles of modern polity (i.e., freedom, equity, solidarity, democracy, etc.); they act as if they believe in them, but at the same time they know that life is a sham. Consequently, they become apologists and generators of a self-perpetuated ostensible democracy (Goldfarb 1991: 20–22, 152) and the world is no more essentially contested (Bauman 1999: 123ff.). Under these conditions, the cynic's psychic apparatus has become elastic enough to incorporate a permanent doubt as a survival factor, being well-off and miserable at the same time (Sloterdijk 1988: 5).

Criticizing Parsons' grand theory, two decades before Goldfarb, Wright Mills had perceived the incapacity of the western publics to feel at home amidst rapid societal transformations and their being driven thereafter to apathy. In his words, 'many people who are disengaged from prevailing allegiances have not acquired new ones, and so are inattentive to political concerns of any kind. They are neither radical nor reactionary. They are inactionary' (Mills 1959/2000: 41). And even when they do not panic, people often experience that 'older ways of feeling and thinking have collapsed and that newer beginnings are ambiguous to the point of moral stasis' (Mills 1959/2000: 4). In these circumstances, engaging emotions toward political authorities is becoming increasingly superfluous; the absence of 'legitimation and the prevalence of mass apathy are surely two of the central political facts about the Western societies today' (Mills 1959/2000: 41). The moral stasis he speaks about is cynicism named otherwise.

TOP DOWN AND BOTTOM UP CYNICISM

Nevertheless, despite being a complex and ubiquitous emotional climate, cynicism is not uniform. Different structural positions in a societal hierarchy may lead to different versions of cynicism, and political cynicism for that matter.

Goldfarb (1991: 14) is crystal-clear: although cynicism is shared by the haves and the have-nots, distinct economic foundations of cynicism exist. In effect, we are dealing with two different versions of the phenomenon. This is a point rarely taken into account by mainstream political sociologists who prefer to analyze political cynicism at a rather abstract and mass level. But, as mocking and amoralistic as it may be in the first place, the cynicism of the elites is quite different from the cynicism of the powerless.

For the postmodern libertarian yuppie anything goes and there is hardly any commitment to any societal objective. S/he can be cynical in an anomic way; i.e., there is nothing to stop her/his ambitions other than those above him/herself, no sympathy, no hope, only calculated success and half-hearted enjoyment. But c'est la vie. This kind of cynicism apologises for the way things are ('they all do it'). On the other hand, for the desperate unemployed and/or the underclass, the absence of value in life is a brute fact and their cynicism is a way of dealing with hardship, a means for getting along with an unacceptable life, most likely an emotional device for bypassing and depressing their anger and shame.

On one hand, the top down cynicism of the powerful is justified by power, because they use their disregard for conventions and ideals to further accumulate power (Goldfarb 1991: 16). Democratic principles become ostensible and are observed for the sake of appearances. Stated otherwise, 'the cynical master lifts the mask, smiles at his weak adversary, and suppresses him' (Sloterdijk 1988: 111). The phrase 'Labour is liberating' (Arbeit macht frei), inscribed above the entrance to the concentration camp at Auschwitz, is one of the most striking examples of this kind of mocking cynicism. Perhaps, Dostoyevsky's Grand Inquisitor in The Brothers Karamasov is the best-ever example of the top down cynicism, a cool facilitator of power.

On the other hand, the bottom up cynicism of the subordinated represents a disillusioned miserable fate, a spiteful stance towards conspicuous consumption, and a withdrawal from a political order that they feel they cannot influence. Their ironic detachment and apathetic cynicism are more or less a defensive mechanism, a strategic resistance in the form of indifference and refuge, a way to maintain distance from a remote and immoral social and political order (Rosenau 1992: 141; Dekker 2005). Furthermore, by making lack of knowledge or power seem intentional, the cynics can gain an appearance of control of something they cannot influence in the first place (Eliasoph 1990: 473; 1998).

In this sense, I could schematically argue that the underclass cynicism is a sentimental flight rather than an emotional basis for fight, as is the case with the cynicism of the powerful. In tandem, however, they contribute to the legitimation of the system through disbelief. Furthermore, it seems to me that both stem from the wider culture of narcissism, yet not in a uniform way (Lash 1978: 31–51). Diffuse feelings of emptiness, defensive introjections to the self, depression, inability to sustain enduring relationships and committal social bonds, pessimism concerning society's future, projection of offensive drives, and inability to mourn are symptoms of narcissistic disorder at personal and societal levels. Hence, it wouldn't be too much to say that narcissistic personality and the culture of narcissism are the socio-psychological foundations of the cynical position.

TWO VECTORS OF CYNICISM

Though the distinction between the cynicism of the haves and the cynicism of the have-nots is ideal typical, it delineates, as it were, a political cultural difference in late modern societies. On this, Sloterdijk (1988: 217ff.) offers his notorious distinction between 'cynicism' and 'kynicism': the first is the mocking and repressing cynicism of the rulers and the ruling culture; the second is the provocative, resisting and self-fulfilling polemic of the servants. As he puts it, 'kynicism and cynicism are constants in our history, typical forms of a polemical consciousness "from below" and "from above"' (1988: 218). This political cultural difference – which under certain circumstances may become division or even a cleavage (Deegan-Krause 2007) – has not been studied properly by political analysts. At any rate, it intersects with another one which has a more normative and evaluative quality. It is the difference between what I would call *constructive* and *destructive* cynicism. It depicts the Janus-face of the cynical position which has evolved throughout western historical and philosophical tradition. At first glance, it could be identified with Sloterdijk's kynicism *versus* cynicism distinction; yet it seems to me that there is a slight but decisive difference between the two disjunctions. Though constructive cynicism draws from the 'kynical' tradition, it is not necessarily the cynicism of servants or the underclass. Conversely, despite its roots in the 'cynical' tradition, destructive cynicism is not the cynicism of the rulers by default. For example, an individual belonging to the underclass may be equally scornful and mocking to a member of his/her own group, as may be the case with someone who belongs to the ruling elite and their attitude towards a member of the underclass, and vice versa.

A few more words to explain my argument are needed at this point: on the one hand, the cynical position, as enlightened false consciousness, can lead to a critical and/or ironic distance from things, to an openness of the historical reasoning and to a toleration of what is different.[12] These aspects are indispensable to the constitution and maintenance of democracy as a way of life and as an organized political system. In that sense cynicism can be seen as a mixture of healthy skepticism and political distrust. Skeptic citizens doubt rather than deny; they criticize responsibly the current performance of politicians and institutional output and are somehow optimistic about the future. Distrusting citizens still believe in the institutional framework of liberal democracy and, like skeptics, are committed to its norms and values. At the same time, they are pessimistic, suspicious and alert as to the political personnel's performance with respect to these norms and values. This combination of healthy skepticism and distrust makes citizens open to innovation and change without driving them to radical relativism and to the nihilistic belief of 'anything goes'. Therefore, this kind of cynicism does not part company with critique, self-criticism and accountability; as enlightened false consciousness it can sustain self-awareness and self-limitation away from megalomaniac, utopian and imaginary representations of subjectivity and political agency. Irrespective of melancholy, sarcasm, guffawing, irony, mistrust, distrust and discontent, these particular aspects of cynicism are conducive to democratic politics and societal dialogue. Hence, cynicism has a constructive character. Cynical individuals who convey these attributes still defend their citizenship in a covert or overt way and keep an eye on public affairs. Constructive cynicism can be seen as an attitude in which the assumptions of democracy are tested and scrutinized, through investigative journalism, public petitions, movements of moral protest, satirical political discourse, profanations of the political sacred, and so on (Rosenau 1992: 141; Offe 1999: 76).[13]

On the other hand, the cynical emotional stance can also easily lead to social and political apathy (Lyon 1994: 72), deep pessimism, cheer mocking and cold nihilism towards the values of democratic polity and civil society. The door is open now to decisionism, and authoritarian and totalitarian forms of social and political behaviour. As the existence of rational or even pragmatic criteria regulating public life is denied beforehand, the very idea of the public good and good life is at stake. Swamped in bad faith and disdain, the cynics find every political argument pointless, they lack reflexivity and exhibit an almost paranoid distrust of political personnel and decision-making processes. At their best, they are Nietzscheans and Karlschmiteans: the will to power and the aesthetization of the Political is all they think of public life (Harvey 1989: 117, 209–210; Rosenau 1992: 143). At their worst, they are bearers of *ressentiment*: withdrawn, inactive and alienated individuals, no discussants, know-nothings, whose political mentality is replete with negativism. Those who follow this kind of cynicism do not just deny politics, they disavow it. Hence, this cynicism is characterised as *destructive*.

METHODOLOGICAL PERSPECTIVES

Arguably, due to its complexity, the concept of political cynicism is a soft one. This is not necessarily detrimental to our understanding of contemporary political culture. As a soft concept, for all its elusiveness, it yearns for adequate qualitative methodologies (in-depth interviews, focus groups, ethnography, discourse analysis and so on) rather than for more sophisticated statistical models and measuring scales alone. In tandem, quantitative and qualitative methodologies will bring to light the nuances and differentiations of meaning hidden under the 'political cynicism' rubric. Deviating from the mainstream theoretical and research agenda, some scholars have attempted alternative research approaches to political cynicism. For example, in her writings, Nina Eliasoph (1990, 1996, 1998) has shown that political cynicism is a style adopted by the less privileged citizens in order to keep distance from the above and remote centres of decision making and to cope with the hardships of their lives. William Gamson (1992: 21, 81–82, 176) documented that citizens' cynicism is not sweeping but situational. Employed as a self-defensive mechanism, cynicism is more common in public places and among familiar acquaintances, such as peer groups, rather than in private environments with close friends. One could argue that the negative self-image of most men and women of the lower rankings is contingent upon the social marginalisation they experience. 'If society has castigated you for much of your life, you normally won't develop the self-confidence needed for political involvement. Rather, you will develop ... a cynicism, that produces withdrawal from social commitments' (Woshinsky 1995: 116–117). Hence, the deployment of the vitriolic and corrosive cynicism of apathy, fatalism and irony. This cynical stance is not just distrust or mistrust which could be made to succumb with the appropriate persuasion techniques. It is much deeper and complicated and it may be analyzed qualitatively by using focus groups and interviews, as Dekker did (2005) in the Dutch case. Needless to say, the tentative interpretation I offered above as to the interplay between political cynicism, guilt and shame begs for verification through grounded research.

Synopsis

A strong case was made in this chapter for an explicit articulation of the political sociology of emotions as a sub-field of 'emotions sociology' or the sociology of emotions. It is not only that many political sociologists and political theorists have carved out the space for such a disciplinary sub-field; it is also the hitherto fast growth of the sociology of emotions itself which has led to an over-diversified thematic agenda that begs for internal formatting. Far from reducing politics to affectivity, the political sociology of emotions is coterminous with political sociology itself plus the emotive angle added in the investigation of its traditional and more recent areas of research. Its raw material is the political emotions defined as lasting affective predispositions supported reciprocally by the political and social norms of a given society, playing a key role in the constitution of its political culture and the authoritative allocation of resources. Political cynicism was exemplified as a case of complex political emotion.

Notes

1 To be sure, these 'traditional' topics had been in the meantime renovated through the advent of life-politics, post-materialism, and cognitive mobilization.
2 It is to be noted, however, that while the rational choice approach exorcized emotion and affectivity from the theorizing of political action altogether, the culturalist-psychological one treated emotion in a more or less metonymic way. 'Affective orientations' and 'party identification' served as sweeping categories which accommodate distinct researchable emotions (for instance, pride, gratitude, joy, solidarity, enthusiasm, devotion, or loyalty).
3 For the characterization of a paradigm as 'normal' a set of institutional prerequisites must take place: relevant academic associations and bodies, conferences, congresses and symposia to be held in different countries worldwide, special journals, handbooks and books series, as well as university courses. All these criteria have been met in the case of the sociology of emotions.

 4 Note that in 1973 Alvin Gouldner drew attention to the sentiments' indispensable role for social theory and the analysis of ideology, kept at bay by the mainstream structural-functionalism of that period (Gouldner 1970: 7–9, 37–40, 205–206).
 5 www.socemot.com/about_the_network/ and www.europeansociologist.org/issue-43-sociology-beyond-europe/rn-reports-%E2%80%93-rn11-sociology-emotions.
 6 Through the establishment of the International Society of Political Psychology founded in 1978, political psychology as a sub-discipline was deeply influenced by the affective turn in social science.
 7 Similarly, scholars in political communication have given attention to the affective dimension of public opinion dynamics, the effects of political advertising, the construction of media events and political rituals, the setting of electoral campaigns, news media consumption, and the impact of social media in identity politics.
 8 Similarly, Wahl-Jorgensen (2019: 6–8) sees emotions as actualizations of affect. Affect is viewed as a superordinate register which make possible the emergence of emotions; the latter are understood as relational embodied interpretations of affect that may be publicly named and circulated. The same happens with Slaby and Mühlhoff (2019: 38) who consider affect as a dynamic reservoir of possibility, a sphere of potential, something that is formative but not yet formed. One wonders, however, why they do not employ herein the notion of virtuality instead of possibility. For an explanation of virtuality, see Chapter 2.
 9 In this respect, the political sociology of emotions perspective decisively informs and enriches political culture study.
10 https://en.wikisource.org/wiki/Fathers_and_Sons/Chapter_25.
11 See Chapter 6.
12 Here 'irony' is understood in the spirit of Vladimir Jankélévitch (1987); namely, as a stance of modest and sensible distrust towards the outside, and as a cool disengagement towards the inside. Irony in this sense is a modality of prudence, a serious gesture of demythologization in tandem with an absence of self-complacency. For Jankélévitch irony is a mode of expression, basically an allegorical discursive strategy according to which the ironist is careful enough not to conflate the spirit and the letter of what is under scrutiny. Rorty's (1989: ch. 4) account is somehow similar but certainly poorer: the ironist doubts the validity of all vocabularies and denounces the possibility of a meta-language. His/her orientation is directed to the private realm and his/her normative objective is to limit, if not to eliminate, pain.
13 It should be remembered at this point that cynicism *qua* distrust is not by definition detrimental to democratic politics; liberal political theory (Hobbes, Locke, etc.) was largely founded on distrust of government (Hardin 1999, 2004).

Part I

The politics of trauma

Chapter 2

On trauma and cultural trauma

The semantic field and the historical-societal context

There is a long and well-known literature record on trauma emanating either from psychoanalysis or from psychiatry and psychology (Alcorn 2019). To summarize the discussion, it suffices to mention Freud's *Studies on Hysteria* (Freud and Breuer 1956) and *Moses* (1986), where he attributes a key position to trauma in order to understand the (dys)functions of the psychic organ. And, of course, there are Post Traumatic Stress Disorder (PTSD), a category in the diagnostic guide of the American Psychiatric Association, included in 1980 for dealing mainly with Vietnam War veterans' psychic diseases, and the concomitant establishment of psychiatric victimology and humanitarian psychiatry. In the meantime, concurrently with memory studies, trauma studies have developed in a steadfast way. Hence, there are heterogeneous resources for dealing with trauma, since its study crosses disciplines and incurs conceptual fluidity (Kansteiner 2004; Leys 2000; Luckhurst 2008), emanating from, among other factors, language and discursive erring and slippages in the attempts to represent the unrepresentable and the unspeakable pain and horror experienced in one's own and others' suffering.

Apart from the US veterans' experiences of the Vietnam War, the world-wide academic and public interest in trauma is due to the legacy of the Holocaust, the other atrocities of World War II, numerous civil wars, racial and ethnic conflicts, crimes committed by military regimes and dictatorships, gender inequalities and the concomitant politics of recognition that contributed to their formation. They have also been fueled by the more recent experiences of post-Communist nationalist conflicts and genocides (in Yugoslavia, Chechnya, Rwanda and elsewhere), and the terrorist attacks in New York, Madrid, London, Paris, and Brussels as well as by the so called war against terrorism, and the refugee crisis caused by the war in Syria after the spectacular collapse of the so-called 'Arab Spring'; also, they emanate from dramatic natural disasters and the abundance of hazardous environmental conditions due to human activity. In a way, from Japan and China to South Africa, and from Northern Ireland and Kosovo to Latin America, the contemporary world seems to be haunted by traumatic memories intrinsically related to the (re)construction of national and local histories, and the (re)formation of collective identities. Due to the horrific collective experiences of various groups and nations in the twentieth and twenty-first centuries, it is not accidental that some scholars speak of our times and culture as 'trauma time' (Edkins 2003: xiv), of 'the culture of trauma' (Miller and Tougaw 2002: 2), 'trauma culture' (Kaplan 2005), or of 'a new century of trauma' (Ward 2012), thereby upgrading 'trauma' into a central imaginative signification and a master signifier of contemporary risk society. To put it otherwise, the idea of trauma is 'established as a commonplace of the contemporary world, a shared truth' (Fassin and Rechtman 2009: 2).

Apart from the world-wide myths of the traumatic genesis of nations as imaginary political communities premised on the predecessors' blood and self-sacrifices, in the post-Cold War period it is not only the perpetuation of wars and violence that generates new victims and traumas, as horrendous as they might be, but a newly formed international relations environment where almost everybody (states, groups, nationalities, ethnicities, movements, etc.) wants to take the position of the victim without currently being one. In a time where victimhood and survivor status have attained substantial symbolic value and victimization itself is pondered as a defining feature of public discourse in late modernity in tandem with self-help therapeutic discourses (Taylor 1992; Illouz 2007: 40–70), they wish that they had been victims in the past, without wishing to be in the present (Todorov 1995; Koulouri 2002; Finkelstein 2003). This helps them to legitimize current acts of violence in the name of a discursively elaborated and highly mythologized past (e.g., the Israeli–Palestinian conflict) or to profit in the international relations power game (Lašas 2016).

On a more general plane, as already implied, the proliferation of trauma studies is called forth by the dynamics of the late-modern global risk society. This is a society which systematically produces threats and as a consequence individuals harbor risk as an involuntary misfortune inflicted by others; in this respect, risk society is a social formation in which the state of emergency threatens to become the normal state (Beck 1992: 24, 79). This dynamic ensures that: a) there is a slow but undisputable spreading of the idea of the subject as 'that-which-can-be-hurt' (Žižek 1997: 136) in the sense that the narcissistic envelope believing that 'it will never happen to us' is ripped apart, and b) there is a loss of trust in persons, institutions, and reference groups. In all likelihood, powerlessness and normlessness render risk society a traumatizing and traumatized society, a society, in other words, which generates a politics of fear (Smith 2006; Richards 2007) and whose self-image is no longer organized around the dominant, optimistic, and reassuring signifier 'progress'. The contemporary acceleration of traumatogenic conditions by no means implies that traumas are systematically found only in late modernity; modernity itself – the dark side of its dialectic, as it were – is inexorably associated with traumas, genocides, wars and mass violence. If anything, Auschwitz and Hiroshima point to the intimate link between trauma and modernity in the long run and in a general and abstract way (Horkheimer and Adorno 2002; Bauman 1989; Eyerman 2012/2017: 5).

But since people have always faced dangers and hazardous conditions, what is qualitatively different in risk society, such that it changes the experience of trauma? All too often in the past people were running particular and isolated risks to be coped with separately and within the optimistic emotional climate of 'progress', 'development', 'modernization', and so on. Nowadays, on the contrary, the very self is in danger; individuals feel all the more unprotected and deprived of resilience (Furedi 2004). As long as the discourse of progress and development has been more or less replaced by the discourse of crisis, anxiety, and fear due to a long list of threatening conditions such as global warming, nuclear winter, biological terrorism, pandemics, loss of biodiversity, and digitalization of surveillance (Posner 2004), 'trauma' turns into a 'dystopia of the spirit, showing us much about our own preoccupations with catastrophe, memory, and the grave difficulties we seem to have in negotiating between the internal and the external worlds' (Roth 2012: 91).

Forty or 50 years ago 'trauma' was rarely discussed outside of the psychiatric fences; under the terms of late-modern risk societies, it has become one of the central imaginary significations. Here Cornelius Castoriadis' terminology is used on purpose in order to illustrate that nowadays reference to trauma, with all its cognitive and emotional implications, is *inter alia* a way in which society as a whole, as well as particular social groups, may think and speak of itself (Castoriadis 1987). Somehow, it is a semiotic means – a dead metaphor (Lakoff and Johnson 1980) extracted from the medical universe of discourse – for the self-representation and self-understanding of societies, groups, and individuals. For Castoriadis' argument – which of course cannot not be given herein the space it deserves – a central imaginative signification is not an image of something or someone, nor is it reduced to, or derivative of something 'objective', external and non-imaginative. Rather, it constitutes and posits social being and social action by giving meaning to, and defining what is and what is not, what is worthwhile and what is not worthwhile. A central imaginative signification reorganizes, reforms, and re-determines retroactively the host of meanings already available in a society or social group; in this way it can offer a new, and adversarial for that matter, interpretative frame of social reality(ies). In a nutshell, the point is not that people did not have any traumatic experiences in the past. It is rather that the moral signifier 'trauma' is acting as a nodal point, a central signification through which many individuals and social groups can imagine themselves retroactively as past, current, or virtual victims. Not accidentally, in the shadow of 9/11 'trauma' was involved in high-brow philosophical debates (Borradori 2004) so that it has irrevocably 'turned into a repertoire of compelling stories about the enigmas of identity, memory and selfhood that have saturated Western cultural life' (Luckhurst 2008: 80). Relatedly, stemming from the historical impact of imperialism and colonialism, a 'culture of victim' and guilty identification with victims has emerged out of a 'duty of repentance'. Hence, the trauma discourse criticizes the Eurocentric assumptions of the superiority of the western liberal and rational thought (Meek 2010: 28); as put in a somewhat cynical aggressive style, 'we Euro-Americans are supposed to have only one obligation: endlessly atoning for what we have inflicted in other parts of humanity' (Bruckner 2010: 34).

Consequently, a vast array of public discourses and academic analyses – comparative or otherwise – is set out with the sociology of emotions being center staged therein. Without endorsing pessimistic statements like 'current debates over trauma are fated to end in an impasse' (Leys 2000: 305), it is possible to navigate the family resemblance among trauma terminologies in order to lay the groundwork for conceptual clarity.

Setting out largely from the Freudian psychoanalysis and Derrida's social philosophy, and based on literary studies, a great deal of work on trauma has been done by the so-called Yale School since the mid-1970s (Cathy Caruth, Geoffrey Hartman, Paul de Man, Shoshana Felman, Dominick LaCapra, Ann Kaplan, etc.). Focusing mostly on the individual and intra-individual levels, the Yale School scholars have understood trauma as an unthinkable occurrence that resists immediate comprehension, the meaning of which is acquired retroactively and is endowed with negative emotionality. Caruth, a widely cited member of the group, stated that a traumatic event 'is not assimilated or experienced fully at the time, but only belatedly, in its repeated *possession* of the one who experiences it' (Caruth 1995a: 4–5). Therefore, in current time an event experienced as *nefas*[1] is apprehended as traumatic as long as it 'cannot be placed within the schemes of prior knowledge' and, in that sense, 'the traumatic event is its future' (Caruth 1995b: 153). This is due to the dynamic process described by Freud as *Nachträglichkeit* (i.e., 'deferred action', 'belatedness', 'afterwardsness'). In the meantime, during the so called latency period, the traumatized person remains silenced and the event cannot be talked out because it is outside the range of normal human experience (Luckhurst 2008: 79). It is manifested, though, in behaviors that originate from the unconscious repetition compulsion, and in this respect the traumatic event is 'fully evident only in connection with another place, and in another time' (Caruth 1995a: 8). Paradoxically, the presence of the trauma is called on by its absence and the ensued speechlessness. Traumas are at first repressed, remain latent, and then are retrieved if and when the right circumstances exist and defense mechanisms begin to be loosened, acquiring a meaning for the subject. Against a common sense lay understanding, trauma is not just any psychic disturbance, but a deep wound caused by unavoidable inflicting situations which cannot be worked out effectively. Therefore, traumas create a void that irrevocably marks the self and one's life narrative. That is why, as 'dystopias of the spirit', they escape immediate understanding and resist representation (Nadal and Calvo 2014: 2; Stavrakakis 1999: 84–85). Oftentimes, then, trauma is considered in terms of the Lacanian notion of the 'Real'; what Lacan (1977b) calls the 'Real' – as opposed and simultaneously linked to the 'Imaginary' and the 'Symbolic' – is an early and deep traumatic kernel in each one of us caused by the dissolution of primary narcissism, i.e., by the entrance of the infant into the reality principle. Actually, trauma is a missed encounter with the Real which has 'an apparently accidental origin' with regards to the subject's life history (Lacan 1977b: 55).

Once social psychologists, historians, psychohistorians, sociologists, and political scientists realized that trauma processes are not confined within an individual psyche or in interatomic dynamics, but ultimately engulf group and societal dimensions, a semantic field emerged consisting of terms like 'mass trauma', 'historical trauma', 'national trauma', 'collective trauma', 'cultural trauma', 'European trauma', and so on. Although there are no dramatic differences among these concepts, taking into account their endurance, evidenced in a number of publications (Alexander 2004a; Eyerman 2001, 2008, 2015; Eyerman *et al.* 2013, 2017), the middle-range social theory of 'cultural trauma' seems to be the most persuasive thematization of trauma at the macro level. Drawing from Alexander's 'strong program' of cultural sociology (Alexander and Smith 2002; Alexander 2012), the theory of cultural trauma advances the task of trauma studies; for the point is not just to discover the 'psychic wounds' in the accounts of traumatic experiences (Hartman 2003: 257), but to weave the experiences into the construction of the social fabric.

The cultural trauma theory

The notion of cultural trauma was first developed during a year-long sojourn at Stanford University's Center for Advanced Study in the Behavioral Sciences. The results of these seminars were compiled as research-based essays in the volume *Cultural Trauma and Collective Identity* (2004) edited by Jeffrey Alexander, Ron Eyerman, Bernhard Giesen, Neil Smelser, and Piotr Sztompka. If it is too much to claim that the theory of cultural trauma has become a new master narrative (Alexander 2003: 94) or a new master paradigm (Aarelaid-Tart 2006: 40), it has certainly evolved into a core research interest within cultural sociology, and ever since several works have been published. For Alexander (2004a: 1; 2012: 6), cultural trauma occurs when members of a collectivity feel that they have suffered a horrendous event that leaves indelible marks on their group consciousness, engraved in their memories forever and changing their future identity in fundamental and irrevocable ways. Accordingly, a cultural trauma is a discursive response to a tear in the social fabric, occurring when the foundations of established collective identity are shaken by one or a series of seemingly interrelated occurrences (Eyerman 2001; 2011a; 2011b).[2] Smelser (2004: 44) offers a similar definition of cultural trauma:

> A memory accepted and publicly given credence by a relevant membership group and evoking an event or situation, which is (a) laden with negative affect, (b) represented as indelible, and (c) regarded as threatening a society's existence or violating one or more of its fundamental cultural presuppositions.

Premised on social constructionism, Smelser's definition assumes that 'cultural traumas are for the most part historically made, not born' (Smelser 2004: 37).

In order to become 'trauma', a shocking and traumatogenic incident – whether a natural disaster or a social dislocation, such as a civil war, mass violence, or genocide – has to undergo a process of social signification; namely, it has to be publicly signified, to get articulated into the paths of public discourse, and to become socially accepted and defined as 'trauma'. For Sztompka (2004: 165–166),

> the cultural traumas generated by major social changes and triggered by traumatizing conditions and situations *interpreted* as threatening, unjust and improper, are expressed by complex social moods, characterized by a number of collective emotions, orientations and attitudes.[3]

Hence, it is not the harmful incident as such that matters here; a lot more important are the modalities by which this incident is transformed into a 'event' that is cognitively and emotionally experienced as a trauma inflicting collective memory and collective identities so that 'people become uncertain about what they should or ought to believe' (Neil 1998: 4). To think that trauma results immediately and invariably from a strike on the body or the mind through a shocking occurrence imposed on a victim is therefore a naturalistic fallacy.[4] Nor is it, however, invented *ex nihilo* or in an arbitrary way (Eyerman 2011a: 152); arguably, it is a historically situated product of discursive (re)constructions. This means that a cultural trauma is discursively mediated and collectively constructed through a process where various interests and carrier groups fight against each other, from the top (elite) as well as bottom up (movements, pressure group initiatives) to make sense of a shocking occurrence and turn it onto an 'event' inscribed into official and non-official collective memory. This is a process, the 'trauma drama', which always unfolds retroactively and it involves rival and alternative conceptualizations and competing discursive strategies. That explains why an event or situation may be realized as trauma at one point in time in a society's or group's history but not in another, and while that dire plight may not be traumatic for other societies whatsoever (Smelser 2004: 36–37). Herein realization is involved in both meanings of the word; namely, as becoming conscious of something and as something becoming real.

Under these terms, it is no surprise that the question 'trauma for whom?' comes to the fore. The inequality of economic, symbolic, social, and political capital largely determines the vulnerability of the particular social groups facing hazardous circumstances. For Sztompka (2004: 166–167), not everyone suffers from trauma in the same way nor does everyone adopt the same strategies for dealing with it. The traumatic potential of an occurrence is situational, i.e., it is largely conditioned by contextual factors and actors' choices.[5] For instance, in war traumas there are manifold personal, political and social factors, as well as domestic and international circumstances, which mediate war experiences and influence whether an individual or a group does or does not become traumatized (Pupavac 2007). It is not to be forgotten in passing that, for Lacan (1977b: 53–54), the individual's trauma, as an encounter with the Real, is a function of *tuché* (chance) and not an automatic result of every single narcissistic frustration.

The contingent nature of cultural traumas is well documented by Eyerman (2011a, 2011b, 2012/2017) while scrutinizing the case of eight political assassinations: John Fitzgerald Kennedy (1963), Martin Luther King, Jr. (1968), Robert Kennedy (1968), and Harvey Milk[6] (1978) in the United States; Olof Palme (1986) and Anna Lindh (2003) in Sweden; and Pim Fortuyn (2002) and Theo van Gogh (2004) in the Netherlands. He analyzed these cases in a totality-wise way by taking into account a number of contextual variables: the timing of the occurrence, the political setting, the authorities' performance, the media coverage, and the presence of influential carrier groups. Accordingly, he concludes that JFK's, Milk's, Palme's and Lindh's assassinations did not trigger a process of cultural trauma; instead, they caused an outpouring of grief that symbolically and emotionally solidified society rather than tore it apart. In contrast, the other political assassinations studied took place in an already polarized political climate and were fueled by antagonistic strategies and interests, and hence initiated respective trauma drama processes.

The in-built contingency of cultural traumas is center-staged by the Nanking Massacre case. In early December 1937 the Japanese Imperial Army invaded China and slaughtered 300,000 civilian Chinese in the Nanking city-area within a six-week period. In addition, the invading troops raped and looted thousands of civilians. This extremity was known to international public opinion from the start. However, it was never constituted as a national 'trauma' for China itself, or for Japan, or even as a trauma for humanity as such, as it was in the case of the Jewish Holocaust (Alexander 2003: 106). As Rui Gao explains, the Maoist silence about the Nanking Massacre and other horrendous war atrocities committed by the invading Japanese army (actually China lost almost twenty-three million people) was a symbolic means of minimizing the national trauma in favor of the class trauma. Pursuing the consolidation of its power, the Communist Party promoted the suffering of the proletariat victims regardless of national origin. It was the myth of class struggle trauma that helped New China to diminish the sores of the War by pointing to the class victimization of both Japanese and Chinese soldiers. The Japanese perpetrators 'appeared as not worthy of consideration in comparison with the ultimate vileness of the Kuomintang regime, and their symbolic significance was largely dwarfed by the dark sacred evilness of true class enemies' (Gao 2013: 72). Thus, the potential national trauma was preempted from the very beginning since Nanking was then the capital of China under the Nationalist government of Chiang Kai-shek. Only later, in the mid-1980s, was the Nanking Massacre endowed with national symbolic meaning in relation to China's foreign affairs with Japan. Drawing attention to the Nanking Massacre has become shorthand for China's efforts to discredit Japan on the international stage. The memory of the massacre sits at the heart of disputes between China and Japan nowadays and the memorial to the massacre has become a prominent part of the commemoration of the war in

China (Mitter 2017). There has been, therefore, a shift from silence to voice and remembrance as far as China is concerned. With regards to Japan, apart from a number of dissident and critical voices, there are a good many official denialist accounts claiming that the atrocities never even happened.

The above having been said, it should be clear that the process of 'trauma drama' entails three fundamental constituent elements: retroactively selective memory, largely negative emotion, and identity. The essentials of this process (or the negotiation of the meaning of the trauma) are, equally: victims, perpetrators, and blame attribution. Retroactively selective memory means that collective memories are formed and reformed through an interplay of oblivion and recollection, through the articulation of a past-in-the-present (Heller 1982) and the concomitant invention of tradition. Resonating with Raymond Williams, I would claim that cultural traumas are formed into the dynamics of the lived culture as this is mediated by the selective tradition against the background of the culture of an entire period (Williams 1965: 65–67, 1987: 320; Demertzis 1985: 113–115). It is through this mediation that, each time, 'the significant past' comes to the fore so that 'from a whole possible area of past and present, certain meanings and practices are chosen for emphasis' (Williams 1980: 39). In our case this is exemplified in the revalued meaning of past hardships experienced retroactively by a society or a group as trauma, and in the absolutely relevant notion of 'chosen trauma' (Volkan 2004) which not infrequently manifests as a sort of obsessive thanatomnesia, not to say thanatomania. In its interaction with the emotional underpinnings of trauma and the identity formation processes, collective memory is by no means a container of images and meanings but 'an active agent in the making of cultural trauma' (Eyerman 2001: 152). For all its intentionality (Alexander 2012: 4), the ensuing meaning struggle, however, should not be deemed merely as a rational strategy for the 'definition of the situation'; that would actually shrink the meaning of the 'meaning struggle' whose cognitive dimension is as strong as its emotional and moral one. The cultural trauma process involves not only strategic and technical interests, and manipulation by agents and entrepreneurs, but entails also an array of short-term and lasting emotions such as shame, pride, compassion, hatred, fear, anxiety, assurance, distrust, envy, resentment, *ressentiment*, dignity, and anger, to name a few. These emotions are integral elements of the identity and will formation all along the process of the trauma drama (Eyerman 2012/2017: 8). Precisely because the hazardous event(s) is deemed to disrupt the social fabric and the routines of societal reproduction, cultural trauma emerges as long as it gives rise to open reflection over the redefinition of who 'we' are vis-à-vis our past selves and others. As nicely put by Alexander, whenever social groups construe events as gravely endangering, suffering becomes a matter of collective concern and a 'we' must be constructed via narrative and coding, and it is this collective identity that experiences and confronts the danger. 'Hundreds and thousands of individuals may have lost their lives, and many more might experience grievous pain. Still, the construction of a shared cultural trauma is not automatically guaranteed. The lives lost and pains experienced are individual facts; shared trauma depends on collective processes of cultural interpretation' (Alexander 2012: 3).

For the most part, the argumentation and counter-argumentation over the victims, the perpetrators, and blame attribution involves one more register: the seeking after of a remedy, the restitution of the traumatic dislocation. This is another aspect of the meaning-making struggle (Eyerman, Madigan, and Ring 2017: 13) and the rebuilding of individual and collective identities. Consequently, apart from fostering group antagonism and conflict, cultural traumas may accord platforms for amelioration of sufferings and reconciliation of rival parties (Alexander 2012: 4). Expressed differently, cultural trauma can be seen as a discursive process where the emotions that are triggered by a traumatic occurrence are worked through and an attempt is made to heal a collective wound (Erikson 1995). In the same vein, Arthur Neil suggested that as soon as it is recognized as a negative experience, cultural trauma helps society to avoid making similar mistakes in the future (Neil 1998: 9–10). Be it noted, though, that there is no guarantee that the healing process, repair or remedy will be consummated. The open accounts and all pending issues that haunt the legacies of the Spanish and the Greek civil wars and the traumatic brutalities of the Latin American dictatorships are a case in point. I shall come back to these issues in subsequent chapters when discussing forgiveness and the commemoration of the civil war in Greece. For the time being, however, and precisely because the trauma drama is on open-ended process, I want to discuss the ontological status of cultural trauma.

As referred to above, relevant literature highlights the situational nature of the trauma process and the importance of contextual factors that make for the potential character of every single case of cultural trauma (Alexander 2012: 29, 113). As historical constructs cultural traumas may or may not emerge out of the discursive mediation of abrupt damages of the societal bond. Had some of the contextual factors played out differently, the outcome might have been different. It could have happened otherwise in virtue of the relative autonomy of the political instance vis-à-vis the economy, to retrieve a standing position held by '70s Marxist Nicos Poulantzas. As it has been neatly put, from an anthropological and semiotic angle, cultural trauma can be seen 'as the rule of uncertainty and the obnubilation of the boundaries of the *Umwelt*, from where a new semiosis with a strong development potential may arise, but which can also turn out to be fatal for the semiosis with its all-destructive power' (Aarelaid-Tart 2006: 53). But is potentiality a plausible term to represent the ontological status of cultural trauma per se?[7] I don't think so. Instead, it seems to me that virtuality is a more accurate linguistic and conceptual designation.

To start with lexical justifications, according to *The Concise Oxford Dictionary of Current English* (1990) the virtual is defined as 'that is such for practical purposes though not in name or according to strict definition'.[8] The *Oxford Advanced Learner's Dictionary of Current English* in 1974 defines the adjective virtual as 'being in fact, acting as what is described, but not accepted openly or in name as such', while in its 1990 edition defines it as something 'being or acting as what is described, but not accepted as such in name or officially'. Finally, in *Chambers English Dictionary* (1990), as an adjective it is defined as 'in effect, though not in fact'. In the medieval Christian outlook, there was a close conceptual affinity between virtue and the virtual due to the unity between the divine mighty and goodness instantiated on the Church that was deemed as the virtual body of the Christ. Hence it has been accepted that even two or three Christians gathered in the name of Jesus virtually form a Church. On this ground, virtualism is a Christian theological assumption that Jesus is virtually present in the mystery of divine thanksgiving.

On a more mundane but no less abstract plane, echoing Aristotle's and Hegel's approaches, the French philosopher of cyberspace Pierre Lévy (1995) argues that 'virtualization' lies at the core of the human condition. Arguably, in his understanding, what potential or possible and virtual share in common is latency and invisibility; on the other side, covert and visible are the real and the actual. In their interplay they make up four interrelated ontological modalities of Being. Specifically, the Real matches up to the potential, and the actual corresponds to the virtual. In this respect, 'realization' is the bringing up of a predetermined possibility and potentiality; what is contained in the real was already latent in the potential. On the contrary, 'actualization' is the transition to the actual in the sense that the actual responds to the virtual which in turn is replete with openness and contingency. For Levy the virtual 'exists' beyond Dasein by inventing new forms and qualities only through the intervention of human agents. Similarly, in Hegel's analysis, actuality consists in two factors: possibility and contingency. The possibility constitutes the inward while contingency is the outward form of actuality and either manifests itself in an immediate unity. For Hegel (1975: 200), the actual (actuality) is premised on will and free choice, and as such it partakes of the essence along its long way towards self-realization. Yet, the added value of Levy's approach is that he pushes the issue a bit further, arguing that in our era what matters is not just the transition from the virtual to the actual, but the opposite move from the actual toward the virtual, a move he calls 'virtualization'. This is a *status nascendi*, an open-ended process of indeterminacy, contingency and risk that is brought forward through will, decisions and agency.

As long as it 'is not the result of an event but the effect of a socio-cultural process … the result of an exercise of human agency, of the successful imposition of a new system of cultural classification' (Alexander 2012: 15), it seems that Smelser's claim that a cultural trauma is not born but historically made does not suffice; it should also be strongly argued that a cultural trauma is not meant to be historically made. This renders it virtual rather than potential. Be it added that insofar as it does not exist outside the system of signification which defines it as such, a cultural trauma stands as an 'institutional fact'. According to John Searle,[9] an institutional fact is ontologically featured by common language, social norms, cultural patterns and social representations. Even if it epistemically assumes an objective reality vis-à-vis particular individuals (e.g., a political party in relation to one of its voters), ontologically it is instituted through the subjective meaning and common culture assumptions. Institutional facts are distinct from 'brute facts' which are realities ontologically independent from subjective states of mind, feelings, norms and linguistic designations (e.g., the distance between the earth and the Moon is 384,395 kilometers). Even if we cannot but use a language to make sense out of them, the (brute) fact stated needs to be distinguished from the statement of it (Searle 1995: 2, 27). Since then, cultural traumas result from collective meaning making process with regards to past pains and affect the identity of an entire collectivity, they come to being as virtual institutional facts which frame the hermeneutic horizon of particular individuals and collectives.[10]

Clinical and cultural trauma: linking the two concepts

That cultural traumas are not just aggregations of individual traumas or the sum total of private sufferings does not mean that the two notions have nothing in common. On the contrary, there is much common ground for thinking of them in tandem. Apart from stark naturalistic accounts according to which traumas are automatically formed by external inflictions, the psychoanalytic understanding of trauma resembles that of cultural sociology considering belatedness and retroactivity. For Freud, the trauma of the earlier, at least childhood, may well be a crack of the protective shield (*Reizschutz*), but still it is always a retrospective experience, whose meaning is always constructed a posteriori (Laplanche and Pontalis 1986: 503–507). Traumas are at first repressed, stay latent and are then retrieved if and when the right circumstances exist, acquiring thereof a meaning for the subject. So, psychic trauma does not follow automatically. Emerging as a symptom (neurosis of traumatic causation) at a later time, when the subject faces circumstances which activate repressed negative feelings and when the defense mechanisms start to be loosened, psychic trauma is always a reconstruction. More than that, Freud asserts (1986: 317) that the psychic trauma does not necessarily result from a bodily experience or a consequence of a specific event (as in the case of cultural trauma). It could well be a completion of an impression or a fantasy. This matches well with Alexander's contention that 'sometimes, in fact, events that are deeply traumatizing may not actually have occurred at all; such imagined events, however, can be as traumatizing as events that have actually occurred' (Alexander 2012: 13; 2004: 8–9). According to Eyerman, cultural traumas can only be known and studied retrospectively.

> It is only after the passing of time – how much exactly is uncertain – that we can know if the affect of a traumatic occurrence is still felt, still alive. In this sense, cultural trauma resembles the trauma experienced by individuals, and its effects remain under the surface and become visible, are revealed, sometimes long after the fact.
>
> (Eyerman 2015: 14)

Similarly, Giesen holds that collective-cultural traumas 'require a time of latency before they can be acted out, spoken about, and worked through' (Giesen 2004b: 116) and from the point of view of an individual life course, one can claim that cultural trauma is a period in which one should rethink the previous life trajectory (Aarelaid-Tart 2006: 48; Fassin and Rechtman 2009: 18).

As already stressed, not every negative and hurtful circumstance leads to cultural or collective trauma. It is the process of social construction (e.g., cultures of revenge are conducive to trauma formation vis-à-vis cultures of forgiveness) that transforms selectively a dreadful and dire condition into 'trauma' as a boundary mark of social memory and collective identity. Likewise, according to Freud (2003: 85–86), a traumatizing condition is converted into trauma when its 'quantity' is such that the dynamics of the pleasure principle cannot master it any more. However, the experience of good motherhood, of a developed ego-ideal, the ability to work through and incorporate, are certain internal pre-traumatic factors that bound the formation of psychic trauma and allow for the individual's adjustment to the post-traumatic environment, reducing in that way the manifestation of symptoms. In both cultural and clinical trauma then, the 'pre-traumatic' conditions selectively determine to a high degree the very formation of trauma, as well as the post-traumatic stage. Hence, either type of trauma is a micro/macro construction.

Another elective affinity between the psychic and the cultural trauma is that both concepts give birth to, and are accompanied by, negative feelings and emotions. Terror, anxiety, fear, shame, humiliation, anger, disgust and guilt are some of the negative feelings stirred by the breaking of the social bond and the normative systems of reference. Smelser is categorical as to the importance of affect in the analysis of cultural trauma: 'if a potentially traumatizing event cannot be endowed with negative affect (e.g., a national tragedy, a national shame, a national catastrophe), then it cannot qualify as being traumatic' (2004: 40). Neil (1998: 3) is also categorical: 'previous feelings of safety and security are replaced by perceptions of danger, chaos, and a crisis of meaning'. One could argue that the 'universal language' of the basic negative emotions as described by Ekman (1993) e.g., fear, anger, sadness, disgust, is necessary in the negotiation of the meaning of a traumatic event.[11] If the impetus of negative affects is not activated, an event cannot be defined and interjected as threatening, disastrous, harmful, and so on.

Finally, an analogy of the defense and working through mechanisms in both kinds of trauma, clinical and cultural, seems to exist. Smelser (2004) underscores the 'displacement' and 'projection' with regards to the attribution of responsibility and the rationalization of trauma. In situations of cultural trauma, moral panics, the demonization of the other, scapegoats, expiatory victims, conspiratorial explanations of history, and so on, constitute defense mechanisms which belong to the same class as those concerning psychic trauma. Another similar defense mechanism is the double tendency of remembering and forgetting. For one and the same traumatic event, precisely because it constitutes a field of competing interpretations and significations, there is, on the one hand, the demand to 'leave everything behind us', and the injunction to 'preserve our historical memory', on the other. By way of analogy, in clinical traumas therapists observe in the same person denial and avoidance (amnesia, emotional paralysis, repression, etc.), but also the reliving of trauma through repetition compulsion.

Clinical and cultural trauma: dismembering the two concepts

We show that what unites clinical and cultural trauma is: (a) both are belated experiences as mnemonic reconstructions of negative encounters; (b) they give birth to, and are accompanied by, negative emotions and sentiments; (c) they activate similar defense mechanisms as far as the attribution of responsibility is concerned; (d) they strongly affect individual and collective identities. In this way cultural traumas straddle individuals and collectives.[12]

There are, however, three fundamental differences between the two concepts. As Eyerman suggests (2001: 3), we can still talk of cultural trauma without it having been necessarily felt by everyone, directly or indirectly. In order for a cultural trauma to exist it does not have to be felt directly by everyone, since some take it up indirectly from the selective social memory (as happens to the next generations of a civil war or a genocide), and it does not have to involve everyone. Obviously, not all Jews were equally affected by the Nazis' 'final solution', plenty of them actually evaded it; yet it does not follow that they have not been affected by the Holocaust trauma (Lev-Wiesel 2007). Cultural-social trauma is not only grounded on group-specific communicative or social memories – i.e., commonly shared bad memories experienced personally and instigating a host of negative emotions – but on cultural or historical memories which are not necessarily lived first hand by everyone. In memory studies, 'cultural' or 'historical' memories are those which exist independent of their carriers, are institutionally shaped via a number of mnemotechniques and mediated by books, commemorating holidays, media, educational systems, popular culture and so on (Levy and Sznaider 2002). Of course, they elicit negative emotions as well. In both ways, therefore, traumas mark collective memory, thus molding the socialization mechanisms and the identity formation processes of the generations to come. Even if some or many people are exempt from this process, the cultural-social trauma does not stop existing and producing pertinent effects. Something like that cannot be said, of course, not even by way of analogy, about the psycho-clinical trauma. From an intergenerational point of view, a cultural trauma is a 'chosen trauma' in Vamik Volkan's sense: a large group's unconscious 'choice' to add to its own identity a past generation's mental representation of a shared event that has caused a large group to face drastic losses, to feel helpless and victimized by another group, and to share a humiliating injury (Volkan 2001).

The second difference has to do with the mechanisms of instituting and sustaining trauma. Clinical trauma is constituted and administered by the inner-psychic mechanisms of repression, denial, adjustment and working through. On the contrary, cultural trauma results from discursive-authoritative mechanisms of defining (and therefore instituting) an event as being traumatic (Smelser 2004: 38–39; Edkins 2003: 44–45). Competing issue claimers, interest groups, the organic and traditional intellectuals of Gramsci or the free-floating intellectuals of Mannheim, and the media contest for: (a) the very existence of the traumatic event itself (e.g., the dispute concerning the truth of the Holocaust), (b) its interpretation (was the 1946–1949 clash in Greece a 'civil war' or an 'insurgence of gangs?'), (c) the proper accompanying emotions (anger, sadness, nostalgia, guilt, shame, disgust, pride, etc.) elicited by concrete structures of feeling or emotional regimes.

The third difference is that the psychic trauma may well not be related to a particular event, but be structured around a fantasy. On the contrary, cultural trauma is always formed by referring to an occurrence or incident whose accuracy, memory and significance is negotiable, i.e., an on-going symbolic-practical accomplishment. Cultural-social trauma is related to an occurrence or occurrences whose significance and meaning may be negotiated and constructed discursively, embellished with imaginative significations; its meaning is not derived out of thin air as there is a factual basis, whatever its exactness, prior to any symbolic mediation, if there is any. What matters here is not its accuracy but the occurrence being a narrative referent for the traumatic meaning- making and mythos-making process. Its 'fate' (i.e., its characterization as trauma or not) depends on the specific regime of signification. So, 'there must be some relation, real or perceived, to some referent, an occurrence, experience or event, which itself appears "always there"' (Eyerman 2015: 7).

Political history is replete with such examples (Eyerman, Alexander, and Breese 2013). By picking up just one we can elucidate the argument above. By and large, nation-building myths are grounded on traumatic narratives; such is the case of the Kosovo trauma, i.e., the defeat of Serbs in the Battle of Kosovo by the Ottoman army 600 years ago. That sorrowful memory retroactively sustains the martyrdom of the Serbian national identity. Embellished with multiple demotic, folklore and religious fantasies, verses, and narratives the real Kosovo has turned into a symbolic referent, a nostalgic-mythical site of self-sacrifiction which, through the centuries, has been 'the most widespread, familiar, habitual and easily usable story Serbs have had to explain things to themselves' (Spasić 2013: 93) vis-à-vis Albanian and Turkish nationalism as well as NATO during the 1999 bombing. Either top-down or bottom-up, the Kosovo Myth has served as a unifying force to give birth to Serbs as a modern nation. Yet, as Spasić claims, the traumatic event of the battle might never have taken place, at least as Serbians have remembered it. What ultimately counts are not the 'reality' of the traumatic event but the vicissitudes of its public representation.

Normative and conceptual prospects of cultural trauma. A very short note

In the main, by replacing 'progress' – the post-World War II central social imaginative signification – as a master signifier of the post-Cold War era, 'trauma' has generated an ambivalent attitude: it can activate self-victimization, a fatalistic culture, and a sense of helplessness which usually lead to the breaking of bonds of trust and confidence at interpersonal, political, and community levels. For one thing, an event is meant to be traumatic when it offends the subject's capabilities to cope with it; further on, however, when interpreted through the grids of public discourse, it culturally implies the betrayal of trust and cooperation, and violates default notions of what it means morally to remain part of a collective (Luckhurst 2008: 10; Zelizer 2002). The meaning-struggle of the drama trauma is not just symbolic; it is also deeply normative and moral and evokes issues about the role of evil in history and society, the quest for justice, the possibility and politics of apology and forgiveness, the interplay of memory and forgetting, the ethics of reconciliation in post-conflict societies, and so on. Culturally speaking, a trauma is carried out paradoxically: it may break and remake the sense of political and moral community and can create links between different cultures (Hutchinson 2016: 6–12; Fassin and Rechtman 2009: 19). To press the point further towards an optimistic version of the cultural trauma theory, traumas have the capacity to widen the field of social understanding and sympathy and to promote a culture of healing and therapy. From an optimistic point of view, the trauma drama process involves remedy of the past sores and wounds. Here, a crucial (although often ambivalent) role is played by the media, since awareness and evaluation of others' traumas are largely accomplished through the means of communication and journalism (Demertzis 2009; Eyerman 2008: 21, 168–169; Tenenboim-Weinblatt 2008).

These are issues to be dealt with in one way or another in the chapters to come; for the present I think an answer should be given to the questions 'what good is the cultural trauma theory?', what is its usefulness for understanding vulnerability in contemporary societies?, and how does it confer added value to the sociological analysis of human pain?

Disciplinary-wise, by and large, cultural trauma is an offspring of the strong program in cultural sociology developed at the Yale Centre for Cultural Sociology by Jeffrey Alexander, Ron Eyerman, Philippe Smith and others (Alexander 2003; Alexander *et al.* 2012). For the most part it naturally intersects with history, historical sociology and memory studies since very many cases of cultural trauma are drawn from past experiences of genocides, nation partitions, war atrocities, and the like.

Theoretically, cultural trauma analysis is mostly done under the auspices of weak social constructionism premised on the assumption that symbolic or cultural structures are absolutely equally important as the 'hard' economic and political structures for the understanding and the thick description of social meaning. Due to its autonomy, culture in cultural sociology accounts is taken as an independent rather than dependent variable as is the case in the sociology of culture for the analysis of the meaning making societal processes. Consequently, any cultural trauma case-study is actually or potentially grounded on a non-essentialist holistic theoretical approach (Mouzelis 2008: 274–284) while scrutinizing the structural determinants of identity formation, memorizations and the emotional energy involved therein. By doing that it enacts also an intellectual enterprise of double hermeneutics (Giddens 1984: 284). To 'uncover layers of meaning' and 'to make deeply buried, culture structures available to the analyst' (Eyerman 2012/2017: 15) while studying collective/cultural traumas is but an effort to interpret feelings and meanings already experienced at first hand by individuals and collectives while carrying on in their daily activities. By the same token, analysts of collective traumas attempt to uncouple causal specificities according to concrete time-space coordinates.

Methodologically, the analysis of cultural traumas is accomplished on three interrelated planes: micro, meso, and macro level. While focusing on the emotions involved in the trauma drama one is not only scrutinizing what is going on within the individual but also, just as much, what is going on *between* the individuals. This is so because individuals are always embedded in social contexts which set 'which emotions are likely to be expressed when and where, on what grounds and for what reasons, by what modes of expression, by whom' (Kemper 2004: 46). At this level of analysis, one is likely to use qualitative methodologies such as in-depth interviews, life narratives, documentary analysis (biographies, autobiographies, etc.). The meso-analytical level focuses on the interaction of carrier groups, the politics of representation, the dynamic of the public sphere, and the impact of the media of communication as they set the stage of the trauma drama. Herein sociological methods such as thematic analysis, case studies, surveys, desk research, and comparative analysis are employed in order to explicate the social environment where past pains are selectively reinterpreted and memorized alongside societal time and identity formation processes.

Under these terms, the cultural trauma theory is a thick description of traumatic experience and is to be seen also as an analytical construct for the study of collective memory (Aarelaid-Tart 2006: 49). Theoretically speaking, it is a multi and inter-disciplinary paradigmatic frame for the systematic understanding of the multilayered process of each particular trauma drama. This seems to me to be of importance because it is not so uncommon for 'trauma' to be used in academic works, let alone in popularized essays, in an a-theoretical way following what Alexander (2004a) calls a 'lay understanding' of trauma. As in many other countries, in Greece the mnemonic legacy of World War II (and the civil war just after that) is an object of both academic and public history, and social anthropology. For all the terminological inflation of suffering, horror, pain, turmoil, and the like, 'trauma' is usually treated on the basis of a common-sense naturalistic account (e.g., Kalyvas and Marantzidis 2015). In a similar mode, analyzing the traumatic experience of 9/11 as framed by the media, Ann Kaplan (2005) rehearses the standard PTSD narrative and proceeds to talk about trauma in a very literal way.

Finally, meta-theoretically or epistemologically, as the cultural trauma analysis allows one to highlight meaning-struggles engaged in the trauma drama, one can understand the 'deeply rooted collective representations that in turn may aid in explaining why the occurrence is powerful or contains the traumatizing potential that it does' (Eyerman 2012/2017: 16). It conforms therefore to the process of societal and hermeneutic reflexivity (Lash 1994). And to the extent that 'cultural sociology is a kind of social psychoanalysis' with its goal being to reveal to men and women the myths they live with (Alexander 2003: 4), it is reasonable to expect that actors, members of the carrier groups, and bystanders alike, may develop agentic qualities and become more knowledgeable and creative with their own traumatic past and that of their ancestors. On this basis, we may ponder the potential of trauma driven agents to make a step from the care of the self toward solicitude for others, from *Sorge* to *Fürsorge* in Heidegger's terminology. Some of the moral derivatives of coping with trauma will be considered in Chapter 5.

Synopsis

Not infrequently, 'trauma' passes as an untheoretical term in socio-historical and political cultural analyses. Stemming from its central staging in contemporary global risk society and post-conflict environments and international affairs, on the one hand, and adopting a weak constructionist perspective, on the other, cultural trauma was construed in this chapter as a retroactive societal meaning-making process initiated by ruptures of the social fabric. These ruptures are usually caused by damaging events like wars, genocides, famines, terrorist attacks, environmental disasters, economic crises, forced migration, etc. As a trauma drama, this process involves three pivotal factors: (a) individual, group, and group-based emotions, (b) controversial policies and politics of memory, and (c) the formation of personal and collective identities. Cultural traumas are publicly narrated collective experiences of past horrendous events by carrier groups who are antagonistic with regards to the nature of the traumatogenic event and the identification of victims and perpetrators. As such they raise normative claims for possible remedy since they may either break or restore the sense of political and moral community. This account takes the discussion of trauma beyond psychological focus on individual psyches into a sociological and politicized framing.

Notes

1 Something contrary to divine law, an impious deed, sin, crime, a wretch, monster (of a person), an impossibility, and horrid! shocking! dreadful! (as interjection). See https://latinlexicon.org/definition.php?p1=1010442 (accessed 7 June 2018).

2 Using instead the term 'historical trauma', Allen Meek defines it as 'disruptions to established forms of identity that are repeated in images and narratives and require continued negotiation and "working through"' (Meek 2010: 39). From a psychoanalytic perspective, Eugene Koh employs 'cultural trauma' in a different, naturalistic, way. He sees trauma as an infliction through which the capacity of someone to make sense of an experience is overwhelmed and therefore one is rendered frozen, or paralyzed, with regard to that particular, and other related, experiences. When this happens to the members of a particular ethnic group or cultural community, he calls it 'cultural trauma' (Koh 2019).

3 For the sake of the argument, the functional equivalent of cultural trauma in terms of historical sociology is the 'historical event'. Sewell defines it as a ramified sequence

of occurrences that is recognized as notable by contemporaries, and that results in a durable transformation of structures (Sewell 1996: 844).

4 A fallacy held by Thomas Elsaesser (2014) who, overwhelmed by the unprecedented force of the attacks on the World Trade Centre in New York, argues that deferral and belatedness are no longer suitable for understanding trauma in the post 9/11 world insofar as the attacks marked a sudden return of referentiality (Elsaesser 2014: 307). For the cinematographic representation of the current Greek crisis in terms of trauma see Sean Homer (2019).

5 Take for instance the horrendous atrocity perpetrated by the Serbs which took place in the Bosnian city of Srebrenica during the war in Yugoslavia in 1995. For more than a decade the Bosnians, as well as other constituents of European public opinion, have been at pains to name that atrocity as genocide, while Serbia was denying it. At the end of the day, in 2007, it was The Hague International Court which characterized that atrocity as genocide without, however, putting the blame on Serbia as a state.

6 Harvey Bernard Milk (1930–1978) was an American politician and the first openly gay elected official in the history of California, where he was elected as City Supervisor for San Francisco. On 27 November 1978, Milk and the Mayor were assassinated by another city supervisor. From that time, Milk became an icon in San Francisco and a martyr in the LGBT community (https://en.wikipedia.org/wiki/Harvey_Milk).

7 This question points to a different notion of ontology than the one Alexander (2012: 14) has in mind when claiming that what is of interest in the cultural trauma theory is not the ontological reality of its basis (i.e., the accuracy of dates, events, and people's moral claims) but 'its epistemology' (i.e., that people come to believe and subjectively ascribe a meaning to hazardous and abrupt situations). My concern is about the ontological character of this belief itself.

8 With regards to informatics, this dictionary describes virtual as something 'not physically existing as such but made by software to appear to do so'.

9 As a philosopher of language, Searle considers the *construction* of social reality, whereas Berger and Luckman (1967), as sociologists, analyze the *social* construction of reality.

10 My understanding of the virtual nature of the cultural trauma construction resembles 'virtual trauma', an idea presented by Allen Meek drawing from Žižek's conception of 'virtual capitalism' and Derrida's 'virtual pace of spectrality'. Commenting on the 9/11 attacks, Meek argues that the initial incomprehensibility of the traumatic event left the space open to the communities of viewers for different interpretations of its images available on television and the internet. So, it is not only the technological nature of the visualized event that matters here but also the range of 'potentialities that may emerge out of a radical disturbance of established social and political structures' (Meek 2010: 172, 186–192).

11 The 'universal language' of the basic emotions does not imply a naturalist conception of emotion in general. To my mind, basic emotions provide a minimum of affective-cultural universals, a thin foundation whereby an infinite array of situationally formed emotions flourish. Between the strong cases of the organic and the extreme constructionist approach, I adopt an intermediate approach of mild constructionism, based on the idea that everything is not a construction or constructable with regard to emotions. A contemporary historian summarizes what Hume, James and other great thinkers of emotions took for granted: 'nearly everyone agrees that there is a biological substratum to emotions that simply cannot be denied, but emotions themselves are extremely plastic' (Rosenwein 2001: 231). See also Kövecses' (2000) formulation of 'body-based social constructionism', which he argues enables us to see anger and its counterparts as both universal and culture specific.

12 This, however, does not mean that there exists a unified psychology of traumatized communities (Koh 2019).

Chapter 3

The civil war(s) trauma

Introduction

I was born in Athens in 1958, a child of a working-class family. Like many thousands of internal migrants throughout the 1950s, my family had moved to the Greek capital, abandoning their village in search of a better life. With limited material resources and social capital, their new urban environment and its promise of upward mobility required social adjustment, emotional energy, and much human cost.

One of my clearest childhood memories is of the many quarrels that took place between my parents and my elder brother. What struck me, or at least, what I now think struck me about those quarrels, especially when the topic was related to mobility or life chances, was my brother's fervent reply to my father, 'the sins of the fathers visit upon their children!' On hearing this biblical reprimand, my father would become speechless, his facial expression showing embarrassment and desperation. All discussion was over. For many years, I was perplexed by these repeated episodes. I could not understand or imagine what the sins of my beloved father might be and how it could ever be possible for that hard-working and honest man to commit any sin at all. Insinuation and silence did not help, but as I entered adolescence, I began to realize that words rarely mentioned by my parents at home, such as 'occupation', 'Resistance', 'Resistance fighter', and 'exile' applied to my father. And with that realization, I felt there was no question of wrongdoing, no sin committed. On the contrary, my father was a Resistance fighter, sent to a detention camp shortly after liberation from the Axis powers. Throughout my formative years, due to my self-identification with my parents, and despite the fact that my father hardly spoke about his past political doings or about politics in general, I found his detention unjust. In time, this sense of injustice drew me almost naturally to the community of the Greek Left, whatever the vagueness of that political description. But those quarrels, my father's shamed face, and the silence visited on him by biblical verdict, stayed with me, an unsettling reminder of a mystery in my mind.

In his later years, my father was somewhat more talkative about his experiences in the 1940s and, from our conversations, two central themes emerged: first, he was emphatic that he 'did no harm to anyone' while in the Resistance; second, the civil war was, for him, a tragic mistake. It did not escape my notice that even though he did not take part in the civil war, he felt that he bore some responsibility for it. His guilt and humiliation had kept him silent. Though he personally committed no wrongdoing, he identified himself with the defeated Left as a whole, a diminished imaginary community. The mystery was almost solved. My father's state of mind reflected the mentality of an entire generation; the generation of the civil war, whose experiences and memories condemned them to shame, fear, and silence. For reasons discussed below, those experiences and memories could never remain generationally specific. While the aftermath of the Greek civil war appeared as an aggregate of personal traumas for those who took an active part in it, it became a cultural trauma that would affect the entire social body for decades. In this chapter I attempt to show when and how this was carried out alongside the tenets of cultural trauma theory discussed in the previous one. In parallel, I will try to indicate commonalities between the drama of the Greek civil war trauma and other cases of civil war. By placing the war in its socio-historical and cultural setting I am using and interpreting evidence from past historical, sociological and ethnographic research, on the one hand and, on the other, original qualitative interviews conducted during 2009–2010 with 11 informants; eight left-wing minded, and three right wing minded. They all took part in the civil war and, since they were interviewed, most of them have passed away.

The event: historical context and prime political cultural consequences

It is officially accepted that the civil war in Greece started in December 1944 and ended in August 1949. Yet, as we will see later, as part of the trauma drama meaning-making, this chronology is controversial. What is not controversial is that the war was waged between the armed forces of the Left, comprised mainly of Greek-speaking Communists joined by a relatively small minority of Slavic-speaking Communists, on the one side, and the Right, composed of paramilitary fighters and the armed forces of the state, backed by British troops and American military aid. The war was won by the latter. The Greek civil war was a defining moment[1] of the Cold War as well as Europe's bloodiest military conflict between the end of World War II and the dissolution of Yugoslavia in 1992–1995. The Great Powers (primarily Britain and the United States but also the Soviet Union) had vested interests, since the outcome would consolidate the post-World War II status quo in the Balkans and thus the Greek civil war was inextricably linked with East–West rivalries from the very beginning.

The Greek civil war was also the first test of the newly formulated American containment theory and of that country's role as a world leader in the fight against Communism. The American agenda was to keep Greece (of all the Balkan nations) out of the Communist bloc and to stop the Soviet sphere of influence from reaching the Mediterranean. The confrontation with Greek Communists served as a model for later American interventions in Guatemala, Lebanon, Cuba, the Dominican Republic, and Vietnam (Kolko 1994: 373–395; Iatrides 2002). On the other side, Stalin and Tito handled the Greek case according to their mutual relations and the cross pressures exerted on them in the UN.

Like every other, the Greek civil war sprang out of a host of socio-historical and politico-cultural roots; a national schism between republicans and royalists, the conflict between refugees who fled the lands of Asia Minor versus the autochtho-nes, and the Metaxas dictatorship (1936–1941) and its anticommunist legislation (Close 1995: 1–7).[2]

The Greek national schism (*Ethnicos Dichasmos*) first manifested in the dif-fering attitudes of King Constantine and Prime Minister Eleftherios Venizelos towards the Entente (Britain, France and Russia) at the outbreak of the First World War. Constantine insisted on neutrality, Venizelos opted for alliance. But Constantine's politics and those of his supporters were deeply conservative whereas the politics of Venizelos were liberal and reformist. Amid the ensuing internal political turmoil between royalists and republicans and pressure from the Entente, Constantine, without abdicating, went into exile (1917). After the defeat of Venizelos in the national elections, Constantine returned to the throne and to a country exhausted by eight years of war (Close 1995: 3–4).

In 1921, under the command of Constantine, the Greek Army undertook a major expedition in Asia Minor, ill-fated due to a Turkish counter-offensive in 1922. Under the lead of Mustafa Kemal, the Turkish army forced Greek troops out of Asia Minor, demolishing Smyrna which was virtually a city of the Greek Diaspora. As a result, 1,500,000 refugees fled to Greece from Asia Minor.[3] This 'catastrophe' resulted in Constantine's abdication and put an end to Greek irre-dentism, but gave rise to a new strain, now between refugee newcomers and autochthones. The frail national economy, burdened by foreign indebtedness and the huge cost of waging wars for a decade, could not sustain the refugees most of whom faced brutality, hardship, and humiliation.

The 1928 national census revealed that the Greek population had risen to 6,204,674 from 5,016,589 in 1920. The large influx of refugees set in motion a significant left-wing labor movement in the urban centers and an acute problem of agrarian reform in the countryside. Along with this there was widespread anti-royalist feeling which resulted in the establishment of the First Greek Republic (1924–1936). Yet, due to the interwar economic depression and the persisting confrontation between royalists and republicans, parliamentary democracy was suspended in 1936 by the dictatorship of the dedicated royalist General Ioannis Metaxas. The new regime, supported if not directed by the King himself, was suc-cessful in destroying the Left through a series of measures such as anticommunist legislation, exile, imprisonment, and repressive surveillance.

From 1941 to 1944, Greece was occupied by the Triple Axis (Germany, Italy, and Bulgaria), which led to a breakdown of state and society (Close 1995: 60–67). To a certain extent it can be argued that the roots and causes of the civil war stem not so much from the aforementioned cleavages but from the dissolution of

Greek society itself during the occupation and the antagonisms, animosities, and hostilities it gave rise to. For instance, the blockade of Greek ports by the British navy, the plunder of natural resources by the occupation forces, and tragic mistakes by the public administration responsible for food distribution precipitated the winter famine of 1941–1942, adding exponentially to the number of those who died due to bombings, assassinations, and guerilla war.[4] Even today the common expression 'occupation syndrome' refers to precautions and proactive consumption based on the assumption that there might be no food in the near future. No doubt, the famine marked the collective memory in a decisive way (Scouras, Chatjidemos, Kaloutsis, Papademetriou 1947).

Very soon, the prewar ruling elites were almost totally discredited in view of their reluctance to undertake any serious Resistance initiatives during the occupation. As elsewhere in Europe, a mass-based liberation movement, called *Ethniko Apeleftherotiko Metopo* (EAM or National Liberation Front), emerged and gained impressive results against the occupying forces. EAM and its own military branch, the *Ethnikos Laikos Apeleftherotikos Stratos* (ELAS [pronounced 'Ellas' – the name of the country itself] or Greek People's Liberation Army), was by far the largest and most powerful organization amongst these forces and was dominated and led mostly by the Communist Party of Greece (KKE).

To be sure, the majority of people were not Communists; EAM enjoyed wide support, 'between a million and two million' (Clogg 1979: 150), including many women who were thus able to participate in forms of social life from which they had previously been excluded. Importantly, many EAM and ELAS members and cadres were strong anti-monarchists who supported Venizelos and were lured by the strategy of the popular fronts (Mavrogordatos 1983: 349). Other Resistance organizations such as the *Ethnikos Dimokratikos Ellinikos Syndesmos* (EDES or National Republican Greek League), though initially liberal, soon developed an anticommunist orientation. Consequently, the Resistance from the beginning was internally divided. Already in 1943, deadly battles took place in the countryside and the Athens area between EAM/ELAS and various non-leftist and anti-leftist organizations. These would be the seeds of the civil war to follow.

In December 1944, three months after liberation, a short but deadly battle occurred between ELA and the British troops that patrolled the Athens area after the evacuation of the German forces and the newly formed Greek gendarmerie EDES (staffed mainly by collaborators) (Close 1995: 137, 141). This battle became known as the Battle of Athens or *Dekemviana* (the December events). At stake were the disarmament of ELAS and the formation of new national armed forces to be controlled by the coalition government of George Papandreou. This government included some ministers appointed by EAM itself. The battle was ignited by the unwarranted shooting of a dozen left-wing demonstrators in the central square of Athens by the gendarmerie. The conflict lasted over a month and in the end the anticommunist camp prevailed and was able to impose its terms on the subsequent Varkiza Agreement, in February 1945. Part of this agreement was to pardon all offenses committed during the *Dekemvriana* except 'common-law crimes against life and property which were not absolutely necessary to the achievement of the political crime concerned'. This clause provided ultra-right wingers and Royalists with legitimacy in launching large-scale violence and terror against members, followers, and sympathizers of EAM/ELAS (Voglis 2000: 74–75).

Left-wingers and ex guerilla fighters found shelter in the mountainous countryside, defending themselves in small and isolated bands. At the same time, the Communist Party was now free to participate in public life but was at pains to put an end to the 'white terror' by peaceful political means. 'White terror' (vis-à-vis the red terror) is the persecution of members of the KKE and other left-leaning citizens by the former members of the collaborationist Security Battalions and the government's paramilitary security services in an attempt by the prewar elites to regain control of the country. This goal was in tandem with the American demand for consolidation of the Yalta Pact (according to which Greece belonged 90 percent to the Western bloc), which meant assuming a tough anticommunist stance. For these reasons, and also due to conflicts within its own leadership (hard-liners versus soft-liners), KKE opted for armed confrontation rather than taking part in the 1946 general election. Much later, in the 1970s, the party would characterize its actions as a mistake.

For the better part of a year, KKE chose not to launch any large-scale military campaign, either because it used military pressure to reach an acceptable political compromise, or simply because it was not ready for a full-blown war. Only after September 1947, when supported by the Eastern bloc (especially Tito's Yugoslavia), did the Party engage in a major military confrontation. This occurred for the most part in the mountainous northwestern region and resulted in the defeat of the Left.

In general, the Greek civil war was a multifaceted phenomenon. First of all, it was a total war that involved military and civilian forces employing conventional and unconventional tactics. Second, it was marked by local particularities and exigencies that have quite recently been recognized by scholars using new methodological tools (Marantzides 2002; Kalyvas 2000; 2002). Often, events in a local context diverged from the large-scale politics of decision-making in government and Communist Party headquarters; therefore, local politics, personal and kinship relations and hostilities were of primary importance in the conflict (Mylonas 2003).

Physical casualties

According to official estimates, by August 1949, there were about 40,000 dead; unofficial accounts place that number as high as 158,000 (Tsoukalas 1969: 89). It is also estimated that up to 60,000 members of *Dimokratikos Stratos Ellados* (DSE or the Democratic Army of Greece), the successor of ELAS, crossed the northern borders of the country and migrated into the surrounding Communist countries, where they remained for several decades. Those of Slavic ethnic origin continue to be prohibited from returning even today. Needless to say, the material disaster was on a much larger scale. One has to add the unprecedented hardship the country faced from the moment it entered World War II until the eve of the liberation.

From 1940 to 1944, almost eight percent of the population was killed, and the value of the national treasury fell by 34 percent (Tsoukalas 1969: 69). According to McVeagh, the American ambassador in Athens, in early 1946, two-thirds of Greece's population survived on only 1,700 calories per day (in comparison to the 2,850 calories of the British); almost 30 percent of the population suffered from malaria, while the incidence of tuberculosis was 15 times higher than that in Britain (Richter 1997: 434). Just after the end of the civil war in 1949, almost ten percent of the population (i.e., 700,000 people), were homeless refugees waiting to re-inhabit their wrecked villages. Cumulatively, World War II and the civil war devastated the Greek economy and ravaged Greek society almost entirely.

All of this exerted a profound impact on the way people became accustomed to violence (Voglis 2002). The civil war rested on a culture of violence inherited from the occupation, a period of collective retaliation, mass execution, deportation of local populations (especially in the region of Eastern Macedonia occupied by Bulgarians[5]), burning of villages, and public exposure of corpses, which was amplified by black (right wing) as well as by red (Communist) terror (Kalyvas 2002). Cruelty and atrocity on both sides marked collective memories and forged personal political identities and life projects in far-reaching ways.

Cleavages and long-term repercussions

For 25 years the most overwhelming consequence of the Greek civil war in political culture was the division between so-called nationally conscious (*ethnikofrones*), healthy, clean, and first-class-citizens on the one side and, on the other, the sick, non-nationally-minded miasma, second-class citizens comprised of defeated Communists, leftists, and non-royalists. This cleavage permeated not only the political realm but every social, economic, and cultural arena; social, political, and economic marginalization was the common plight for the defeated Communist Left. As a result, the public sector was purged of any non-nationally-minded civil servant. This cleft intersected with the previously mentioned interwar division between royalists and republicans.

Until the end of the 50s, the space for any strongly-worded discourse challenging the post-civil war establishment was extremely narrow. But from the beginning of the crucial decade of the '60s that space widened as the 'Union of the Center' party (*Enosi Kentrou*) challenged the dominance of ERE (the right-wing dominant party). In addition, economic development in the tertiary and manufacturing sectors allowed for the massive and very fast accession of domestic migrants to the labor market. There existed, however, an unbridgeable contradiction: while economic incorporation continued and created the conditions for social consensus and the gradual de-EAMification of the petit bourgeois masses

(Charalambis 1989: 196), the structure of the post-civil war state (monarchy, army, national-mindedness, etc.) did not allow for the lifting of their political exclusion. The petit bourgeoisified who were defeated in the civil war, already incorporated in the market and the consumerist way of life, demanded moral recognition and political representation. In the new socio-economic environment their fear gradually gave way to resentment and indignation and accumulated emotional energy which led to massive social rallies between 1963 and 1964.

Seemingly, it was a period where the post-civil war regime was about to lose its grip. Yet, under the prospect of losing control in the parliamentary elections scheduled for May 1967, April's *coup d'etat* in effect blocked every outlet for the democratic incorporation of the not-nationally-minded in the political system and cancelled the mood for further massive protests. On top of the traumatic memory of the civil war, there now came the stroke of the imposition of dictatorship, bringing a fatalistic belief in the inevitability of political inequality and marginalization. Their humiliation of the defeated in the civil war was only accentuated by this course of events. The main consequence of the dictatorship was that the contradiction between the demand for political and moral recognition, and the powerlessness to impose it – combined with the chronic reliving of endless vindictiveness, hostility and indignation – shaped a deep-rooted *ressentiment*. Either as an individual or a collective complex emotion, as will be shown in Chapter 7, *ressentiment* played a major role in Greek populism for years to come.

The civil war as a collective injury

No one celebrated in the streets when the civil war ended, as they had the liberation from German occupation in October 1944. Nor was its end celebrated in later years in any massive or popular way. A few years after its end, the civil war was almost purged from the official discourse of its primary actors – victors and defeated alike – which is something quite uncommon in the history of civil wars. With respect to historical/official memory, 29 August 1949 – the date of the end of the civil war – has never been elevated to the status of a National Holiday as was the case in Franco's Spain, apart from in the dictatorship years (1967–1974) which were in any case discredited. Nevertheless, the memory of the civil war was conveyed through a number of local celebrations and memorial services, for example, for the victims of the Communists in emblematic sites where deadly battles occurred (e.g., Meligalas, Makrygiannis, Vitsi), but the end of the civil war as such was not commemorated in any substantial way. Similarly, for all the devastation incurred in World War II, 8 May is not celebrated in Greece to commemorate the end of that war.[6] Instead, it is 28 October, the first day of the victorious Greek-Italian war, which is celebrated as a National Holiday, symbolizing thereby the unity of the nation. As the end of World War II almost coincides with the beginning of the civil war, the victorious national elite did not want to connect the two events (Voglis 2008). From 1950 to 1967, the anticommunist discourse did not much refer to the then recently ended war or to the red violence; rather its focus was on the threat of the so-called international Communism in tandem with the Cold War emotional climate.

A politics of oblivion gradually came into effect, generating public as well as private silence around the civil war, the same painful silence my father had been enduring for two and a half decades. The post-civil-war Left embraced the catchword 'forgetting' (the past) to help alleviate its marginalization. Similarly, the official discourse of the right-wing in the period between 1950 and 1967 was built around forgetting in an effort to gain greater legitimacy over multiple constituencies. It is indicative that during the 1950s the emergency measures and censorship prevented the Greek film industry from actually producing a single film that directly addressed the violence of the civil war. Instead, in a number of successful folklore films (e.g., *Astero, Sarakatsanissa, Gerakina*) one can observe allusions to the struggles of the civil war as a filial conflict, as well as a desire for reconciliation and social cohesion (Potamitis 2008: 132–134). Even during the dictatorship, which promoted a great number of propaganda films about the Communist atrocities and treason in an attempt to blame the Left and define accordingly the public memory of the war, there were no explicit and open references to the civil war; in public speeches and propaganda such references were more indirect and metaphoric than straight and polemic. Why the silence then?

Horrendous as its impact on the body social might be, I would argue that during the period from 1950–1974 the civil war was primarily experienced as a collection of private injuries that could not be accommodated into the societal symbolic universe. The tragic occurrence of the war was experienced as a perpetual shock inflicting numbness and silence, a defense mechanism of collectives and individuals alike. Had this aggregate of individual suffering been openly discussed, recognized, and signified in the public sphere, it would then have resulted in a cultural trauma. I shall deal with this later on. For the time being, some further explanation is needed regarding the veil of silence about the war.

Affective casualties

For the defeated Left, the emotional injuries were as widespread as the physical ones referred to above; in addition to the battlefield, a moral and emotional war was waged. All detainees in prisons or places of exile were pressed to sign declarations of repentance, through which they recanted their political ideas and the Communist Party itself. This method of demoralization during the civil war developed into an industry of recantation. In several thousand cases, these declarations were signed after a long and painful process of physical and psychological torture which my father did not managed to avoid. What is more, these declarations were widely publicized in the local and national press, as well as in the village communities; those who signed were forced to prove their true repentance by informing on comrades, sending public letters repudiating Communism, by joining the military police to arrest and torture their former comrades and friends. As one of my informants said

> *violence was immense, psychological mainly but physical as well ... the moment I was forced to sign they commanded us to take an oath and to write three letters ... one to the priest of our parish, one to the gendarmerie of our region ... and another one addressed to the newspapers ...*

(man, aged 83)

Through this mechanism people were 'reformed', transformed into good citizens who denounced their past identity (Voglis 2000: 76–77). Some could not stand such humiliation and committed suicide; a very tough emotional cross-pressure was exerted on all those who signed but did not alter their beliefs about Communism or the Left in general, as they were stigmatized by both the authorities and the Communist Party itself. Activating a reflex syndrome of suspicion, the party organization treated such people not as politically defeated and physically exhausted subjects but as sinful and compromised individuals who would not defend the moral superiority of the party.

For those who signed such declarations, my father included, the emotional cost cannot be fully understood except by reference to the political culture of Greek Communism, as well as the emotional culture of the Left in the decades before the war. The political socialization of the Greek Communists had been carried out in a *milieu* of self-asserted marginalization. Following the Bolshevik revolution, the KKE was founded in 1918. From early on, the party, as an institution and through its individual members, in line with the Marxist-Leninist creed and the Third International, was at pains to comply with the official policies of the Greek state and public opinion. In 1920, KKE strongly opposed the irredentist war in Asia Minor denouncing it as imperialist and adventurous. In 1924, it supported a 'unified and independent Macedonia and Thrace'[7] propagating the idea of a working-class revolution in Greece and the Balkans. During 1929, the opening year of the Great Depression, in tandem with popular sentiment, it organized militant rallies and strikes which resulted in many casualties. In 1930, it unsuccessfully declared a general strike and advocated the establishment of the Soviet regime in Greece.

The repressive state apparatus responded harshly to these political projects by prosecuting hundreds of party members. What is more, the 1936 Metaxas dictatorship declared the Communist Party illegal; almost 2,000 of its members were arrested or exiled, and they were forced to sign declarations of repentance. The entire network of its organization was for the most part demolished by the secret police. Those who remained free had to be very secretive in their contact with each other and in their private everyday lives.

Given the quasi-religious adherence to the Communist utopia, these experiences and practices contributed to the shaping of an emotional culture and an emotional climate, as well as a *habitus* of strong group-mindedness, suspiciousness against real or alleged police agents, and against traitors and revisionists within the party ranks; in other words, a disciplinary solidarity as well as a sense of being a righteous or expiatory victim. By and large, this emotional *habitus* was reactivated during the years of Resistance (1941–1944) and afterward. In effect, it was a defense mechanism for coping with disappointment and humiliation. This inherited emotional climate of mutual suspicion in tandem with the process of humiliating self-negation imposed by the postwar regime (Voglis 2000) contributed to the silencing of the civil war in the period between 1950 and 1974.

Designating the war

Hitherto, I have been referring to the Greek civil war through the use of English words. Yet, one should bear in mind that in Greek there is no semantic equivalent to *civil war*; in fact, the Greek word which usually substitutes for *civil war* is *emphylios polemos* (internecine or filial war, war within the same race). The idea of a civil war is premised on the notion of civil society and civil sphere; it presupposes individualized citizens who are organized according to collective goals and/or interests, who are in conflict over the definition of a society's historicity. On the contrary, the Greek political discourse cannot linguistically support the idea of an inner-state war *qua* civil war, precisely because it is endowed with a variety of pre-modern and anti-modern social significations. This is a result of the absence of a deeply rooted bourgeois culture (Mouzelis 1986; Charalambis and Demertzis 1993; Demertzis 1997). Under these terms, Greek economic capital at large has been commercial rather than productive, and socio-economic development has been thoroughly carried out by the state and not by a robust capitalist market. This model had made for the prominence of loose party structures, it has been conducive to clientelistic electoral politics, and it contributes to an atrophic civil society (Demertzis 1997). For more than a century, Greek society evolved within the tenets of cultural nationalism and traditionalism rather than political and socioeconomic modernity, resembling what Riggs (1964) defines as a 'prismatic society', that is a society with limited differentiation and highly mixed structural functions. In other words, despite the modernization processes experienced since the last quarter of the nineteenth century and the emergence of a stillborn class politics and interest intermediation in the interwar period, the hegemonic political cultural setting within which the civil war took place was of a *Gemeinschaft* rather than a *Gesellschaft*.

As in other Balkan and Eastern European countries, the nation-state in Greece, as a post-traditional mode of domination, is supported by what has been called cultural nationalism, that is an ideological discourse according to which the nation is far from being an association premised on modernity's civic liberties, but is rather a particularistic, *qua* horizontal brotherhood, an ethno-cultural community of language, religion, tradition, race, and habits, with romanticized historical memories (Kohn 1961: 329–330, 457; Sugar 1969: 19–20, 34–35). With an absent strong civil sphere and with a communal *habitus* prevailing, it follows that the armed conflict between Greeks in the mid-1940s was designated as internecine or filial conflict rather than a proper civil war. It was understood as a conflict within the same national family, between men and women of the same blood, namely between brothers and sisters, rather than between opposing life projects and mutually exclusive societal interests.

No wonder that although in 1943–1944 and 1945–1946 both the EAM and the anti-EAM bloc warned of an imminent civil war (*emphylios*), during the period of the long and large-scale fighting (1946–1949) both sides were cautious enough not to use the term *emphylios polemos* to describe what was happening. Had they employed such a semantic designation they would have discredited themselves as violators of the transcendental racial/national unity. To put it in another way, the constitutive civiclessness of the Greek civil war can be explained by the moral, if not sacred, character of national community, which by necessity precluded the actors from defining their actions according to the only available codification the universe of Greek political discourse could offer them: an 'internecine war'.

Since the 'civil war' *qua emphylios polemos* was excluded from public discourse early on, both sides spoke metaphorically about what was happening and, in the process, demonized each other. Demonization created symbolically the enemy it was legitimate to kill. As long as it was morally unbearable to take responsibility for the waging of an internecine war, each rival symbolically struggled for the de-humanization of the other, drawing legitimacy from the consensual myth of national-communitarian unity. For the Right, *Dekemvriana* were considered a rebellion, and their opponents were rebels against the legal national government. For them, the 1946–1949 conflict was a war against bandits and outlaws (*symmoritopolemos*), a war against Communist bandits who betrayed their country by pursuing a path of partial annexation to the Soviet bloc or the Slavs. Among others, this was premised on the anticipation that the DSE and its government (*Prosorini Democratiki Kivernisi* – Interim Democratic Government) would serve Bulgaria's geostrategic ambitions, after the Soviet Union's proposal at the Peace Conference of 1946 that western Thrace, actually a Greek territory, should be conceded to the then-socialist Bulgaria. For the Left, the 1946–1949 internal conflict as well as the December events were described as a 'people's liberation war', a 'people's democratic struggle', 'people's self-defense', 'armed struggle of DSE', 'armed struggle', or simply 'struggle'. The opponents of the Left were identified as monarchists-fascists and reactionaries who gave up the country to British and American troops, whose presence was no less than a 'second occupation'. This is why, in the countryside, those on the Left referred to the civil war as 'second guerilla war' (*deftero andartiko*). Apparently, these semantic designations on both sides were the necessary symbolic arms which prepared the actors to hate and kill the enemy and to be ready to get killed themselves. Via an adversarial meaning-making process, both sides were defending the nation, albeit in a different fashion; namely, the Right was defending the restoration of national unity whereas the Left was concerned for the liberation and the reconstitution of the nation. For their advocates, both projects awarded legitimacy to violent and brutal actions and moral superiority to each side's own self-contained political outlook.

However, with the ceasefire in 1949, these metaphors were no longer in public use. To be sure, in the subsequent years, until the 1980s, *symmoritopolemos* has been the typical right-wing designation in public speeches and political documents, and also in legislation and jurisprudence, to the extent that any reference to the war was made at all. After the mid-1950s, apart from the seven years of dictatorship, the ruling elites refrained from bringing the war onto the public agenda when they could. The prevailing stance was silence. However, the veil of silence was socially imposed but not directly enforced by a repressive state apparatus.[8] It was imposed as a win-win political choice as well as a spontaneous collective response to a horrendous event. Within such a complex situation, it has not always been easy to conduct academic research on the Greek civil war; for the long period of silence (1950–1974), the topic was a taboo and, in addition, scholars were reluctant to deal with it because of the difficulty in accessing archives prior to the early 1980s. Systematic scholarly work started after the political changeover in 1974.

Last but not least, it is necessary to take heed of the fact that in many other civil wars one can observe the same terminological civiclessness premised on the different dependency path of each country with respect to political modernity. To use a few examples, the word in Serbian/Croatian is *bratocibilački rat*, which means internecine war or war between brothers (armed conflict between Yugoslav people during the WWII and conflict within national communities later on). The Polish equivalent of the civil war is *Wojna Domowa* which means 'domestic war'; in the feudal era, civil war in Japanese was 内乱 = Nairan (internal unrest) whereas in the present era it is 内戦 = Naisen (internal fight, internal war). Civil war is also understood in Chinese as 'internal fight' or 'domestic battle' or even 'domestic war' civil war (内战). In Czech there are two terminological versions of the concept: the colloquial version is *občanská válka*; when it comes to the struggle with the collaborators in World War II, *Građanski rat* is the literary translation of 'civil war'. Similar but not identical is the terminological designation of the Finnish civil war; until the 1970s, when the consensus politics had prevailed and sharp memories of the war were alleviated, civil war (*kansalaissota*) was very much in use by the defeated Social Democrats, and class war (*luokkasota*) was the currency used by left-wing socialists and Communists. The victors in the war named it 'war for Freedom' (*vapaussota*), or 'war for liberation' (against Russian imperialism). Contrary to the Greek case, the Reds in Finland used 'civil' or 'class war' during the actual fighting in 1918 (Alapuro, 2002).

From collective injury to cultural trauma

According to cultural trauma theory, an event, as destructive as it may be, will become or produce cultural trauma only when connected to the hermeneutic horizons of social action. A cultural trauma involves the realization (in both senses of the term) of a common plight. As such, it must be collectively defined to influence and/or change the systems of reference, established roles, rules, *habitus*, and narratives of an entire society or, at least, of a significant part of it. In other words, a cultural trauma is a total social event and not just an aggregate of individual experiences. A dislocating event like a civil war does not in itself constitute a 'cultural trauma'. To become one, such an event must undergo a process of social signification; namely, it has to be signified and become socially accepted and constructed as 'trauma'.

Keeping in mind that the fundamental elements of cultural trauma theory are memory, emotion, and identity, I uphold that, as a totalizing social event, the Greek civil war can be described as a cultural trauma because it affected collective memories, group consciousness, and the organizational principles of Greek society, redirecting its orientation for several decades. Yet, as already mentioned, this was a process which could not have happened in a straightforward fashion; as it is the process of social construction that transforms selectively a painful condition of many individuals into cultural trauma of a collective, the Greek civil war is transfigured into an exemplary case of cultural trauma through a 'trauma process', which fills the gap between event and representation (Alexander 2004a: 11). The remainder of this section is about this process.

The seven-year dictatorship collapsed in July 1974. This occurred through a combination of grassroots activism and international pressure, but mostly due to its own inefficacy in securing economic development after the 1973 oil crisis and the collapse of its nationalistic foreign policy on the Cyprus issue (which led to the invasion and military occupation of the northern part of the island by Turkish troops). The political changeover (the so-called *Metapolitefsi*) put an end to the post-civil-war regime and constituted a major turn in the political opportunity structure of the country, and, retroactively, set in motion the Greek civil war trauma process (or trauma drama). KKE was legalized, new parties were formed, all civil rights and liberties were reinstated, and the monarchy was overthrown; at the same time, the country prepared itself to join the European Union. Given the absolute discrediting of *ethnikofrosini* (national-mindedness) and its advocates, an entirely new context for political culture opened up. This allowed for the re-interpretation of the official and collective memory and the experience(s) of the civil war itself.

The carving out of a democratic public space removed the veil of silence and made it possible for various carrier groups such as parties, intellectuals, political refugees and prisoners, media, journalists, academics, and artists to bring the civil war into focus as an issue whose meaning was publicly negotiated and symbolically processed. Although right and center-right claim-makers were active, it was ultimately the left and center-left advocates that gained hegemony in this trauma process. To put it differently, the subsequent transformation of the civil war from collective injury or group pain into cultural trauma was primarily launched as an internal affair of the Left.

Analytically, I would divide this trauma process into two phases: the phase of selective construction from 1974–1990, and the reflexive construction period from 1990 on. As in any other cultural trauma (Alexander 2004a: 10–24), in both phases the trauma process was conveyed in different and yet interrelated instances, whose impact on the final outcome may vary: official and collective memory, group identity, institutional arenas (aesthetic, media, scientific, legal), and the attribution of responsibility endeavors.

Selective construction phase (1974–1990)

In the era of *Metapolitefsi* (officially, the Third Greek Republic) another catchword was keenly embraced by the Left and the Right alike: 'reconciliation'.[9] The cleavage between first-class and second-class citizens was to be suspended once and for all, here and now. Such an almost universal demand was premised on a paradoxical act of 'remembering to forget' (Bhabha 1991: 93); namely, on a highly selective process of restructuring the official and the collective memory of the 1940s. With some exceptions, the mnemonic community of the defeated Left is built around the Resistance rather than the civil war. Now, however, the talk of the civil war was frankly and deliberately put aside, in accordance with the consensual reconciliation imperative. Hence, for the first phase of the trauma process another kind of politics of oblivion was carried through by both sides, while, as is usual in such cases, at the level of unofficial memory, the civil war was antagonistically recollected so that a divided or even multifaceted collective memory emerged (Halbwachs 1992: 172, 182; Connerton 1989: 38–99; Aguilar 1996; Eyerman 2001: 5–22; Voglis 2008). At the level of official memory, the civil war was obliterated, disappearing into the shadows of the Resistance'.[10] The newly constituted Third Hellenic Republic had to consolidate itself and was searching for a founding myth capable of cementing emotional energy. Among others, in the host of non-academic bibliographies concerning the 1940s comprised from propaganda material, veterans' memoirs, autobiographies, biographies, diaries, illustrated texts, congresses, and convention minutes which appeared after 1974 (Marantzides and Antoniou 2004), as well as in party documents and political discourse, one could detect the emergence of this myth: through bypassing the civil war, resistance against the Axis powers became the master narrative.

A defining moment in this process came shortly after the 1981 general elections, when Socialists (PASOK), led by Andreas Papandreou, took office for the first time in the country's history and supported the legal recognition of the Resistance (Law 1285/1982) turning it into 'National Resistance', a term never before used. Without much enthusiasm, National Resistance has ever since been officially commemorated on 25 November. On that day in 1942, a major offensive against the Axis was carried out by combined ELAS and EDES forces. Typically, recognition was given to individual Resistance fighters, who were awarded an appropriate title and pension. Accordingly, the names of hundreds of avenues, streets, and squares across the country were changed overnight to 'National

Resistance'.[11] In a prominent and historic square in the center of Athens a statue was erected and dedicated to 'national reconciliation.[12] Clearly, a myth-making process had been taking place top-down, mainly to the extent that the period of Resistance (1941–1944) was cleansed of any disturbing stains of internecine conflict and radical, if not virtually all, revolutionary projects subjugated entirely to the nationalist discourse.[13] In Greece and many other countries, a mythic all-embracing national Resistance movement promoted a postwar collective admiration that expunged any taint of toleration and collaboration with the occupiers. Actually, unlike other European countries (Fleischer, 2008: 235), in Greece the punishment for collaborators was extremely mild, and many survived precisely because they served the anticommunist cause (Haidia, 2000). In addition, large parts of the newly formed postwar ruling class had been collaborators and black-market dealers during the occupation or usurpers of Jewish property. Not surprisingly these facts have been suppressed. In effect, the *resistancialist myth* did not so much glorify the Resistance, as celebrate a people *in resistance*, a people symbolized without intermediaries such as political parties, movements, or clandestine leaders (Rousso 1991: 18; Giesen 2004a: 148). In this vein, sometimes, the retroactive recall of the Resistance period is accomplished through metaphysical and almost religious metaphors; one of our informants said: '... *whoever lived through that era, whether today is in the right or left side, is like receiving the Holy Communion, as if one drinks fresh water from a source, it is something you get power from ...*' (woman, aged 84).

With some exceptions, even nowadays the mnemonic community of the defeated Left is built around the Resistance rather than the civil war. This is repeatedly observed in numerous testimonies and narratives. Half of the left minded of my informants were more than reluctant to use the word *emphylios polemos* (civil war) a decade ago; as one of them mentioned: '... *well the civil, I do not want to say the word war, between the Democratic Army and Governmental Army from '46 to '49 is the continuation of the national Resistance ...*' According to another interviewee, '... *gradually, after 1946 the resistance against the British in the beginning and against the Americans afterwards has been named civil war*'.

Nevertheless, of the former political and social outcasts, many must have experienced a kind of bitter and halfhearted satisfaction, when the Resistance was officially recognized. What two of our interviewees said about it might reflect a more widespread mood:

> ... *they got me in trial for high treason ... us who fought the Germans ... they give me a 320 Euros pension ... this is the way they compensate us ... can you imagine? ... here we had the greatest Resistance throughout Europe....*
> (man, aged 85)

> ... *we risked our lives and this makes us proud ... those memories are great but painful, it is the most patriotic and heroic I've ever lived. Well, nowadays*

I am looking at myself, my house etc., I've become a petty bourgeois like others.

(man, aged 83)

Another defining moment in the memorization of the civil war came seven years later when a ritual of great significance for the 'nationalization' of the Resistance and the promotion of 'national reconciliation' took place on the fortieth anniversary of the end of the civil war. Hundreds of thousands of secret police files of so-called non-nationally minded citizens were ceremonially burned despite the objections of Greek historians. As a result, a precious and rich corpus of documents was lost forever. That ritual was performed according to the 1863/1989 law under which the term *symmoritopolemos* was officially replaced by the term *emphylios polemos*, and its duration defined as from 1944 to 1949. Also, the word 'bandits' (*symmotites*) was replaced by 'Democratic Army' (*Dimokratikos Stratos*). That law was enacted by the three-month coalition government of the right-wing party of *Nea Democratia* (New Democracy) and the left-wing and the originally Communist *Synaspismos* (Left Coalition) formed on the eve of the post–Cold War era and it apparently had great symbolic impact. In contradistinction, public commemoration of the civil war has been rare and sporadic, inflicting sometimes embarrassment and bitterness.[14] Within this context, the mnemonic rituals of the Right and the commemoration of its victory over the Reds were denounced by the so-called democratic camp as 'hate rites'.

The portrayal of the 1941–1944 Resistance as a 'chosen glory' (Volkan 2005), mingled imaginatively with the far less massive resistance against the military *hunta*, was supported by the performances of left-leaning artists, especially composers like Mikis Theodorakis and Yannis Markopoulos, who delivered concerts in stadiums all over Greece, attended by large crowds and broadcasted over state-owned TV channels. This artistic work focused on the heroic, enduring, and victorious people who defended the motherland despite their mistakes and misgivings. A prominent position in these performances was afforded to *andartika tragoudia* (guerilla songs), which have been extremely popular for more than a decade.

This cultural politics of oblivion bears witness to two defense mechanisms: 'displacement' and 'projection' with regards to the attribution of responsibility and the rationalization of trauma. The 1946–1949 conflict is here interpreted as the straightforward outcome of British and U.S. intervention in Greek political life.[15] The December 1944 events, let alone the 1943–1944 clashes between ELAS, EDES, and the Greek Security Battalions which were formed in 1943 under the direction of the German occupation forces (Mazower 1993), are scarcely mentioned at all. Scapegoats, expiatory victims, and conspiratorial explanations of history have been more than frequently employed to identify the causes and consequences of the civil war. For instance, hundreds of KKE executive members who disagreed with official decisions were physically exterminated by the party's death squads *Organosi Perifrourisis Laikou Agona* (OPLA or Organization for the

Protection of the People's Struggle); they were accused of spying and exhibiting a 'reactionary' petty bourgeois mentality. As an aside, the acronym OPLA is also the Greek word for 'arms'; here we have a case of a living metaphor designating death.

Ever since the late 1950s, the self-representation of the Left has been that of an expiatory or pious victim. The public memory of the Left – not just of the KKE-, selected the innumerable atrocities it suffered in the so called 'white terror' of the Right between 1945 and 1946 as well as the unbearable prosecutions they endured during and after the war (executions, exile, imprisonments, rapes, tortures, social marginalization, etc.). To this end, films like *The Man with the Carnation* (1980) and *Stony Years* (1985) visualized retroactively the suffering of the Left, rendering it thus a pious victim. At the same time, however, numerous malpractices, atrocities, and responsibilities were repressed or even disavowed in the public memory of the Left.

Likewise, the 1974–1990 mnemonic community of the victorious Right has been equally embarrassed by the civil war since it is more than reluctant to remember the approximately 5,000 of its opponents executed by the extraordinary court-martials and the nearly 70,000 prisoners and exiles who were convicted between 1947 and the early 1950s. This is nicely depicted in the 1977 Theo Angelopoulos' film *I Kynigi* (The Hunters), which is about 'non memory', that is, the refusal of the Right to take responsibility for its victims. Certainly, the anticommunist victory conferred contentment, security, and, nevertheless, a self-censored silenced celebration because of the hecatombs on each side and the overwhelmed demand for 'forgetting'.[16] Yet, the public memory of the Right selectively retains the 'red terror' and communist crimes, putting aside the issue of the Security Battalions which during the December 1944 events were the backbone of the anti-EAM forces. Most of their members joined the National Army against DSE soon after, in order to escape legal prosecution for war crimes.

The attribution of responsibility for the civil war to the British and the Americans is a classic example of a conspiratorial view of history, quite common in the Greek populist political discourse, either left or right.[17] Under these terms, Greeks *in toto* are the expiatory victims of the foreigners, an interpretive motive supported by the famous 1975 movie *O Thiasos* (Traveling Players) of Angelopoulos. Another similar defense mechanism is the double tendency of remembering and forgetting. For one and the same traumatic event, such as the Greek civil war, there is, on the one hand, the demand to 'leave everything behind us' in the name of national reconciliation; on the other hand, however, there is the injunction to 'preserve our historical memory'.

Either option leads to unsuccessful mourning and, paradoxically, in spite of being profound political options, they depoliticize the civil war itself by subsuming it into the nationalistic discourse. The adversarial political identities of the opponents of the 1940s are symbolically transfigured to the extent that (a) any revolutionary or counter-revolutionary potential of the civil war has been systematically suppressed from public discourse and popular memory; and (b) the opponents were discursively endowed with nonpolitical, nearly metaphysical, traits – noble defenders of the race and the nation on the one side and selfless and benign patriots who hunt the reactionary servants of imperialism on the other. Even the bridging of opposites described as 'reconciliation' instead of 'compromise' has its own significance; actually, 'reconciliation' is the counterpart of filial war as it presupposes two formerly homogenous parts that have only temporarily parted. On the contrary, devoid of any moralistic overtones, 'compromise' is a political concept premised on power relations and the convergence of strategic projects in a public sphere. All in all, the politics of oblivion did not contribute to a sort of *Vergangenheitsbewältigung*, i.e., a systematic reappraisal and collective process of coming to terms with the past, something which is not, of course, a Greek peculiarity (Fleischer 2008: 196, 209, 234, 246–247).

Similarly, in the phase of selective construction of the cultural trauma, the Left was entangled in a symbolic antinomy as to its collective identity: by claiming the glory of the Resistance, it was no longer morally defeated while, at the same time, it could claim the role of the victim of the post-civil-war and the post-dictatorship polity. This antinomy has been grounded on two mutually exclusive vectors: the silencing of its strong opposition to class society (that is, the ultimate stake of the Communists in the civil war) and the affirmation of the national society they wanted to be part of. Effectively, this antinomy was premised on a particular sort of emotional reflexivity. 'Emotional reflexivity' in this context is meant as a sort of emotional dynamics, a capacity to negotiate relationships by changing the structure of feeling and, therefore, how others feel within these relationships. It is a process in which social actors have feelings about and try to understand and alter their lives in relation to others (Holmes 2010). Accordingly, defense of the Resistance was meant to de-stigmatize the Greek Communists and the Left in general from being labeled traitors and national outcasts, while at the same time, it was a symbolic means for transforming the trauma of humiliation they experienced after the defeat into the bestowed pride of fully-fledged citizens. This de-stigmatizing emotional reflexivity has been carried through a legitimating discourse guided by militant testimonial zeal (Panagiotopoulos 1994) which, as a rule, framed the Greek Communists as martyrs for the homeland during the occupation and as the innocent victims of a revengeful state. It is not accidental that some of my informants used the word 'Golgotha' when describing their experience as victims of the war, that is a religious metaphor through which their activities acquire a nonpolitical status. Only recently, very few intellectuals of the Left would deviate from this justifying discourse by discreetly referring to the civil war as a strategic political option carried out by the Greek Communist Party in the 1940s and not as a fatal tragedy of the Greek people as a whole.

The 'nationalization' of the Resistance was an equally crucial signpost for the collective identity of the Right; its effect has been quite the opposite in relation to what it conferred on the losers of the war. It was argued before that, through the nationalization of the Resistance, the Greek Communists felt that they were no longer morally defeated although they remained losers. The opposite seems to be the case with members of the Right involved in the civil war: though they are still the victors on the battlefield, they have been morally degraded. The change of the regime of signification of the civil war after 1982 gave rise to embarrassment, frustration and anger as the followers of the Right felt deprived of the certainties bestowed by the anticommunist discourse and the *ethnikofrosini*. It is somehow a post-victory trauma when people deem retroactively that their past military deeds and subsequent political attitudes are not socially recognized or even frequently denounced: '*I ask you, when the state did wrong? When it was sending out its soldiers to kill themselves or now that it does not honor them?*' This is the way one of the subjects interviewed by Antoniou (2007) expresses himself in view of the redefinition of the Resistance. One of our right-oriented informants expressed his regret for the 'psychological domination of the Communist Party' so that 'historical memory and historical facts are ignored'. This post-victory trauma is likely to react to the official strategy of forgetting and 'reconciliation' (Antoniou 2007).

Reflexive construction phase (1990–2018)

During the last 25 post-Cold War years or so, a number of significant and interrelated changes in all fields of social life have been taking place in Greece, adding some new qualities to the trauma process regarding the civil war. Economically, up to 2008–2009 a rapid growth in the GNP, mainly concentrated in tertiary sector was achieved, backed by massive revenues from EU funds and stock market investments directed by 'casino capitalism' logic and heavy consumerism. Politically, this period was characterized by the gradual sedimentation of the party system from polarized pluralism into a two-party system to the extent that the two main catch-all parties, PASOK and *Nea Democratia*, had managed to minimize or eliminate other significant political forces; thus, in the main, the total left vote, comprised of a number of parliamentary and non-parliamentary parties, barely exceeded 12 percent in the various elections. In spite of the strong turnout fostered by compulsory electoral participation, a waning partisanship and disenchantment with public affairs have been systematically documented. In the realm of cultural values, a privatized atomism versus reflexive individualization antinomy can be observed, so that politically adiaphoric cohorts or instrumentally oriented publics coexist with agents of post-materialist libertarian orientations and advocates of identity politics.

Albeit differentially, this political cultural *milieu* affected the way in which the civil war has been culturally interpreted by the members of the first, second, and third generation. After 60 years (1949–2009), due to the cohort effect and the gradual disengagement from party politics, the left-wing parties included, the civil war was seen in a more distanced way; it might be true that for many apathetic youngsters (the third generation) the civil war was a thing of the past, but it can be reasonably argued that for the bearers of libertarian post-materialist values of the second generation the civil war was a present/past reality. For the carrier groups of this generation, my generation, it is not just that the war's effects are still pertinent or that the civil war injuries hibernate in collective moods in spite of the emotional inoculation of 'reconciliation'. Rather, at stake had been a coming to terms with the past of the country (*Vergangenheitsbewaltigung*), a more or less disillusioned stance toward the politics of oblivion either of the 1950–1974 or the 1974–1990 period, which joins hands with similar undertakings with regards to other countries' traumatic experiences (Fleischer 2008: 196, 209, 234, and 246–247) after World War II and the Cold War. This stance was facilitated by the explosion of memory and the rise of public history vis-à-vis postmodern consumerist lures that subvert historical consciousness. The issue gaining currency is a quest of reflexive historicity with regards to the civil war inheritance and the demise of the hegemonic politics of oblivion. It would not be much to say therefore that the master narrative of reconciliation has lost its unquestionable grip, and new sensibilities were to mediate the construction of cultural trauma in all instances referred to above.

Indicative is the great controversy that took place in late 2009 over the aesthetic value and the political cultural significance of the widely watched and strongly marketed movie of Pantelis Voulgaris *Psychi Vathia* (*Soul Deep*). The movie has been viewed by almost 200,000 people, quite an impressive audience for a production of the local film industry.[18] This is a film about the vicissitudes of two young brothers who almost accidentally get involved in the opposite camps of the 1946–1949 filial war. The last scene of the film depicts the hugging of a devastated couple just after the very last battle of the war; the couple is made up of one of the two brothers, soldiers of the National Army, and a young girl who fought for the Democratic Army. In 1985, Voulgaris' *Stony Years*, referred to above as emblematic of the (self) victimization of the Greek Left, was very favorably received; his more current *Soul Deep* had an ambivalent reception in many newspaper articles and blogs: on the one side, many applaud it as the ultimate symbol of national reconciliation; on the other side, however, critics and viewers alike question the simplistic attribution of the causes of the war to the 'foreigners' and the director's reluctance to delve deeply into its issues.

The questioning or the bypassing of reconciliation was even better depicted through the unprecedented success of the 1997 biography of Aris Velouchiotes (1905–1945), the founder and leader of ELAS, by the novelist and essayist Dionysis Charitopoulos: *Aris. O archigos ton atakton* (Aris. Leader of the Irregulars). In Greece's small book market, this 800-page, well-written biography went through four editions (the last in 2009); it has been reviewed widely and circulated in more than 200,000 copies. In 2012 it was translated and published into English with a slightly different title (Charitopoulos, 2012). Aris Velouchiotes is a mythic figure of Greece's recent history, something like the local version of Che Guevara, part and parcel of the symbolic repertoire of the left political culture. A distinguished member of KKE, he did not accept the Varkiza Agreement and soon after (March 1945), he founded the 'new ELAS' by recruiting a handful of comrades who were against the 'new occupation by the British'. Totally isolated and denounced by the Communist Party, which at that time opted for peaceful and mass-scale political activity in the cities, he was bound to fail. In June 1945, trapped by military troops, Velouchiotes committed suicide. His head was hung up for public display. The paradox is that a year later, KKE itself started a civil war based on the premises he had advocated.

For many on the Left, this thrilling personality is a symbol of a virtual revolution, a signpost of what the Communist Left could have done in order to avoid defeat. As I understand it, the great success of the book, at least among the left-leaning readers, is due not just to the storytelling of the deeds of a chosen hero; it is based on a contra-factual attitude: things could have happened otherwise; we could have won the war, if the party had adopted the appropriate strategy in time. Therefore, the reception of the book is about retroactive (self) critique and reconstitution of the collective memory and identity of the Left well beyond the imperative of reconciliation.[19]

An additional instance resonating a more reflexive stance towards the civil war is the 'when it started versus when it ended' controversy that is taking place among political sociologists and historians in Greece, as well as among public intellectuals; it is premised on the criteria that should be used to demarcate the transition of an armed conflict within a state into a civil war. This is a controversy that draws on the literature of comparative studies of civil wars (Fearon and Laitin 2003; Sambanis 2002). Thus, while the mainstream position until the late 1980s was 30 March 1946 to 29 August 1949, for some it has been more appropriate to speak of civil war when one refers to the armed confrontations between ELAS, EDES, and the Security Battalions in 1943, in spite of the fact that a large-scale military mobilization was not involved. Andreas Papandreou himself argued that conflicts between Greeks during the Resistance period can be regarded as civil war irrespective of their sporadic or intermittent nature, provided that contradictory and mutually exclusive socio-political projects lay behind them (Papandreou, 1969).[20] In equal measure, the December 1944 Battle of Athens is seen as part of the civil war. Casting the blame on the Reds, the historical narrative of the right-side recounts 'three rounds' in the Communists' plan to seize power: 1943, 1944, and 1946–1949. This is however strongly disputed by left-minded authors and commentators (e.g., Margarites 1989).

Another issue, reiterating the Communist historical narrative, is whether or not the 1945–1946 right-wing violence against the defeated EAM/ELAS, which erupted after the Varkiza Agreement, signals the beginning of the civil war (Mazower 2000: 6–7 and 31–32) or whether this was a 'unilateral civil war' – which was to be followed soon after by a 'bilateral civil war'. Nowadays, however, after the legal settlement of the issue in the late 1980s, most Greeks hold the view that the civil war occurred from 1944 until 1949.

With regard to the attribution of responsibility, another illustration of the demand for an appraisal of the civil war as a respectable academic and scholarly activity is underpinned by the need to redirect attention away from the question 'Whose fault was it?' toward the question 'How did the civil war take place?'

(Marantzides and Antoniou 2004; Mazower 2000: 8). As carrier groups of the trauma-drama, scholars not only dispute details of the civil war but also disagree as to how it should be studied. A heated debate exists between traditional historians who base their work chiefly, if not exclusively, on the study of archives and a group of younger and post-revisionist scholars who apply oral history, memory and local studies, and clinical approaches. The thrust of the dispute concerns the appropriateness of oral and local history methods and the possible disintegration of the field through topical approaches and piecemeal studies. Be it noted that this dispute straddles the fence between academic institutions and the public space by means of newspapers, journals, magazines, and artistic creation.

This trend was further promoted from 2010 onwards as a very large number of books, novels, conferences, public speeches, articles, theatrical performances, university courses, and social media focusing on the civil war and the tormented decade of the 40s came forcefully to the fore. But it was in the realm of political rhetoric where this acme close to reaching its peak in a dramatic way. For one thing, the 40s and the civil war are now narrated in multiple and pluralistic ways and there seems not to be any dominant reading whatsoever. On top of that, however, the advent of the sovereign debt crisis provided a new political and discursive opportunities structure for the articulation of political meaning making processes through the selective memorization of the past. In a country which (a) lost more than one-quarter of its GDP, (b) saw unemployment reach a record 27 percent, (c) experienced soaring poverty levels, (d) ranks last among the EU28 in the social justice index,[21] and where (e) sovereign debt-to-gross domestic product (GDP) ratio went from 120 percent in 2010 to an estimated 170 percent by 2019, the highest among the EU28, the likelihood is that citizens and political forces of the radical left and right will drastically de-legitimize the political system *in toto* and the political personnel for that matter. As the recession was growing and the steering capacity of the politico-administrative system increasingly fell short, a profound representational crisis emerged (Karyotis and Gerodimos 2015). The 2012 earthquake-like elections put an end to the traditional by-partism and the emotional energy and its moldings through propaganda before, during, and after the electoral campaigns gave rise to a novel political division intersecting the left-right cleavage: the pro- and anti-bailout division. The two principal actors who have gained most from the landslide of bi-partism were the radical left oriented SYRIZA and the Nazi party, Golden Dawn. At the January 2015 national elections, SYRIZA gained a relative majority and was able to lead a coalition government with the far-right party of Independent Greeks. This bizarre coalition has been premised – among other things – on the anti-bailout (or anti-memorandum) strong stance of both parties.

The division over the memorandums which enacted the bail-out agreements was sustained by a political emotional climate replete with negative affectivity (Davou and Demertzis 2013). It is poignant to note that many populists and media personalities described the bailout treatment between Greece, the European

Central Bank and the International Monetary Fund as a new 'occupation', bondage, colonization, and subjugation. Anachronistically, images and labels of the 40s were selectively used to discredit the bailout proponents: they were denounced as traitors, collaborators, Quislings, and the like. In return, anti-memorandum followers such as the Greek *Indignatos* were self-characterized as fighters, bearers of a new Resistance, subjects of a revolutionary multitude and so on. Many from the pro-memorandum 'camp' were prepared to denounce every single grievance and mobilization against austerity measures as populist and demagogic. During all five years until 2015 the wounds of the civil war reopened through a symbolic interplay between past and present. This is indicative of the virtual nature of cultural trauma we referred to in Chapter 2. The trauma drama of the Greek civil war is still an open-ended process.

Nevertheless, the bellicose rhetoric shrank after 2015; the new government took office and started optimistic negotiations with international lenders. Negotiations went awry and, before its collapse, the government organized a referendum in the summer of 2015 on whether Greece should accept the bailout conditions. The voters rejected the bailout conditions, but the SYRIZA-led government accepted a bailout package containing larger pension cuts and tax increases than the one rejected by voters. Paradoxically, however, SYRIZA won the autumn 2015 election, ratifying a 'realistic turn' towards mainstream politics again, in coalition with the Independent Greeks, launching the implementation of an unprecedented austerity program.

Meta-theoretical repercussions

Based on personal interest, in this chapter, I have attempted to show when and how the collective suffering of the Greek civil war was socially constructed, and thus transformed into cultural trauma. This has been a long process which started, as already analyzed, with a protracted period of silence. It would not be inaccurate to say that, among other things, this silence was product of a cryptic, as it were, memory and not merely the natural consequence of unacknowledged shame, embarrassment, and fear. Suffering was not just, or totally repressed or disavowed, nor was it simply forgotten; it was also ciphered, disguised and hidden. This is the way the crypt of memory works as it holds past experiences of pain: disguises the act of hiding and hides the disguise itself. According to Derrida (1986: xiv, xv) this confers safety on the subject because it operates as a sealed interior within an external environment and has the 'ability to isolate, to protect, to shelter from any penetration ... the crypt hides as it holds'. This hermetic function is a defense of the self against violence. If this holds true, the likelihood is that my father's silence was not only grounded in shame but was also a gesture to protect his family.

Mark Mazower (2018) describes a similar stance held by his grandfather, Max, who was a cosmopolitan and well-educated revolutionary socialist in Tsarist Russia and who took part in the Bolshevik Revolution. He married

Frouma, the sister of Nikolai Krylenko, a prominent figure of the new regime, who served as People's Commissar for Justice and Prosecutor General of the Russian Soviet Federated Socialist Republic. Nevertheless, Krylenko was arrested himself during the Great Purge and sentenced to death. In the meantime, Max and Frouma escaped and settled in London where they flourished and gave birth to Mark's dad, Bill Mazower. However, the price Max had to pay for immigrating to England and building a new home there was to denounce his passionate activism and keep his wife's kinship to Krylenko secret. Max realized that the movement for which he had fought 'had lost out and languished in oblivion'. Thus Mark explained to himself why his grandfather was surrounded by a 'melancholy aura of dashed hopes' which ended up in a persistent silence about the past. But that silence – thinks Mark – had a constructive and positive influence: it was a way to tell his kids that they were free to follow their own course in life with no fetters to bind them. Thinking of how I grew up in the 70s and the '80s, a period of upward social mobility in Greece, I do feel that my own father's silence had moved somewhat along the same track as that of Mark Mazower's grandfather.

Nowadays there is little or no silence about the civil war in Greece and the fact is that after 70 years, this war still fires the Greek political psyche. This is so because the master narrative of national reconciliation promoted from 1974 onward lags behind a systematic reappraisal and coming to terms with the past which steers clear of political correctness. As a result, the issue of forgiving or its functional equivalent has not yet been seriously raised. Forgiveness is crucial to cultural trauma theory, at least to its optimistic version, because a consistent concept of trauma, as a living metaphor, refers to a dynamic process that includes both the traumatic element itself and the process of its healing. Forgiveness is part of the healing process, an integral element of mourning and remedy. Certainly, to forgive is not to forget, nor is it denial or disavowal. I will be commenting on this in a subsequent chapter. For the time being a question is raised as to who exactly is the victim and who is the perpetrator in the Greek civil war. Who is supposed to forgive whom? This question was unthinkable 20, or even ten, years ago.

Today, in a period where the cultural trauma of that war is reflexively constructed through transcendences and regressions, whoever utters this question has to avoid negationism and historical revisionism (which is not identical with revisionist historiography). It may be true that nowadays scholars and students approach the civil war in a more distanced way, but it is equally true that it lurks in collective moods, political and national stereotypes, and social memory(ies).[22] This became apparent during the 2010–2015 period. What ultimately matters are the reconstruction of the past and the rewriting of history in a progressive way. Certainly, it will be a long process which ultimately concerns the next generation, as the real participants in the civil war have almost passed away. Yet, it might be less painful and it may leave room for mutual recognition.

Synopsis

An interpretation of the Greek civil war as an exemplary case of cultural trauma, according to the relevant middle-range theory, was attempted in this chapter. This was done through illuminating comparisons with other civil wars and by show-casing the politics of memory, the antagonistic regime of signification the war has been designated by, and the emotional dynamics during the war and after the war was over. It was argued that after a quarter of a century period of silence, the trauma drama of the Greek civil war underwent two phases: the phase of selective construction (1974–1990), and the reflexive construction phase (1990 onwards). Currently, the trauma drama of that war molds much of the political identity formation processes in Greece as its legacies are evidenced in public memory and emotionality.

Notes

1 On 5 March 1946 Churchill demarcated East versus West with his notorious 'Iron Curtain' statement; the Truman Doctrine and Marshall Plan followed some months later. Their timing chimed with two crucial incidents of the Greek civil war. 30 March 1946 is conventionally understood as the date on which the conflict began, and in September 1947, with USSR backing, the Greek Communist Party led the war to its climax (the *limnes* [lakes] plan which aimed at the secession of a region in the northern part of the country which was to be declared a popular democracy).

2 These divisions were much less intense than the intersecting cleavages which fueled the Spanish civil war (regional/ethnic differences, Catholics versus anticlerical groups, class conflicts), or the Finnish civil war which had a much more solid class basis due to the strong organizational unity of the Finnish workers' movement (Alapuro 2002).

3 See also Chapter 5 on this.

4 Though exact numbers cannot be defined, it is estimated that approximately 100,000 people died due to the famine (Fleischer 1986; Hondros 1983) To be sure however, one should not exaggerate with the long last destructive consequences of the occupation as these were used in the construction of the post war victimization of the nations involved. After the war, almost each country constructed a powerful myth about the 'unprecedented destructions' it suffered by the enemy, suppressing thus from the public memory the thorns of collaboration, the toleration of the Jewish genocide and ethnic cleansings executed either by the Axis forces or the Resistance, as well as alignment with the occupation forces.

5 With one of her elder sisters, in early 1942 at the age of 14, my own mother was forced by the Bulgarian occupation authorities to move from the northern Aegean island of Thassos to Athens where she had a first-hand experience of the famine.

6 For quite different reasons, the Baltic countries do not celebrate 8 May either; for them the real liberation came in 1991 when the Soviet empire collapsed.

7 Large parts of these geographical areas which had been under the Ottoman rule from the sixteenth until the early twentieth centuries belong now to the Greek nation-state as a result of the Balkan Wars (1912–1913) and the World War I resolutions. KKE's policy for an independent Macedonia and Thrace state met public outrage because, if for no other reason, it was just a few years beforehand that these territories had been annexed to the Greek state after bloody wars supported by a widespread irredentist ideology. Nationalist sensibilities over the so-called Macedonian Question challenge not only the Greek public opinion of today, but also Bulgarian

foreign policy, and – of course – the citizens and the governments of the Republic of North Macedonia. This country's population of approximately two million people consists first of inhabitants of Slavic and, second, of Albanian ethnic origin, with Romani, Serbian, Turkish and Bulgarian minorities making up a much smaller part of the population. An overarching Macedonian nationality is supposed to politically bridge these ethnic differences. See among others Demertzis *et al.* (1999) and https://en.wikipedia.org/wiki/Macedonia_naming_dispute https://en.wikipedia.org/wiki/Republic_of_Macedonia.

8 However, that was not the case with the Katyn massacre in 1939. For almost 50 years the socialist regime forcibly imposed silence concerning the fact that the crime was committed by the Soviets rather than the Nazis. See Bartmanski and Eyerman (2013).

9 This discursive gesture, however, was not followed by any sort of formal or informal reconciliation committee, body, tribune, dialogic process, etc.

10 A similar binary remembrance was put into effect in Poland with regards to the Katyn massacre and in Mao's China with regard to the massive atrocities of the Japanese army (Bartmanski and Eyerman 2013; Gao 2013). Yet, it seems to me to be misleading to draw too sharp a line between the official/public and the collective/popular memory; in the on-going hegemony process, bridges are built and various kinds of overlapping are formed.

11 Implementing an 'integrated plan of commemoration', a similar effort has been taking place in Spain over the last ten years or so. The names of high-profile streets and squares, dedicated to Francisco Franco, the dictator who ruled Spain from 1939 to 1975, as well as to generals (e.g., Yagüe, Mola, Varela, Fanjul) who took part in the military uprising against the Spanish republic in 1936, and who served under Franco in the bloody civil war, have been changed as a result of the so-called historical memory law that was passed under the socialist government of José Luis Zapatero in 2007 (www.ft.com/content/b7bb4a8a-a8d2-11e5-955c-1e1d6de94879).

12 This was not initiated from below; it was implemented top-down by public authorities. Monuments and memorials offer official recognition and create sites of memory and commemoration. Yet, since monument and memorial are not designated by two different words in Greek, it is not clear whether this statue is to commemorate the occurrence of the civil war or remember those who died in it. As Arthur Danto maintained, 'we erect monuments so that we shall always remember, and build memorials so that we shall never forget' (Danto 1985: 152 – cited by Eyerman *et al.* 2017: 23).

13 Similarly, in the trauma drama of 1974 in Cyprus, the Turkish military aggression overshadows the memory of the Greek-Cypriot coup against the legal government of the island, which actually triggered the Turkish invasion and sparked a short but harsh filial conflict just before the invasion. See Roudometof and Christou (2013).

14 I guess the same kind of bitterness is harboured by the followers of the US Confederacy who have been recently witnessed the removal of many monuments and memorials from a good number of American cities in the wake of the Charleston church shooting in June 2015. These removals – not destructions – as spectacular as they may be, including statues of prominent figures like Robert E. Lee and Nathan Bedford Forrest, are a clear sign against white supremacy and nostalgia for the good old days of the slavery that was defended by the big farmers and troops of the South during the American Civil War. In a divided emotional atmosphere, the public in the United Stated seems to be still tormented by the civil war. See among others https://en.wikipedia.org/wiki/Removal_of_Confederate_monuments_and_memorials#Florida.

15 Clearly, the civil war is not unique in terms of the British and US involvement; one can find similar stories in Palestine and Israel, India, Cyprus and so on, according to the divide and conquer politics. The point is that this involvement has been imaginarily overstated.

16 A note of caution is needed at this point. I do not aim to counterbalance the suffering of the two major actors, nor do I wish to make up for each one's political responsibilities. If anything, I am principally interested in how and not why the war was carried out, and not whose fault it was.

17 Likewise, the dominant account of the civil war in today's Finland resides in a seemingly inconceivable revolt by a part of the people 'against itself', broadcasting the cause of the war outside the nation – Reds were 'infected' or 'misled' by the Russians to betray their own fatherland (Alapuro 2002).

18 www.myfilm.gr/6361.html. The success of the film in generating public dialogue about the civil war can be compared to both Wajda's movie *Katyn*, which in Poland narrated more than anything else in the past the need for the memorization of the massacred victims (Bartmanski and Eyerman 2013) and the performance of Beckett's play *Waiting for Godot* in Sarajevo and New Orleans as a symbolic means for the construction of vicarious trauma (Breese 2013).

19 Lately, very few intellectuals of the Left have deviated from this justifying discourse by discreetly referring to the civil war as a strategic political option carried out by the Greek Communist Party in the 1940s and not as a fatal tragedy of the Greek people as a whole. In June 2018, 73 years after his death, KKE officially restored Velouchiotis, admitting that his deletion was a mistake and that his political choices then were in the right direction.

20 This has been recently witnessed in the cases of Iraq and Syria where civil and inter-ethnic conflicts coexisted with resistance to the American, British, and Russian troops.

21 www.bertelsmann-stiftung.de/en/publications/publication/did/social-justice-in-the-eu-index-report-2017-1/

22 This holds true for contemporary Spain. In July 2018 the Socialist prime minister Pedro Sánchez ordered by decree the immediate removal of General Franco's remains from his tomb in the grandiose basilica at the Valle de los Caídos (Valley of the Fallen) to a more modest burial place. This decision, 80 years after the end of the civil war, ignited heated disputes over its meaning and political usage. (www.nytimes.com/2018/07/07/world/europe/spain-franco.html). In the 28 April 2019 election, profiting, among others, from those disputes, the newly formed far-right party Vox took 10.26 percent of the vote and 24 seats in Parliament and managed to become the first far-right grouping to win more than a single seat in Congress since Spain returned to democracy after the fall of Franco in 1975. Eventually, Franco's remains were exhumed on 24 October 2019; the exhumation took place ahead of a repeat national election on 10 November the same year. There were voices against that and voices protesting that real reparations for the civil war and the dictatorship never happened (www.nytimes.com/2019/10/24/world/europe/franco-exhumed.html).

Chapter 4

Mediatizing traumatic experience and the emotions

Introduction

In Chapter 2 it was argued that contemporary risk society is at the same time a traumatizing and a traumatized society; namely, it does not only systematically produce threats about hazardous situations in the future (e.g., nuclear risks), but entails the politics of fear (Anselmi and Gouliamos 1998). In this respect, it has been described also as an 'angst society' (Scott 2000) linked to fatalism and victimization processes which, in turn, cultivate a fatalistic culture and a concomitant feeling of helplessness. Victimization and self-victimization are propped up by the breaking of bonds of social trust and confidence to institutions. As Edkins (2003: 4) proposes, an event is called traumatic not only when it offends the subject's capabilities and wellbeing, but when at the same time it implies the betrayal and breaching of relations of trust. Similarly, Beck (1992: 28, 61) held that 'risks experienced presume a normative horizon of lost security and broken trust' and that risk is lived nowadays as a condensation of 'wounded images of a life worth living' and its side effects have 'voices, faces, eyes and tears'. Therefore, the insistence on cynicism, nihilism, anti-party sentiments, and distrust, apparent in social theory and political sociology literature of the last three decades (Sloterdijk 1988; Goldfarb 1991; Bewes 1997), should be apprehended in connection with the sociology of trauma and emotions.

In Chapter 2 it was also argued that cultural traumas have the capacity to widen the field of social understanding and sympathy. Their institutionalization necessarily implies the designation of victims, the attribution of responsibility through the trauma drama process which is embellished by the media discourse. In this chapter I intend to discuss the media's impact on the way media users moralize other's traumas as they are depicted by the means of communication. As will be shown, an interesting academic and public discussion on online forums about the possible moral consequences of the mediatization of traumas has been developing over the last ten years or so. Regarding the media and cultural studies, this discussion has not gained much currency yet; however, when seen in conjunction with cognate deliberations in media psychology it seems that a substantive body of knowledge is already in place.

Emotions and the media

We all know by now that for a number of societal, cultural and epistemic reasons constitutive of late modernity (the demise of logocentric knowledge, feminist studies, body studies, consumer culture, individualism, informalization of social manners) emotions have reemerged as a fully legitimate and much demanding topic of inquiry and that an 'emotional turn' has marked much of the sociological stuff. For all their links and conceptual contiguities with sociology, it seems that nothing similar has ever taken place in the field of communication and media studies. Unlike sociological literature, emotions and affect have not reemerged as a normal and legitimate subject matter; from the very beginning, in one way or another, they were to be found in the earliest of human communication and mass communication studies.

As early as the first quarter of the twentieth century, American scholars conducted psychological research on film and radio broadcasting which was moving beyond audience ratings and coverage and extended to emotional aspects of reception, emotional gratifications, and the impact of film and radio usage. With the advent of television, emotional-psychological research expanded and flourished so that, gradually, from the 1960s onwards, a sustainable theoretical body with different conceptualizations and methodologies emerged under the rubric 'media and emotions' (Wirth and Schramm 2005).[1]

The topic of the media and emotions is not restricted to media psychology where an impressive body of theoretical and empirical research has been produced over the last thirty years or so; as I shall try to show in this chapter, the media-emotions nexus can be traced in many other areas of academic and public concern.[2] Moreover, not all fields and sub-fields in communication studies have treated emotions equally; in fact, there are areas of importance, such as communication policy and media economics, where the analysis of emotions is virtually absent. To take two more examples from political communication, much remains to be done in the areas of agenda setting and media framing. Traditionally, the agenda building of an issue has been studied as comprising three main components: (a) the media agenda, which influences (b) the public agenda, which in turn may influence (c) the policy agenda, i.e., political elites' issue priorities (Dearing and Rogers 1996). In this tradition, the emphasis has been given to cognitions, issue attention and awareness with the affective component being effectively absent from the researchers' priorities. Nevertheless, the media do not only prioritize certain issues but call attention to some emotions while ignoring others (Döveling 2009).

By the same token, the research tradition of news framing has been cognitively oriented as well, to the extent that it focuses on interpretation and schemes of perception conveyed by the news media. According to Entman (1993: 52)

> to frame is to select some aspects of a perceived reality and make them more
> salient in a communicating text, in such a way as to promote a particular

problem definition, causal interpretation, moral evaluation, and/or treatment recommendation for the item described.

Studies leaning on the framing hypothesis have shown that the way news is framed may dramatically alter opinions among readers, and news frame studies have also shown that certain standard frames tend to appear in political reporting (e.g., thematic, episodic, conflict orientation, game, and so on). Moral evaluations have been present as an undercurrent in earlier research; yet, their affective pair, i.e., emotions, have either, in the best case, been inferred indirectly or, in the worst case, totally ignored. Nonetheless taking emotions seriously would enhance agenda-setting and frame analysis in tandem with qualitative audience research by understanding not only what news consumers think but what they feel about news items and consequently under which frame of reference they may form political judgments and take decisions over disputed public issues. It is indicative that although Kinder (2007: 159) holds that frames include metaphors, exemplars, catchphrases, visual images, rhetorical flourishes, and justifications, which in politics bring about emotional arousal, he emphatically claims that from this point of view, 'research on framing has so far explored just a small patch of the whole territory'. From the standpoint of media psychology, Nabi (2003) maintains that rather than considering how emotions function within traditional paradigms of attitude change, communication scholars should explore the possibility that emotions serve as frames for issues, privileging certain information in terms of accessibility and thus guiding subsequent decision making.

In a nutshell and at the risk of oversimplifying, I would argue that the areas of study in which emotions have been (or should be) taken into consideration include: propaganda, persuasion, commercial and political advertising, risk communication, uses and gratification, media violence, news analysis, journalism, film studies, audience ethnography, and new media. Some of these areas are quite over-researched (e.g., media violence, advertising, persuasion) while others are under-researched (e.g., media framing, agenda setting), as happens in exactly the same way for the political sociology of emotions discussed in the first chapter.

Conceptualizing emotions in communication and media studies

By and large, and given the absence of any major affective turn in communication and media studies, scholars have tended to deal with emotions in three different ways: metaphorically, metonymically, and denotatively. Below I provide a few examples of these conceptualizations.

Metaphoric conceptualizations

In many cases, communication scholars have treated emotions evoked by or imputed to the media by placing them metaphorically at a different level of abstraction; in the first place, metaphor is commonly classified as a trope in which one entity is described by comparing it to another but without directly asserting a comparison. For Lacan, who was heavily influenced by Jacobson in this respect, metaphor is the substitution of one signifier for another. In an analogous parlance, metaphor corresponds to Saussure's paradigmatic relations which hold *in absentia* (Evans 1996: 111). Ever since advocates of the uses and gratification theory(ies) referred to 'escapism' they have treated emotions *in absentia* because under this signifier a number of different emotions could be accommodated: calmness, tranquility, joy, hope, delight, contentment, elation, worship, mindfulness, fascination, amazement, astonishment or even gloominess. More than this, the very rubric 'gratifications' in the name of this theory serves as a metaphor of multiple and concurrent emotions experienced by media users in real time. The most that scholars in this tradition are willing to admit is that the 'affective needs' the users seek to gratify are needs relating to aesthetic experiences, love, and friendship (Rayburn 1996). The same holds true for 'sensationalism' and 'infotainment' in news analysis (Schudson 1997; Postman 1985: 87), 'the symbolic power of the media' in the sociology of communication (Thompson 1995), and 'meaning' and 'identification' in audience ethnography (Wilson 1993). No doubt, all these narrative tropes are indices of emotions and affective states of mind which, nonetheless, remain vague; or, to be more accurate, these tropes are substitutes for emotional terms due to the metaphoric discursive context in use.

Metonymic conceptualizations

Not infrequently, scholars refer to emotions metonymically by using a combinatorial (syntagmatic) axis of theoretical language (Evans 1996: 113–114). In this respect, they do not substitute (as in metaphoric conceptions) but link emotional terms to one another. Yet this combinatorial link involves a perpetual deferral of meaning in the sense that emotional terms are not equated but connected to each other according to the logic of the parts-whole relationship. Hence, elements of vagueness and imprecision come to the fore. An exemplary metonymic use of emotions can be found in Fiske's analysis (1987: 224–239) of television's discourse where he emphatically registers 'pleasures' instead of more distinct emotions like joy, happiness, surprise, astonishment, love and so on. Similarly, for Massumi (2002) 'affect' is not only seen as the primary feature for the understanding of the user's experience in the new media environments but as a central 'medium' for the understanding of our information- and image-based late capitalist culture. It is what makes for 'synesthesia', i.e., the connections between the senses, and 'kinaesthesia', i.e., feelings of movement through moving images and icons on the screen in our computer. As already mentioned in the first chapter, Massumi's 'affect' serves as a whole and overarching thymotic-ontological web linked metonymically to particular emotions that can spring from it. In a similar vein, much of the advertising literature sticks with such metonymic notions as 'desires', 'emotional appeals' and 'emotional conditioning', that are built into many ads, or with the 'emotional responses' of consumers (Leiss *et al.* 1986: 225, 233, 289; Bocock 1993: 76ff.). Clearly, general terms like 'emotional appeals' or 'emotional responses' are conducive to more specific emotions such as guilt, shame, hope, sadness, trust, confidence, cheerfulness, loyalty, greediness and so on.

Denotative approaches

In contrast to metaphoric and metonymic uses, some communication scholars focus on specific and discrete emotions. This trend does not necessarily coincide with the emotional/affective turn in sociology and the humanities although it gets certain feedback from this; most notably, the denotative approach is commensurable with the sociology of emotions' preference for the analysis of particular emotions within concrete socio-cultural and relational settings. In this context, one of the most representative cases is the analysis of shame and guilt in public communication, social influence and persuasion processes; in this analysis, strong or weak guilt appeal of messages is related to differential levels of aroused guilt in the recipients and persuasive outcomes (O'Keefe 2000; Planalp *et al.* 2000). Similarly, Breakwell (2007: 109–172) discusses particular emotions involved in risk communication from a psychological point of view. On the basis that risk communication usually takes places in forums that are highly emotionally charged, she offers ample research evidence as to the role of particular emotions in enhancing media messages about risks and hazards. Some of the most prominent emotions in her analysis include: (i) induced fear in communication messages for risk awareness and risk aversion; (ii) trust and distrust towards public institutions regarding willingness to undertake a specific risk such as hosting a nuclear waste repository in one's community; and, (iii) worry, regret, anger, outrage, terror, and panic as to the understanding of the public's perception of risk. More generally, in media psychology, there has been an on-going tendency to see emotions as explanatory mechanisms within established media effects theories and as outcomes of audience's interactions with media content. In this context, Robi Nabi (1999) has been giving particular attention to discrete, message-induced negative emotions that influence attitudes in persuasion processes in the sense that emotion type, emotional intensity, and emotion placement within a message are expected to mediate information processing depth, message acceptance or rejection, and information recall. She has also developed a model of coping with discrete emotions in media use, such as regret, jealousy, sadness and anxiety, so as to predict selection and perception of media content based on the individual's need to reframe the person-environment situation (Nabi 2002). Within the context of media psychology, Vorderer and Hartmann (2009) have delved into the research and theoretical agenda of media entertainment, pointing out particular positive and negative emotions experienced by the users of media content.

All three approaches have their pros and cons as they build on different levels of abstraction. Metaphoric approaches offer broad-spectrum explanations about people's reactions to and people's uses of media, but sweep aside proper analysis of particular emotions. Metonymic approaches to emotions bring the affective dimension to the fore much more effectively, yet they still lack specificity and concreteness. Finally, administrative or critical research on discrete emotions (such as fear, cynicism, shame and so on), albeit more concrete, offers a compartmentalized account and fails to recognize that emotions are experienced in a complex and/or flow-like manner rather than one at a time.

By and large, in the media effects literature the media are seen as causes of emotions. Less research has been done the other way around; namely, to see specific emotions as predictors of the selection of media content, particularly in entertainment television. Another prominent feature in media and emotions research is its almost one-sided emphasis on negative emotions and the overlooking of positive ones in analyzing various aspects of human and mass communication like social influence, persuasion, decision-making, attitude formation, watching television, consuming news, vote preference and so on.

As noted earlier, there are a number of extensively researched, and other less researched, areas of investigation in communication studies with regard to media and the emotions. In the rest of this chapter, I shall deal with an issue that broadens the theoretical agenda and straddles media studies, cultural studies and media ethics on the one hand, and the sociology and, to a lesser degree, the psychology of emotions on the other; I am referring to the mediatization of traumas.[3] As I shall show, analyzing this issue entails the analysis of both negative and positive emotions, and a double understanding of the media as both causes and instances of audience's affective responses.

Media, trauma, and morality

Media and the 'politics of pity'

The strong feelings that accompany cultural-social trauma entail identification with the victims, not only by those who suffer from it in the first place (the inner group), but also by the wider public, the bystanders. Here, a crucial (although often ambivalent) role is played by the media. In the information society, awareness and evaluation of others' traumas are largely accomplished through the means of communication and journalism (Eyerman 2008: 21, 168–169; Zelizer and Allan 2002; Andén-Papadopoulos 2003; Tenenboim-Weinblatt 2008).

A common wisdom in the critical media debate is that suffering is commodified chiefly by the electronic media and members of the audience become passive spectators or onlookers of distant death and pain endowed with no moral commitment. Thus mediatisation of traumas leads to quasi-emotions; namely, emotions, which do not motivate, do not endure, and mortify or dump our sensitivities (Meštrović 1997), contributing to the 'social production of indifference' (Robins 2001: 536), 'premised on the capacity of the television to screen the suffering of the world and at the same time the capacity to screen out the brutal reality of that suffering. Spectators see the faces of those who are suffering, but they do not necessarily feel any moral obligation. Therefore, a mediatised trauma is seen as mere entertainment and detached curiosity (Kansteiner 2004).

Yet, it is also argued that due to the extended availability of their messages about people's suffering, the media allow forms of sympathy with distant others (Thompson 1995: 258ff.; Baer 2001; Sontag 2003). The mediated participation in the pain of others can lead to new forms of social interaction (Alexander 2004b: 22, 24) and it may initiate what Luc Boltanski (1999: 3–19), drawing from Hannah Arendt (1973: 66–98), calls 'the politics of pity', premised on strong moral sentiments and public action in favour of the unfortunates.[4] Among others, this is possible because media content is often linked to memory and offers opportunities for bearing witness which may establish moral accountability by moving individuals from the personal act of seeing to the adoption of a public point of view (Zelizer 2002; Coonfield 2007; Tenenboim-Weinblatt 2008). The mediated participation in the pain of others (especially when the trauma is acute and hard to face), in other words the spectacle of suffering at a distance by people who do not suffer, may possibly induce moral obligation and responsibility for the distant unfortunate. For in more than one-way, different media genres perform the fine tuning of audiences' sense of morality by offering opportunities for discussions about good and evil (Schwab and Schwender 2011: 22). The politics of pity is premised on commitment, on strong moral sentiments like indignation and compassion, as well as on specific forms of individual and collective public action in favor of the unfortunates, such as accusation of the persecutors, petitions, demonstrations, humanitarian action, fund raising and so on. In contrast to Meštrović's thesis on quasi-emotions and Baudrillard's analysis of simulacra, other scholars discern a strong possibility of effective political action in the present triggered by the spectacle of suffering, by carving out new public domains for the defense of the unfortunate victims (Chouliaraki 2004, 2006).

Luc Boltanski differentiates (1999: 96–101) the presence of moral sentiments as a necessary condition for the articulation of the politics of pity, as distinct from sentimentalism, that is, an essentially aesthetic stance towards the pain of the other. Investing in excessive affectivity, the sentimental person seeks out the spectacle of suffering not in order to relieve it, but in order to get pleasure from the fact that aroused sensibility confirms his/her humanity. For Boltanski (1999: 99) it becomes crucial 'to separate real emotions, the externalization of the inner going back directly to the roots of the heart, from purely external, imitated or depicted emotions with no inner reference'. I would claim that sentimentalism presents itself as a tear-wet apathy; most notably, it is the counterpart of the vicarious and conspicuous indignation described by Meštrović (1997: 26, 62–63) as components of emotional sterility and postemotionalism.

To be sure, participation in the pain of others, new forms of solidarity and the emotional armor of the politics of pity is a very complex issue that touches on the essentials of moral philosophy. A whole array of phenomena of fellow-feeling is involved in the politics of pity and the visualization of others' suffering is similar to those analyzed insuperably by Max Scheler (1954: 8–36). Vicarious feelings of sympathy, pity, understanding, imitation and emotional identification are evoked and it is certainly too difficult, if not impossible, to separate pity from compassion, as Boltanski does. Compassion is supposed to be premised on a face-to-face basis, on immediacy and a community of feeling not conducive to political action proper. Pity, on the contrary, is said to be premised on the distance between the spectator and the sufferer and it is this distance that permits the development of political strategies and the enactment of impersonal rules (Boltanski 1999: 6). In Scheler's terms (1954: 13), to whom Boltanski scarcely makes any reference, pity is fellow-feeling which involves intentional reference to the other person's experience in the sense that '*my* commiseration and *his* suffering are phenomenologically *two different facts*'.

Ambivalent political emotionality

Yet, in actual postmodern mediatized politics it would be too much to expect crystal clear differences in the emotional responses to the visualization of the distant others' sufferings and traumas. Besides, the difference between compassion and pity is not self-evident for everyone in the academic community. For instance, Sznaider (1998) speaks of 'public compassion' which originates in an abstract, theoretical, and rational idea of humanity rather than in a face-to-face encounter with suffering persons. He holds that public compassion draws from the humanitarian movements that arose in the eighteenth and nineteenth centuries, such as movements to abolish slavery and child labor, rather than from the tradition of religious charity. For Nussbaum (2001: 301) compassion is 'a painful emotion occasioned by the awareness of another person's undeserved misfortune' and it seems that in her account 'awareness' does not necessarily entail physical proximity and interpersonal communication. I guess that Scheler himself would steer clear of Boltanski's 'pity' to the extent that 'modern humanitarianism' or 'humanitarian love' is only interested in the sum total of human individuals, it is a quantitative equalitarian force that 'does not command and value the personal act of love from man to man, but primarily the impersonal "institution" of welfare' (Scheler 1961: 116, 120–121). Also, we could take notice of Höijer's (2003: 20) point that

> compassion has to do with perceiving the suffering and the needs of distant others through media images and reports. Global compassion is then a moral sensibility or concern for remote strangers from different continents, cultures and societies.

Nevertheless, in fairness to arguments *a la* Boltanski and Lilie Chouliaraki, it needs to be said that the crux of the matter is not of a terminological nature; rather, the crucial moral question is whether the media's presentation of the others' traumas undermines the nature of sympathy in the human form as the 'taking the attitude of the other when one is assisting the other' (Mead 1934: 299) or as 'a function of the whole mind' according to which 'what a person is and what he can understand' is accomplished 'through the life of others' (Cooley 1964: 140). This question could be posited in another way as well: to what extent can the media of communication eradicate the natural disposition of the human being to empathize with other people, a disposition much praised by moral philosophers like David Hume and Adam Smith?

The response to this question oscillates between two extremes: On one hand, estimating the representation of others' traumas through the logic of the 'culture industry', simulation and 'hyper-reality' leads to quite different conclusions from those put forward by Alexander, Boltanski, Chouliaraki, and Sontag.[5] For the 'apocalyptic' followers of Adorno and Jameson the commodity aesthetics of the electronic media have blocked all possible critical reflection, as false accounts of collective memory and social-cultural traumas have colonized the audience's historical imagination instead of liberating and broadening it. In this vein, it is thought that media aesthetics and consumerism are detrimental to the development of vigorous solidarities and active trust and it would be naïve to undercut the legacy of the Frankfurt School altogether. In fact, following the beats of war journalism by offering dramatic coverage of civilian populations in pain (Luostarinen 2002), the media not only sell human tragedies in a global market place, they also cultivate the numbing effect, if only because 'the spectacle of suffering becomes domesticated by the experience of watching television' (Chouliaraki 2004: 189). Within the semiotic chain of equivalences, mediatized traumatic events become, as it were, de-evented since they lose their uniqueness and acquire a moral banality due to viewers' denial in the face of human suffering (Cohen 2001) or due to their 'compassion fatigue' and their blasé 'over-familiarity' with distant suffering through the info-tainment framing of news coverage (Silverstone 2007: 62–63; Tester 1997: 39; Moeller 1999).[6] On top of this, one should take into account particular results of psychological research which point to the fact that mediatized and visualized traumas and the exposure to others' suffering bring about vicarious helplessness since the subject feels that control of his/her environment is out of hand (Johnson and Davey 1997). A good part of Ann Kaplan's analysis of trauma culture refers to vicarious traumas caused by mediatized catastrophes where spectators do not feel the protagonist's trauma itself but 'feel the pain evoked by empathy' (Kaplan 2005: 90).

On the other side, however, it would be too harsh to exclude moral sensibility from the mediated quasi-interaction (Thompson 1995: 87–118). The electronic mass media does help the politics of pity and global compassion to emerge as the immediate speed in the transmission of distant others' traumas and suffering facilitates recipients' identification, somehow, with the visualized victims. That was the case, for instance, with the so called 'Kosovocaust', an aftermath of the 'CNN effect'; i.e., footage and news photos articulated in reference to the 'lessons of the Holocaust' provoked intense moral outcries among western public opinion, thus affecting to a considerable degree international decision making.

Yet, identification with the victims can be accomplished retroactively as well; for instance, the film *Schindler's List* and, almost two decades earlier, the TV drama *The Holocaust* (not to mention a host of other products of popular culture on the Jewish cultural trauma) greatly contributed to the formation of a global awareness and a strong moral stance. Either through real time transmission or retroactively, or both, the media may build up a sort of 'cosmopolitan memory' sustained by the visualization of others' pain – say, for instance, the Rwanda genocide, the atrocities in Somalia, the Khmer Rouge's extermination of one-third of Cambodia's population, the famine in Darfur, 11 September, the Syrian 'refugee crisis' , the COVID-19 pandemic – triggering in the spectators the expression of some of the most basic emotions recognizable by everyone: disgust and anger for the perpetrators, sadness and fear for the victims.

Given the above, a notice of caution is necessary at this point. Among others, the vector of oscillation is contingent upon the kind of transmission of media content as well as upon the media genre. Live coverage of distant traumas may orchestrate an emotional reaction at a distance on the spot vis-à-vis recorded images of past events which enact processes of selective remembering and/or forgetting. Also, news outlets about others' suffering may have different impact on viewers to non-news media content such as documentaries or drama (Kyriakidou 2014, 2017; Scott 2014).[7] For example, documentaries might invite more elaborate and mindful approaches whereas news may provoke spontaneous and less thoughtful emotional reactions. But images, in turn, may stimulate intense emotional engagement, sometimes bringing about a sort of 'emotional politics' in the sense that news reporting of disaster and trauma motivates people to charitable giving and pressing governments to intervene. Yet, it is doubtful whether anything serious was done to stop the Syrian refugee flows after the world-wide spreading of the image of the dead three-year-old Syrian boy – Aylan Kurdi – lying face down on a beach in Bodrum in September 2015. Admittedly much depends on the particular conditions of discursive and emotional appropriation of media content, which is very often de-codified in moral terms. Even if 'moral assessments have become an integral, pivotal part of the disposition theory of emotion that has been employed to explain the enjoyment of drama' (Zillmann 2011: 109) and even if 'the role of moral judgments and empathy have traditionally been considered integral' to media entertainment (Nabi *et al.* 2011: 124), we should keep in mind that the audiences of others' pain are embedded in concrete social-cultural contexts of reception. These contexts posit the condition of possibility for the experience of empathy or compassion, as well as for the decision to act or not in response to distant others.

Moral and emotional interpellations mediatized

Willy-nilly the spectator is addressed as a witness of the evil and while viewing it s/he is interpellated as a moral subject; as long as this interpellation takes place, new locuses of global solidarity and ethical universality are carved out

(Levy and Sznaider 2002: 88), fueled, at the least, by the previously mentioned emotions. It is precisely through these emotions that television becomes 'an agent of moral responsibility' (Chouliaraki 2004: 186) and, consequently, a facilitator of the 'democratization of responsibility' (Thompson 1995: 263–264) and the democratization of compassion (Wahl-Jorgensen 2019: 81). What is more, these emotions are triggered by a sort of 'economy of witnessing' consisting in two functions: being an eye-witness of and bearing witness to an atrocity or wrongdoing.

> Being an eye-witness ... entails watching the event as it happens and engages with the objective depiction of historical truth; bearing witness entails watching the event as a universal truth ... and engages with a traumatic moment that borders the unrepresentable.
>
> (Chouliaraki 2010: 524)

> Needless to say, much depends on whether the drama of the depicted traumas is represented in the media as tragedy or as melodrama.
>
> (Eyerman 2008: 17)[8]

It can be argued that time-space compression is a sufficient condition for the rising of cosmopolitan memories and the global spreading of responsibility; the ultimate, though necessary, condition is the feeling of guilt. It is guilt that allows the spectators to absorb the suffering of the distant others and their traumatic history; but why is this so? Attempting an interpretation – and here I am roughly following a sort of psychoanalytic argumentation, though not so closely as it deserves – I would claim that this is so because every normal or average person is endowed with unconscious guilt due to the Superego's imperatives (Freud 2001). The paradox Freud underscores is that the more ethical the subject is, the more guilty s/he feels. Ambivalence towards the father or everyone who assumes the role of the father, and the subsequent repressed aggression come back to the Ego through the Superego. This is so, not only because the subject has repressed the forbidden drives before, as it were, an external parental authority; what counts more is that the subject feels anxious in front of the internal authority, the Superego. This internal authority monitors all forbidden desires so that intention becomes equivalent to wrongdoing. That is why many people feel guilty without prior wrongdoing.

Yet, it is not only the severe and punishing Superego that elicits guilt; it is also the symbolic Law underlying all social relations, i.e., the Law of the signifier, which according to Lacan commands that not everything is possible in human affairs. The prohibition of incest is an example of the symbolic Law which actually 'superimposes the kingdom of culture on that of nature' (Lacan 1977a: 66). It seems to me that somehow the Lacanian account of the Law, closely related to Kant's Categorical Imperative, is linked to the normative vulnerability Velleman (2003) speaks about. Normative vulnerability is the sense of being unjustified and defenseless against negative reactions and responses addressed by the other(s) in a warranted way. Thus, even if one commits no wrongdoing one may feel guilty upon the imaginary anticipation that there is somewhere someone else suffering who resents or envies one's good fortune. So, whenever a spectator is in front of a horrendous mediatized event, e.g., the collapse of the twin towers on 11 September or the Rwanda genocide or the Hurricanes Katrina and Harvey, not only does s/he feel that the symbolic order, represented by the Law, is violated by the intrusion of pure negativity (or evil, one could say) which evades any discursive intermediation; what is more, s/he experiences that his/her secure state and wellbeing is unacceptable and unjustified before the victims' tragic plight. In virtue of the unconscious guilt, the spectator feels that the violation of the Law is somehow her/his responsibility. Besides, I would also claim that the spectator of the pain of Third World distant others on western television and the internet may also experience a preconscious guilt in line with the following logic: although I as a person have done nothing to bring about their suffering, somehow I am guilty because I enjoy the goods of the capitalist center which exploits and dominates countries in the periphery.[9]

In one way or another, therefore, guilt is an ontogenetic moral ground for the development of 'emotional politics', precisely because it is rooted deep in the human psyche. It is not an accidental or contingent but an immanent moral stance. Yet, the crucial point in the information age is the degree of its universalizability against the grammar of the media; apart from redefining the interplay between distance and proximity, the latter systematically promote particularity over universality through personalization, dramatization and episodic framing of traumatic situations.

As ambivalent as their moral impact might be, and irrespective of the 'compassion fatigue' and the routinization of the others' traumas in the media (Tester 2001: 13; Alexander 2003: 103), observing the pain of others through the media cannot totally shield spectators from moral interpellation, from their direct or indirect moralization (Wahl-Jorgersen 2019: 77–83). It is certainly true that media reporting on distant suffering serves cynical commercial interests; telethons dedicated to the alleviation of Third World suffering and misfortunes are part of entertainment programming and offer ample opportunities for humanitarian sponsoring and image making. It is not inaccurate to say that viewers insulate themselves against the impact of 'traumatic' images, or even receive a sort of voyeuristic pleasure from what is called 'trauma porn' (Meek 2010: 31). It is true that, frequently, the politics of pity or compassion are reduced to giving money for charity just in order to keep the distant other at arm's length. It is also true that mediatized cosmopolitan memories buttress the post-democratic ideological discourse on 'human rights' which provides moral grounds to international interventions described euphemistically as 'humanitarian interventions' which of course create new victims, as was in fact the case when NATO and the U.S. – prompted by what was called in Tony Blair's Britain 'ethical foreign policy' – dropped bombs over Kosovo and Serbia in Spring 1999.

All this is true, but I want to believe that it is not the whole truth. It seems to me that there is always a moral reminder, a sort of unconscious or preconscious guilt, which escapes the commercial logic of the medium (Bennet 2003) and under certain circumstances overwhelms quasi-emotions and leads to agentic moral public action. This virtual moral stance has a neurobiological origin that actually sets the conditions of possibility for its emergence; neuro-scientific evidence suggests that clusters of 92 neurons in the premotor cortex have the ability to represent the action or experience of another and produce the same emotional energy as if this action or experience was performed by oneself (Williams 2009: 255). Therefore, 'the same neurons are activated when I feel fear as when I observe you feeling fear' and 'when I am performing the action and when I am observing you performing the action' (Schreiber 2007: 52–53). Far from endorsing a sort of neuro-scientific determinism, it seems to me that herein lies a non-reversible neurobiological substratum of empathy and sympathy precisely because mirror neurons facilitate mental representations of how other people think and feel. Different from emotional contagion, in this context I define empathy as an other-oriented emotion that stems from the apprehension of another's condition, and that is identical or almost identical to what the other person feels or would be expected to feel. Rather than feeling the same emotion as the other person, sympathy is defined as an involving other-oriented affective response that consists of feeling sorrow for the needy other (Eisenberg 2004). Well before the discovery of mirror neurons, research in twins evidenced a biological basis of empathy.

Luckily, the function of mirror neurons provides a firm basis for the validation of Adam Smith's programmatic statement:

> How selfish so ever man may be supposed, there are evidently some principles in his nature, which interest him in the fortune of others, and render their happiness necessary for him, though he derives nothing from it except the pleasure of seeing it. Of this kind is pity or compassion, the emotion which we feel for the misery of others, when we either see it, or are made to conceive it in a very lively manner.
>
> (Smith 1976: 9)

For Smith, this ability to put ourselves in the place of the other is elementary in everyone, not confined to the virtuous and the humane: 'the greatest ruffian, the most hardened violator of the laws of society, is not altogether without it' (ibid).[10] Be it noted, however, that the fragmented and inflated way the media depict distant suffering may add up to an 'empty empathy' in the sense that episodic and out-of-the-context media framing elicits a narrow sentimentality on the spot, without turning empathy into a more powerful means of effectively responding to others' misfortunes (Kaplan 2005: 93).[11]

Though he had not anticipated this link between classic moral philosophy and modern neuro-science, it is actually this which provides ground for the surfacing of those rare cases where the public media assume the role of the 'mediapolis'

Roger Silverstone (2007) spoke so passionately about, i.e., as a space of socio-political dialogue and deliberation of moral significance with remote others in pain through and within the media of communication. This dialogue is expected to contribute to keeping a proper distance from the victims, bringing them neither too close in nor too far away so as to enable someone to adopt their position.[12] Here I would claim that by virtue of mediapolis, though fragile and precarious, the feelings of guilt, indignation and sadness aroused while watching the unpleasant plight of distant others can be a stimulating condition for alternative moral-practical reflection. Despite the reluctance people feel to interpret their concerns and sentiments into determinate courses of action, the mediatized trauma of the distant other could give rise to a sense of responsibility for his/her life and to a sense of dignity. It prompts also what Hans Jonas (1984) regards as the attribute that differentiates humans as a species *par excellence*: the undertaking of substantive responsibility towards the Being in its entirety and the other human beings.[13] Perhaps the mediatization of traumas is unable to mobilize the Levinasian ethics of being *for* the other instead of being simply *with* the other. The likelihood is that time-space compression is conducive to the moral stance of being with the other, due to the disguised proximity of the sufferers in the screen. Yet, as soon as this takes place, it is already too much; one can maintain, therefore, that, serving as carrier groups, the media make possible an ethics of care and responsibility in our age where care seems impossible. This is accomplished, *inter alia*, through the social construction of 'moral universals', i.e., generalized symbols of human suffering and moral evil (Alexander 2003: 27–84).

By the same token, I would even argue that the media may make our direct or indirect encounter with the suffering of the others easier and mobilize that sort of moral minimalism that Michael Walzer (1994) was writing about: a moral minimum, a 'thin morality', which does not serve any particular interest but instead regulates everyone's behavior in a mutually beneficial way. For if we only have a contextual thick morality, then how are we to object to powers annihilating shared life-worlds and public spaces? In a contemporary cosmopolitan context counteracted by atavistic nationalisms, how else is an encompassing cosmopolitan democracy and a global civil sphere going to take roots if not through mature human subjects 'capable of formulating problems in terms of universal rights and justice, as well as having the capacity to feel sympathy and empathy?' (Stevenson 1997: 74). Fully aware of the ambivalence permeating the media of communication and fully undertaking the task of 'bringing democracy back in', Jeffrey Alexander defends a new open-ended notion of civil society marked by universalistic solidarity, justice, and the secular faith that democracy is a way of life. In this context, he sees the media as a platform of sentimental education which may help people empathize and sympathize with socially relegated groups (e.g., the African-American community) widening thus the limits of social inclusion and civility. In this respect, the media establish the communicative boundaries between civil and uncivil domains of social life. As long as they promote solidarity, they help us to imaginatively take the place of the other and, in this way, we are simultaneously self-willed autonomous individuals and members of an extra-individual solidarity with universalistic claims (Alexander 2006: 75, 403, 307). Wasn't this, after all, the meaning of the highly mediatized international mobilizations against the war in Iraq in 2003 – which for Jürgen Habermas, Jacques Derrida, Umberto Eco, and Richard Rorty, heralded the 'Renaissance of Europe' – as well as the solidarity that was expressed for the people in South East Asia after the earthquake of 26 December, 2004? Yes, that was the meaning then; the question though is whether it still holds strong for most of us in the age of 'post-truth' and growing presentism.

Trends of the media-moral emotions nexus

As trauma becomes a central imaginative signification of our era, changing social bonds and the terms of social remembering, the media of communication become all the more crucial for the articulation of traumatic historical memory. The media are not the only instruments for the (re)construction of collective and historical memory in view of traumatic experiences; they are one of the most powerful mnemotechnical apparatuses because they stretch time and space, thereby redefining the interplay between distance and proximity. Effectively, as an arena of the cultural trauma drama process, they may make for the possibility of trauma not to be experienced directly by anyone. Yet, even more important is the fact that the media conveys moral implications when they render traumas as part of the spectacle and the emotional public sphere.

Academic wisdom suggests that in the media saturated postmodern society 'the grand narrative has lost its credibility' bringing on de-legitimation and radical suspicion of 'pre-established rules' (Lyotard 1984: 37, 81; Rosenau 1992: 133–137). As individuals become all the more aporetic and distrustful of the offered truth of the 'mutual spying of ideologies' and the fragmentation of societal and political reality, there is no reasonable basis to support their choices. Consequently, in the absence of moral parameters, the moral situation of postmodernity is that 'yesterday's idealists have become pragmatic' (Bauman 1993: 2) and that 'liars call liars liars' (Sloterdijk 1988: xxvii).

Another common wisdom, both academic and mundane, which is nonetheless related to the first, maintains that due to the aesthetization of public discourse conferred by the hyper-mediation and 'compulsive inflation' of still and moving images, most people get satiated when encountering mediatised traumas. In a media saturated society, the intimate spectacle of others' suffering, it is said, makes for denial and unfeelingness and/or quasi-emotionality; to this end, traumas become anaesthetized banalities, and the elicited compassion is absorbed in the immediacy of the media moments, it is experienced alongside the tacit feeling rules of presentism (Revault d'Allonnes 2008). Technologies of spectacle 'permeate the body and are articulated as forms of memory, experience, belonging, emotion, and ethics' in such a way that 'the fundamental link between life and death and conceptions of ethics and justice are progressively eroded' (Lafleur 2007: 225).

To a certain extent, the argument I articulated in this chapter attempted to challenge these wisdoms by suggesting that the media may assume the role of a quasi-transcendental moral claimer when it comes to the representation of others' traumas and suffering. Representation of others' traumas brings forth emotional shock and a process of interpretation as to who the victims and the perpetrators are, initiating, by the same token, moral judgment and active solidarity premised on the universal feeling of guilt and the neuro-biological ground of sympathy and empathy. Also, one should take note that when representing our own traumas, the media may contribute to the healing process and working through; this is accomplished through the definition of the trauma (i.e., transforming via signification a traumatogenic event into a cultural trauma), its narration, and its articulation into collective memory. In a way, this threefold process contributes to successful mourning and reconnection to ordinary life (Herman 1992; Zelizer 2002; Wastell 2005). It is perfectly accepted that media outlets are variously negotiated and perceived and appropriated in social-cultural contexts which often leave room for alternative de-codifications (Hall 1980; Stevenson 1997: 63) that do not always create representations that contribute to ideological views of what is socially and culturally acceptable and desirable; consequently, affect elicited from observing others' trauma as well as ours may not necessarily serve as a commodity (emotional capital) in the media market, and may not unavoidably lead to sentimentalism.

Arguably, ever since the publication of Silverstone's *Mediapolis*, a lot has been done in the areas of humanitarian communication and moral media studies; more emphasis has been given to theoretical conceptualizations and less to empirical evidence concerning the moral agency exhibited by different audiences in view of distant human suffering[14] (Kyriakidou 2015). A further push forward was made by bringing together the sociology of emotions, media psychology, and media ethics (Döveling, von Scheve, and Konijn 2011). Yet more work is to be done when it comes to the mediatization of collective traumas in more national settings beyond, say, 9/11 or the Holocaust. To this end, trauma studies, and cultural trauma analyses for that matter, have to align more closely with media morality studies; as difficult and multi-layered as this may be, qualitative or quantitative research on mediatised distant human pain – either through textual/visual or audience analysis – is needed. As the media is one among many arenas in which the trauma drama is taking place, its role is to be put into a much wider context where moral claims are formed. Research on distant suffering may contribute to the articulation of a cultural trauma that is in the process of becoming, moving thus beyond individual moral responses exposed in audience media researches by contributing to societal reflexivity *in situ*.

Synopsis

On the basis that the mediatization of socio-political life and social consciousness is a fundamental aspect of post-modern culture, this chapter has focused on the ways traumatic experiences are depicted and triggered by the media of communication. It also accounted for the ways emotions have been treated in media scholarship. It was argued that, for the most part, the mediatization of others' traumas gives rise to 'quasi-emotions' which are part and parcel of the spectacularization of suffering; however, it was also argued that there is space for full-fledged emotions like sympathy and compassion for the victims and anger, resentment, or disgust for the wrongdoers to give rise to support thereafter for a possible 'politics of pity'. While it has strong unintended consequences in foreign affairs, the politics of pity at the grassroots may occasionally inform decision making in a more reflexive way. This moral-emotional function of the media is feasible through the emergence of 'moral universals' despite widespread cynicism and apathy.

Notes

1 In an attempt to delineate the sub-field of the sociology of emotions, four chapters of the reader edited by Greco and Stenner (2008) focus on the relations between the media and emotions.

2 This has been systematically documented by Döveling, von Scheve and Konijn (2011), where a multilayered and multi-disciplinary approach of the media-emotions nexus is accomplished with media psychology staging high in the analyses of different topics.

3 From a different perspective directed by the premises of war studies, Nico Carpentier's edited volume (2007) touches upon a number of issues raised in this chapter.

4 These issues refer to the essentials of moral philosophy and entail a host of conceptual controversies as to the meaning of compassion, pity, fellow-feeling and the like. See: Scheler (1954), Sznaider (1998), Hoggett (2006), Nussbaum (2001) and Höijer (2003).

5 For the latter, exposure to traumatic images can function as a 'challenge to turn our attention to, to think, to learn, to check the explanations invoked by those in power in order to justify collective pain' (Sontag 2003: 121–122).

6 In this respect, not only the members of the audience but journalists themselves often become desensitized in the face of mass suffering (Milka and Warfield 2017). In any case, their work requires extensive emotional labor (Wahl-Jorgensen 2019: 39).

7 Anyhow, the emotional underpinnings of news media political communication of any kind whatsoever have only recently started to be empirically scrutinized (Otto *et al.* 2019).

8 September 11 made clear that when it comes to 'our' trauma, American media and journalists adopt a much less melodramatic and entertaining stance (Rosen 2002).

9 Setting out from a perpetual remorse for modernity's dark side (colonialism, world wars, environmental degradation, etc.) westerners over-react, Pascal Bruckner claims, with regards to guilty conscience and sympathy expressed for the less fortunate of this world, as if the latter are relieved of all responsibility for their situation. But self-denigration is a kind of indirect and no less megalomaniac self-glorification: the West is the master and destroyer of the planet *in toto* and in this respect the western citizen cannot or should not feel guilty for his/her mighty position (Bruckner 2010: 35, 42).

10 This feeling Smith calls 'sympathy'. A more extended discussion on Smith's moral sentiments is to follow in Chapter 6.

11 Arguably, empty empathy, as understood by Kaplan, is cause and effect of 'quasi-emotions', as described by Meštrović (1997).

12 Although both were much influenced by Levinasian ethics, Silverstone's optimistic idea of 'mediapolis' stands opposite to Bauman's 'telecity' where moral relationships of responsibility are superseded by aesthetics and entertainment. 'In the telecity, the

others appear solely as objects of enjoyment.... Offering amusement is their only right to exist' (Bauman 1993: 178).

13 Silverstone actually adopts Jonas' idea of being someone responsible even for the actions of one's enemies within the mediapolis, inviting thereby Daniel Dayan's critical insinuation that this is ultimately a heavenly and not an earthy stance, characterized by moral maximalism (Dayan 2007: 117, 119).

14 To take an example, stemming from film and photography studies and post-structuralist theorizations, Meek (2010) scrutinizes the blurring of media logic and traumatic logic in the case of 9/11 mainly without, however, coping with spectators' own reception of trauma images.

Chapter 5

Trauma and the politics of forgiveness

Entering the past

On 21 to 24 April 2016, the famous Berlin-based documentary theater group, Rimini Protokoll,[1] delivered a much-debated performance in the sparkling and avant-garde Onassis Cultural Center, Athens.[2] The performance was about Hitler's *Mein Kampf*, on the occasion of its reprinting in 2016 in Germany, where it had been banned for 70 years. The artistic group's intention was to deconstruct the notorious 90-year-old text in an effort to counteract its appeal among potential publics by showcasing how the Bible of Nazism is replete with paranoiac truisms. The performance of 22 April was preceded by an address by the activist Rainer Höss,[3] grandson of Auschwitz Commander Rudolf Höss. The 22 April event was part of a series of similar events initiated by Rainer Höss, who set himself the task of activating European collective memory against Nazism and the negation of the Holocaust.[4] Höss stated a case of apology and repentance: 'You can't choose your family, or your nationality', and 'Exempt me from my grandfather's generation', he said. Notwithstanding its boldness, the performance gave rise to some critical commentaries on the congruence and timing of the topic. For instance, doubts were cast[5] on whether the German theater company could play *Mein Kampf*, not at the dazzling Onassis Cultural Center, but at the Rimini military cemetery, where 146 Greek soldiers are buried. In the Italian city of Rimini, a six-day deadly battle against Nazi troops took place at the end of World War II, with the Greek Third Mountain Brigade involved on the side of the Allies. So the issue at hand was whether the Rimini Protokoll's artistic gesture could overcome the mnemonic inertia of the Rimini cemetery.

Earlier, a previously scheduled meeting on 29 January 2016 between Adriana Faranda and Agnese Moro was cancelled after fierce protests were raised by the Red Brigade's victims' relatives, by political associations, by judges, and by justice servants. The meeting was to take place in a seminar for the service and varieties of justice organized by the Istituto Superiore di Scienze Religiose,[6] Syracuse. The topic of the seminar was justice's function in different branches of juridical systems. Faranda is a repentant Red Brigades member who took part in the 1978 assassination of Italian Prime Minister Aldo Moro, the Christian Democrat who had planned to form a 'great coalition' with the Italian Communist Party in the 1970s, and Agnese Moro is the daughter of the murdered politician. After the cancellation, both women, joined by another repentant terrorist, Manglio Milani, and a relative of a victim killed in a bombing incident caused by neo-fascists in 1978, sent an open letter to the Administrative Board of the Istituto, saying, among other things: 'Dear Judges, pushing us into silence is an injustice to memory'.

These examples bear witness to how societies remember a hurtful past, to who is entitled to apologize for whom, and to how reconciliation between traumatized subjects and wrongdoers is possible. These are open-ended issues full of aporetic traces insofar as trauma is extracted from individual pathologies and forms, as we have already argued, a central feature of the experience of our time. Some of the hidden and overt dimensions of these issues can be explored against the background of the political sociology of emotions. In Chapter 2 it was maintained that cultural traumas straddle individuals and collectives and that what unites clinical and cultural trauma is that, among other things, (a) both are belated experiences as mnemonic reconstructions of negative encounters; (b) they give birth to, and are accompanied by, negative emotional energy; and (c) they activate similar defense mechanisms. We also saw that, from an optimistic point of view, trauma and cultural trauma theory involves and evokes remedial action and healing processes. Arguably, there are different kinds of remedies and healing practices contingent upon the nature of the trauma, the range of the afflicted groups, the carrying capacity of the public arenas wherein the trauma drama is taking place, and the concomitant resources of the carrier groups. However, it is certainly true that remedial initiatives are deeply and strongly intertwined with moral claims.

Typically speaking, one could point to four individual and/or collective moral responses to trauma: revenge, demand for justice, forgiveness, and non-forgiveness.[7] Each response involves a number of different as well as interrelated sentiments. The discussion to follow considers possible emotional and moral reactions to trauma, with a particular focus on forgiveness and *ressentiment*. Forgiveness is deemed a moral virtue and *ressentiment* is considered a complex negative emotion. Both have received much attention, not only for their inherent importance but also for their robust relevance to contemporary trauma and therapy culture as well as to the philology of resilience against failures inflicted by neo-liberal economics and social policies.

Reactions to traumas

Revenge

Revenge is an immediate reaction motivated mainly by anger or wrath, hatred, and resentment. Resentment here – and in the chapter to come – is understood as righteous moral anger, a negative reactive attitude that a person develops upon experiencing indifference, insult, or injury from another person, and it implies strong disapproval of the injurer, who is reasonably considered responsible for his/her actions (Strawson 1974: 7, 14; Murphy 2005). Stated differently, resentment is a legitimate and valuable form of anger or rage in response to perceived moral wrongs (Brudholm 2008: 9). Generally speaking, anger and cognate emotions like disgust, hatred, and hostility generate aversive affectivity, which signifies repugnance for something or someone. In its most elementary form, anger is caused by impediments to one's aim in the midst of a task, a disturbance of one's course of action (Schieman 2006). Whenever anger, and political anger for that matter, goes beyond the removal of impediments for goal achievement and involves blame attribution it acquires a moral pointer. According to Kennedy's translation, in his *Rhetoric*, Aristotle (1992: 1378a) described anger as a 'desire, accompanied by [mental and physical] distress, for apparent retaliation because of an apparent slight that was directed, without justification, against oneself or those near to one'. According to more popular translation, it is 'an impulse, accompanied by pain, to a conspicuous revenge for a conspicuous slight directed without justification towards what concerns oneself or towards what concerns one's friends'.[8] In an alternative translation provided by Martha Nussbaum (2016: 17): anger 'is a desire accompanied by pain for an imagined retribution on account of an imagined slighting inflicted by people who have no legitimate reason to slight oneself or one's own'. It is the lack of 'right justification' and the concomitant violation of fairness plus the transgression of a sense of justice that render anger more than an egoistic, self-targeted, emotion. Its appraisal basis draws from a subjective sense of broken justice or moral order. This is perhaps a strong reason to better translate his notions of *orgē* and *thumos* as resentment rather than as anger. Yet, the issue begs for some further elaboration.

Anger may be well grounded and well established as long as it defends one's self-respect and dignity, protests against injustice as it signals that a wrongdoing has been occurred, and discourages others' aggression. However, in Aristotle anger is intrinsically linked to retributive pay back, it is about a conspicuous revenge for a conspicuous slight concerning oneself or one's own. Revenge-driven anger may be evoked either by down-ranking or by some narcissistic ego-injury. And this is precisely what, according to Nussbaum (2016), renders anger morally problematic. As anger is targeted at the particular person or the persons who are regarded to have insulted or down-ranked someone, one is oriented to the past rather than 'toward more productive forward-looking thoughts' (Nussbaum 2016: 6) to rectify individual and social well-being for the time to come. The moral flaw of anger consists in its retaliatory and retributive nature vis-à-vis a 'rational' stance toward restitution advocated by Nussbaum herself (2016: 29, 93, 100, 129).[9] Likewise, it also consists in the compensatory illusion that the pain of the offender will pay off the damage done in the first place.

Nevertheless, as indicated in many Greek tragedies, the tendency to retaliation and revenge 'is deeply rooted in human psychology' (Nussbaum 2016: 40) and under conditions of moral atrocities, vengeance comes forth as a moral necessity (Peterson 2018: 152). In terms of a sequence, vengeance seems to be first underpinned by anger and hatred, followed shortly after by resentment; the latter requires extended cognitive assessment of an unjust or harmful situation and develops later, while anger and hatred explode immediately. These three emotions account for the victim's damaged dignity and self-respect. Minow (1998: 10) postulated that whoever is not resentful of injuries done to him/her is a person who lacks self-respect. Retaliation, an eye for an eye, and giving what's coming to the wrongdoer are the most usual expressions of vengeance. This stance springs from action readiness toward unforeseen offenses like terrorist attacks (e.g., New York, 11 September 2001; London, 7 July 2005; Paris, 13 November 2015; Brussels, 22 March 2016, etc.). Yet, there is always the risk of a disproportionate response, in which the victim violates domestic and/or international law out of malicious spitefulness or hurt arrogance (Badiou 2016). In this case, a vicious cycle of vendetta and excessive violence is initiated that fails to restore what was damaged in the first place. A question then is how vengeance can be distinguished from the pursuit of justice.

A possible remedy is retribution, i.e., a public punishment that stands as a fair functional equivalent of vengeance. For Minow (1998: 12), retribution is seen as 'vengeance curbed by the intervention of someone other than the victim and by principles of proportionality and individual rights'. According to Schwartz (1978), organized retribution is an impartial infliction of punishment in accordance with prescribed and fixed rules of proportionality that steer clear of the possible excesses of *lex talionis*.[10] Organized retribution cannot but take place within a political realm supported by an operational juridical system. There is no functional equivalent, though, of the organized retribution within the intimate sphere; from a Stoic point of view, Nussbaum (2016: 31) claims that a rational and responsible stand against insults and wrongdoings is to go through anger and come to a compassionate hope state of mind, a really difficult moral task indeed. With organized retribution, perpetrators get their deserts in decency, i.e., they are punished according to the rule of law and norms of civility. In this way, reversal of the roles of perpetrator and victim is avoided, while the victim retains his/her self-respect. Of course, retribution cannot purge horrible memories, nor does it preclude regression to overt or covert actions of revenge. In primitive societies the infinite perpetuation of bloody revenge, which occasionally destroyed the entire social fabric, was prevented through the ceremonial sacrifice of someone innocent from the perpetrator's group. At a later stage the sacrificial victim was chosen from the animal realm. That ritual functioned as a moderate settlement showing respect for the injury inflicted to the victim's family or group; the perpetrator himself was not to be sanctified because that would not symbolically save society from the vicious cycle of revenge. In this sense, 'sacrifice is primarily an act of violence without the risk of vengeance' (Girard 2005: 14–15).

Demand for justice

In contemporary societies, it is hard to imagine retribution's being accomplished outside formal retributive justice procedures; in a way, retribution is legally contained vengeance through prosecutorial justice. In René Girard's words, 'the juridical system never hesitates to confront violence head on, because it possesses a monopoly on the means of revenge' (Girard 2005: 23). For some individual victims, this may be satisfactory. For other victims, and, moreover, when societal relations are at stake, demand for retributive justice may be deemed not appropriate. With Truth and Reconciliation Committees in South Africa and elsewhere, the point was not retribution but recognition and reparation, and in this respect restorative justice was enacted (Elster 2006; Flam 2013). Because the goal was national rebuilding and reconciliation, it was argued that full prosecution of crimes would perpetuate social divisions and make it impossible for the body politic and its citizens of all colors to be healed (Hutchinson 2016: 251, 254–255).

Demand for restorative justice presupposes a higher level of abstraction than retribution; it seeks to repair past injustices while supporting the humanity of both offenders and victims by delivering corrective changes in the record, in relationships, and in future actions (Minow 1998: 91–92). It does not just deliver legal punishment to wrongdoers. Like revenge, reparation through restorative justice is underpinned by resentment but not by hatred or wrath; instead, grief, sorrow, and hope play important roles.[11] When traumatized victims are given the chance to narrate their suffering and pain publicly, they are able to regain their lost dignity; in addition, perpetrators are also heard, and through their true remorse and apology – when and if they are delivered – they are reconnected with their humanness. Justice done in this respect is not only restorative but therapeutic as well (Campbell 2004).

Whenever material and symbolic restitution is secured, reparative procedures may confer greater comfort on the victims. Nevertheless, whenever intergenerational claims are involved, restitution processes cannot ever go far enough (Cowen 2006). Restitution is not to be carried out in center-stage courts or committees, but within informal *milieus* or community-based commissions, where cleansing rituals and truth telling in front of familiar people and relatives is easier (Kaindaneh and Rigby 2012: 170–171). Truth telling does not exonerate wrongdoers of their crimes; what it ultimately does is to reaffirm the wrongness of the violation of traumatized victims' humanity (Minow 1998: 146) as they are entitled to ask for dignity and justice (Alcorn 2019). Witnessing of suffering, memorializing, restitution, remorse, and acknowledgment of the humanity and subjectivity of both victims and perpetrators are essentials of the reparative demand for justice. They do not eliminate sorrow, grief, resentment, and haunting by horrific memories, nor do they clean the slate, but they do contribute to the remedy of societal dislocation and to the reconciliation of traumatic past animosities. In addition, demand for justice may, under certain circumstances, prepare for another moral stance toward traumatic experiences, i.e., forgiveness.

Forgiveness

Forgiveness comes forth as a response to the question of whether legal resolutions are sufficient to suture individual and collective traumas. From the outset, it should be stated that forgiveness is not forgetting, nor is it equated with justice. If anything, 'the law does not always offer justice to the victim' as the court 'may refuse to hear a particular case, to accept the evidence of the victim, or to find the perpetrator guilty beyond reasonable doubt'. Apart from that, what would the meaning of restorative justice be in the face of crimes against humanity? What would the institution of law eventually mean for victims who are physically and emotionally demolished? Even if justice is done, injustice remains (Campbell 2004: 340–341). This is where forgiveness comes in and that is why demand for justice may or may not be contingent upon forgiveness, nor is forgiveness a prerequisite for the service of justice in relation to traumatic events like crimes against humanity.

In approaching forgiveness within the politics of trauma, it is necessary to keep in mind that the constituents of cultural trauma are memory, emotion, and identity, and that cultural trauma itself is substantiated through the trauma drama process. Schematically speaking, this process may conclude – if it ever does – with one of two different outcomes: either self-perpetuating victimization[12] and negativity, or healing by moving ahead toward a post-traumatic future. The pessimistic and optimistic versions of the trauma drama correlate with unfinished and successful mourning, respectively. It is through lack of successful mourning that historical/cultural trauma continues to excite people and stir up violence (Volkan 2005). Without procedures of individual and institutional reflexivity upon historical memory, vindictiveness is inevitably perpetuated and the blood of the forefathers keeps asking for vengeance, while shame is disguised as anger and hatred against the national, racial, ethnic, and religious other (Scheff 2000). In contrast, the work of mourning integrates the traumatic experience into both the life-world and the system of social reproduction, eliminating or at least reducing aversive affectivity. At the individual level and according to the psychoanalytic tradition, trauma is a dynamic process, encompassing both the rupture of the psychic tissue and its healing. At the level of collective trauma, the functional equivalent of clinical-individual healing is mourning (as opposed to melancholia) and forgiveness.

Forgiveness, now, is far from simple; it is not just reduced unforgiveness, and dealing with it involves a fascinating, complex, and contradictory thought landscape that has theological, philosophical, psychological, juridical-political, and sociological stepping-stones (Hughes 2016). Examples indicating the wide range of issues involved are the subfield 'Forgiveness Studies' and the existence of institutes like the International Forgiveness Institute,[13] study centers, and other relevant institutions and establishments (e.g., International Forgiveness Day, Worldwide Forgiveness Alliance, etc.). There are voices, however, that are reluctant to admit that forgiveness is such a hot topic and an essential factor of moral public life. For instance, Nussbaum (2016: 59) discerns a tendency to use the word forgiveness 'for whatever attitudes one thinks good in the management of anger'; further, she claims that 'forgiving' is simply 'an all-purpose term of commendation in the general neighborhood of dealing with wrongdoing'. Similarly,

Murphy (2005) casts some doubt on the 'forgiveness movement' as it seems to gloss over uncritical sentimental boosterism and robust moral thinking about human dignity. However, one should not ignore the fact that, currently, over 2,500 empirical studies on forgiveness can be identified and that, more or less, scholars have arrived at some consensus on the meaning of fundamental terms regarding the semantic field of forgiveness (Worthington and Cowden 2017).

The constituency of forgiveness

Provided that forgiving is not conflated with excusing, pardoning, or condonation[14] and keeping in mind that in the English language as well as in other languages there is no easy demarcation between forgiveness and cognates like acquittal, absolution, mercy, and forbearance – a typical case of forgiveness involves a victim, a victimizer, memory, will, and responsibility. Forgiveness is a moral act of will – literally a speech act ('I forgive you') – by which the victim relieves the victimizer of her/his responsibility. Certainly, forgiveness is not an obligation; rather, it is a mysterious and unpredictable faculty (Tavuchis 1991), a moral virtue or an imperfect duty. Imperfect duties allow for agential discretion about when and with respect to whom to discharge the duty (Hughes, 2016). In this sense, forgiveness is a purposeful redirection of thought and feeling, as if time could be turned back, and the victim grants the victimizer (and himself/herself) the prospect of a fresh start. The memory of the wrongdoing is essential, since forgiveness is never decoupled from the unbearable burden of time and the ineradicable trace of pain. If, on the contrary, simple passing of time is expected to bring remedy to traumatic experiences, the likelihood is that forgetting would turn into a precarious solution because, as Vladimir Jankélevitch warrants,

> the flame still smolders in the room where memory is and it can reawaken the fire. What tells us that rancor will not be reborn from the cold cinders of forgetting that the flame of anger will not wake up from the embers of rancor? No, nothing tells us this.
>
> (Jankélevitch 2005: 36)

The emotionality of forgiveness

As a moral virtue for coping with personal and collective wrongdoings and insults, forgiveness entails strong will and emotional reflexivity. As Larocco put it (2011), it metabolizes the shame inflicted by a wound; it is the silent assault of the self against the power of malice and injustice. In addition, forgiveness eliminates, or at least reduces, negative emotions (hostility, hatred, malevolence, envy, anger, resentment, etc.); it acquits and liberates the guilty party/victimizer, although her/his actions are never forgotten. Because forgiveness is performed through an exercise of free will, it liberates both victim and perpetrator from the vicious circle of revenge and negativity (Arendt 1958: 236–241) and in this respect it alters the meaning of the past. Furthermore, forgiveness reverses the balance of power between traumatizer and traumatized: the victimizer's absolution rests on the will of the victim.

It is because of this reversal, although it encapsulates the moral leverage of forgiveness, that Nusbaum denounces it as a private and public virtue. Instead, she opts for a public emotional climate and mental disposition free from rancor, anger, and resentment which are supposed to guide policy making and the pursuit of the common good (Nussbaum 2016: 217). This can be accomplished, she argues, through the cultivation of unconditional generosity, empathy, trust, and acting love in the spirit of the father of the Prodigal Son. After reviewing Jewish and Christian versions of forgiveness, where the former is assessed as punitive and the other as morally sadistic and shame inflicting (Nussbaum 2016: 60–74), she bases her argument on the assumption that there is an in-built transactional element in forgiveness characterized by 'score keeping' memory and an inquisitorial record of apologies given. The person to be forgiven must express regret, acknowledge the pain inflicted by her/his actions, commit to becoming a better person, and offer a narrative about how he/she came to harm someone (Griswold 2007: 149–150). In this respect, forgiveness brings anger's wish for payback back again, as it is ultimately a sort of subtle revenge. Yet, one could counter-argue that in an era of generalized cynicism and 'post-truth' this is already too much; it could also be said that Nussbaum is offering an *etic* interpretation of the forgiving experience which is at odds with an *emic* hermeneutic perspective.[15]

How is it delivered?

Forgiveness can be given, asked for, or even felt – but given to and asked for by whom, when, and under what circumstances? Can all wrongdoings and harms be forgiven? Does forgiveness, like apology, have to be communicated? A common assumption among scholars is that forgiveness must be given in a proper way under the right circumstances (Griswold 2007); otherwise, a request for forgiveness may prove to be a further insult to the victim and victimize her/him yet again (Brudholm 2008: 2). Thinking of resentment as moral anger linked to self-respect, Jeffry Murphy (1998: 17) contends that 'a too ready tendency to forgive may properly be regarded as a vice because it may be a sign that one lacks respect for oneself'. In the post-Cold-War period (and in many post-civil-war countries), forgiveness/reconciliation is considered to be of utmost importance for rebuilding community; however, it is often achieved to the detriment of other values, such as justice and dignity. Yet not every transgression can be forgiven and excused. As Tavuchis (1991: 123) says, 'there are acts in the moral spectrum that are beyond forgiveness ... [and] individual and collective actors impervious to sorrow'.

Several questions pertain to the difference between 'personal' and 'group' forgiveness – and apology for that matter: is it possible for a group to forgive another group? And, in this case, what does an apology mean? Is it possible for a person to forgive a collectivity, a group, a nation, or a state? If so, does this not presuppose – in many instances – the idea of collective responsibility, which is incompatible with the western conception of justice? Is forgiveness an individual, and not a collective, virtue? At any rate, 'the ability of assorted individuals to remember and forgive does not translate easily into the claim that forgiveness is pertinent to the collective realities of political life' (Shriver 1998: 71).

Further considerations pertain to distinction between 'private' and 'public' forgiveness and apology (Sarat and Hussain 2007). Does forgiveness necessarily take place backstage, away from the spotlight of political spectacle, or could it also be a public undertaking, a ritual-performative act? For example, do the granting of amnesty or the public apology of the guilty parties and the equally public absolution of sin qualify? As a rule, public forgiveness is contingent upon public apology, something common in contemporary international politics (Tavuchis 1991; Cunningham 2012; Brooks 1999). Not infrequently, especially in countries where there is urgent need for reconciliation and national rebuilding or for restoration of a state's international relations, forgiveness may assume the form of hasty or pseudo-forgiveness, a rhetorical soothing, a ritualistic instance of the postmodern 'politics of regret' (Olick 2007), while its normative foundations have nothing to do with the height of the spirit of forgiveness (in contradistinction to the depth of fault and wrongdoing), as conceived by Ricoeur (2004b: 457–506), who has raised these questions. In this case, pseudo-forgiveness does not lead to a modification of negative feelings, which presupposes a strong will to start anew.[16] Rather, pseudo-forgiveness involves a mixture of prudish hypocrisy and forced magnanimity, leading to a ritualistic evaporation of guilt and responsibility in the name of social and national progress. In this way emotional healing, directed by elites, becomes more about retribution and revenge, rather than emotional catharsis and – as an unintended consequence – the emotions elicited by trauma perpetuate existing antagonisms (Hutchinson and Bleiker 2008). At any rate, however, apology in politics should not be understated, as it may provide justifications for reforms and changes in international relations that are grounded in acknowledgment of historical injustices (Nobles 2008).

Public and official apologies raise a couple of additional questions: should the one who forgives be the one who was directly traumatized? Can someone forgive on behalf of another? Can one apologize for deeds for which one is not responsible? Is the initial wrongdoer the only one bound to apologize? These questions resurrect issues raised by the public gesture of Rainer Höss referred to earlier in this chapter. Forgiveness and apologies in politics are a relatively new and sensitive subject and there is no unequivocal or cut-and-dried answer to these questions since they refer to complex sociopolitical issues such as, for example: post-apartheid South Africa; post-Khmer Cambodia; the position of African-American and Japanese-American citizens in the United States; Sino-Japanese relations in light of the Nanking Massacre; German relations with Israel and Greece; United States' relations with Vietnam; Japan and the Philippines' issues regarding the World War II atrocities committed by the former's army; white Australia's relation to the Aborigines; Italian-Libyan relations after Italy's invasion during the Great War; Israel's attitude toward Switzerland for its policy toward Jewish refugees during the Nazi period. The international campaign 'War

against Impunity' and the international movement for Truth Justice and Reconciliation have been successful enough in raising awareness and mobilizing stake-
holders (e.g., International Crime Tribunals, the International Criminal Court) for
the defense of victims suffering from all kind of atrocities and for the persecution
of victimizers worldwide. Although the proliferation of Truth and Reconciliation Committees backed by the 'noble heritage' of the UN Universal Declaration
of Human Rights has affected the moral codes and the emotional regime(s) of
international community, one should bear in mind the power asymmetries and the
Realpolitik involved therein (Flam 2013).

At any rate, in social and international affairs it is not always that easy to define
exactly what is an apologizable offense. Yet the offender's apology should not
be mistaken for account, excuse, disclaimer or justification. For to apologize
means voluntary declaration by the wrongdoer that s/he has no excuse, justification, or explanation for the harm inflicted. On the contrary, accounts, excuses,
explanations or justifications entail detachment, and ask the offended to be reasonable enough toward the wrong action or inaction, and to release thereafter the
offender upon this basis. Authentic apology, however, 'requires not detachment
but acknowledgment and painful embracement of our deeds, coupled with a declaration of regret' (Tavuchis 1991: 19). If public apologies and forgiveness, for
that matter, are not contingent upon individual and collective guilt and atonement,
they end up in sheer politics (Tollefsen 2006).

Easy versus difficult forgiveness, conditioned versus absolute forgiveness

As a rule, emotionally speaking, apology is triggered by shame, guilt, sorrow,
remorse, regret, embarrassment, sadness, or even self-pity. Yet, it is possible for
these emotions not to be really felt when an apologizer asks for forgiveness, particularly with official intrastate apologies, which may be seen as cheap gestures
if not coupled with concrete reparations (Cunningham 2012: 152). There is no
ultimate way to check the apologizer's sincerity and true repentance. As Hughes
argued (2016),

> ... we all too often do *not* know much at all about wrongdoers' intentions,
> motives, desires, and thoughts to confidently pass judgment on whether we
> can reasonably forgive them, and so the connection between understanding
> wrongdoers and forgiving them in the light of that understanding remains
> contentious.

Hence, albeit at the interpersonal level at least, apology is not just a speech act
but requires oral admission and the wrongdoer's sorrow, embarrassment and
shame (Tavuchis 1991; Scheff 2000: 135) and as long as 'successful forgiveness and political apology depend on truth telling' (Griswold 2007: xxiv), there
is always a risk in forgiving upon apology. Differentiating forgiveness from

justice, Arendt (1958) and Jankélévitch (2005) endorsed a conception of forgiveness upon receipt of apology or repentance. Through repentance, the transgressor acknowledges responsibility for her/his action, and through forgiveness, both victim and victimizer are released from the past by undoing the consequences of the original trespass. In this respect, forgiveness is a special sort of remembrance and forgetting. For Jankélévitch (2005: 34) true forgiveness of the offender is an act of freedom of a *causa sui* and a person-to-person relationship where the forgiver behaves toward the perpetrator as if he or she had never committed the action, rather than merely forgetting or rationalizing it. But then again, he insists that only repentance can make the possibility of forgiveness 'sensible and right'. And not only that: prior to repentance is the desire for it. On the contrary, in the Old Testament tradition forgiveness has its limits in the sense that even if God's forgiveness is not contingent upon man's repentance, those who do not forgive others are not going to be forgiven (Reimer 2007: 271, 277).

Similarly, in Ricoeur's understanding, forgiveness is unilateral, not transactional. It is a gift out of love (Ricoeur 2004b: 460–461). Thus, it cannot be demanded or ordered; it is offered out of singular free will. In parallel, though, it requires continued condemnation of the past action, in the sense that it redeems the actor while rejecting the action. Emotionally, this means that the abused agent is able to empathize – not sympathize – with the abuser;[17] it also means that the abused empathizer is a person of insight whose understanding surpasses mere technical knowledge. For Gadamer (1979: 288), this person 'is prepared to accept the particular situation of the other person, and hence he is also most inclined to be forbearing or to forgive'. Nussbaum's argumentation on generosity comes quite close to Ricoeur's reasoning although she makes no reference to him at all, nor to Arendt or Jankélévitch for that matter. Essentializing, as it were, forgiveness as an inherently crypto-vengeance kind of transaction, she is after generosity, justice, and love as values to follow both in private and in the public sphere.

Not infrequently, however, in post-conflict societies, uncritical, unthoughtful, or easy forgiveness is offered whenever decision makers urge reconciliation and present forgiveness as a moral imperative and political duty, overlooking the victim's right of denial. Easy and uncritical forgiveness is inauthentic and at odds with progress in spirituality; it is hasty and adopts a stance of 'let bygones be bygones', equating forgiving and commanded forgetting. As Shriver (1998: 220) noted regarding his own country, commercial interests urge the U.S. government toward a policy of 'forgiving and forgetting and getting on with the business of making money' in new trade relations with former national enemies. In contrast, 'difficult forgiveness' (Ricoeur 2004b) is directly inscribed on a core of moral, social, aesthetic, and emotional reflexivity. It cannot be predicted or demanded. Difficult forgiveness is the result of recollection, of an effort to recall and reconstruct the traumatic past, to re-inscribe it in a personal history, and to reposition it on the emotional plane, instead of reliving it through repetition compulsion. Psychoanalytically speaking, repetition compulsion (the eternal return of the evil) represents a kind of dramatization and is analogous to reminiscence, while recollection refers to the subject's recognition of his/her history and realization of its relation to the future (Evans 1996: 162). In historiographic terms, difficult forgiveness leads to a novel rewriting of history (Nobles 2008).

The distinction between reminiscence and recollection could be said to correspond to Todorov's distinction between 'literal' and 'exemplary' memory. In his analysis, both sorts of memory are selective, with the difference being that the first type subordinates the present to the past, while the second allows the past to be renegotiated in the present (Todorov 1995: 31–32). Both represent ways of using and recalling the traumatic past. In the first case, a painful event is kept in memory intransitively and its recall becomes a central reference point for the individual's or the group's total existence. Personal and social identities are causally connected to the traumatic primal event, which represents a privileged – if not the main – source of meaning (for example, the 'Holocaust' defines Jewish memory and identity, etc.). Under these terms one can speak pessimistically of 'fixed' cultural trauma, i.e., a trauma drama marked by unstoppable grief and self-perpetuating anger (Arcel 2014: 31, 97). In the second case, the painful-traumatic event is retrieved into memory and experienced in its correct proportions, becoming a transitory event that allows us to feel the pain of the generalized other through a combination of reflective internal dialogue and empathy with the others, either intimate or distant. Our own or a former generation's experience can then become universalized and function as an example. Alexander (2004b) moved in much the same spirit when he treated the Holocaust as a universal and universalizing metaphor for evil. Literal memory is melancholic, sad, and resentful. Exemplary memory, on the other hand, leads to successful mourning and allows an opportunity to utilize the lessons of suffered injustice to fight current injustices, to abandon the narcissistically victimized self and reach toward the other. The United Nations' cosmopolitan programs like 'Remembrance and Education'[18] are structured accordingly, and are aimed at motivating people not to forget the potential for mass violence, atrocity, and genocide. They also aim at forging 'imagined communities of cosmopolitan belonging' grounded on 'practical enlightenment' and a 'notion of humanity as a singular, trans-historical collectivity plagued by an unfortunate appetite for barbarity' (Skillington 2015: 179, 182, 194). To this end, it is necessary for individuals and collectives to get involved in a process of *Aufarbeitung* (Adorno) and *Vergangenheitsbewältigung* (Habermas), i.e., to come to terms with and to master the traumatic past in sublimating ways through 'dialogic remembering' (Assmann 2009).

Effectively, 'difficult forgiveness' follows the path of exemplary memory, reconstructing the past in an emancipating way, with emancipation in this instance denoting the loosening of the grip of myth, the abandonment of the self-perpetuating narcissism of victimization, and the promotion of critical self-awareness on a personal and collective level. Ricoeur's 'difficult forgiveness' echoes Hegel's (admittedly obscure) conceptualization of forgiveness as the reunification of a self who hitherto has been acting according to its own inner self-referential law and conscience without any mediation with the other. In this respect, evil is understood as the consciousness' lack of correspondence with the universal, wherein someone refuses to 'throw himself away for someone else'. Forgiveness is posited as 'the reconciling affirmation, the "yes", with which both egos desist from their existence in opposition,' and thus 'the existence of the ego [is] expanded into a duality' (Hegel 1967: 668, 670, 674–679).

Yet, Ricoeur's approach is more akin to Derrida's 'unconditional forgiveness' than to Hegel's account. Much philosophical and theological literature on forgiveness focuses on its conditionality or absoluteness. While recognizing the importance of apology, reparation, and reconciliation in post-conflict societies (i.e., Nussbaum's transactional forgiveness), Derrida (2003) favored an 'absolute' and 'exceptional' conception: forgiveness should move beyond economic transaction and not be conditional on expectations, restitutions, demands, or declarations of repentance.[19] He advocated an 'unconditional, gracious, infinite, an economic forgiveness granted *to the guilty as guilty*, without counterpart, even to those who do not repent or ask forgiveness' (Derrida 2003: 34).[20] In his words, 'the most problematic ... is that *forgiveness must have a meaning*. And this meaning must determine itself on the ground of salvation, of reconciliation, redemption, atonement, I would say even sacrifice' (2003: 36). Therefore, forgiveness has nothing to do with the imperative of social and political health, nor does forgiveness amount to a therapy of reconciliation (Derrida 2003: 41). True forgiveness, according to Derrida, forgives the unforgivable.[21] Only then can it be deemed as springing from authentic free will, as an exceptional event or, in theological terms, as a miracle. On the one hand, Derrida's concept, stemming partly from the Abrahamic tradition[22] on forgiveness, remains at the personal level – absolute and unconditional forgiveness can only be an interpersonal moral gesture: I forgive someone, not something, and I do it in a radical way, because I grant forgiveness while wanting nothing in return. On the other hand, Derrida had to have realized the importance of cognate stances like mercy, amnesty, forbearance, pardon, and condonation for the regulation of collective life and institutions. He was torn between his radical idea of pure forgiveness and the pragmatic processes of reconciliation (Derrida 2003: 51).

A good case of unconditional forgiveness is described by Libby Tata Arcel, a Copenhagen University professor of psychology who was Director of the Rehabilitation Center for Torture Victims at the Institute of Clinical Psychology. Arcel was of Greek origin and her last book (2014), written in Greek, is about the intergenerational trauma of the so-called Asia Minor Disaster, exemplified among her own family members when she systematically interviewed them over a considerable period of time. The Asia Minor Disaster or Catastrophe (Μικρασιατική Καταστροφή) is the name Greeks use for the traumatic defeat of the Greek Army in the 1919–1922 Greco-Turkish War referred to already in Chapter 3. They fought that war against the Turkish National Movement, grown out of the partitioning of the Ottoman Empire after World War I. For the Turks themselves that war is known as the Western Front (*Kurtuluş Savaşı, Batı Cephesi*) of the Turkish War of Independence. By the Treaty of Sevres the area around Smyrna was awarded to Greece as an occupation zone for five years. Greek troops landed in Smyrna harbor in May 1919, and were greeted with enthusiasm by the Greek population which had been rooted there ever since ancient times. The occupation zone soon spread beyond what the Sevres Treaty envisioned and this initiated Turkish resistance. Initially, the Greek army had success in their efforts to extend their hold over western and northwestern Asia Minor. However, their advance was checked in 1921 by Turkish forces; in August 1922 the front collapsed and the war effectively ended with the recapture of Smyrna by Turkish troops and irregulars, resulting in the destruction of the city and the slaughter of many ethnic Greeks therein. That defeat put an end to Greek irredentism and led to the expulsion of the Greek presence from Asia Minor as 1,500,000 refugees fled to Greece (and almost 700,000 moved to Turkey). Among them was Arcel's mother, then a young girl, who managed to escape the disaster together with her mother, Arcel's grandmother, and her four siblings. In the meantime, her father, Arcel's grandfather, was burnt alive by Turkish irregulars in his own village church with a number of Greek fellows. The image of the burning husband and father were to haunt the traumatic refugee identity of Arcel's antecedents for the years to come. Nevertheless, the family managed to acclimatise and make a living. Arcel's mother made a decent marriage and gave birth to four children who thereafter made their own families.

As a psychologist, Arcel decided to scrutinize the traumatic memories of the refugee experience down to the third-generation offspring. The entire process of interviewing her relatives took a while as she had to do it mostly while in Greece during the summer breaks. Although she hadn't previously expressed any such wish, a couple of years before she passed away Arcel's mother asked her daughters to make a trip back to the village of her childhood in Turkey. They actually made that trip and the old woman visited her old house, now inhabited by a second-generation Turkish family who welcomed her warmly and were friendly. Several months after the event Arcel heard her mother sitting in the yard of her house quietly uttering three words: 'I forgive them'. Arcel, surprised, asked her: 'mother, whom do you forgive?' 'The Turks', the mother said, 'the Turks who burnt my beloved father alive' (Arcel 2014: 14). She did that for all the traumatic memory, without any apology, contrition, or reparation. She couldn't live with resentment and hatred any more.

The notion of unconditional forgiveness, though, is not appealing to everyone. Echoing the Levinasian imperative to be for the other rather than being with the other, instead of unconditional forgiveness Nussbaum advocates what she calls 'unconditional love'. Of course, unconditional forgiveness has the lead over transactional forgiveness, 'but it is not free of moral danger' which consists in the wronged party having 'angry feelings first, and then choos[ing] to waive them', sticking thus to 'some type of payback wish'. Second, 'unconditional forgiveness remains backward-looking ... it says nothing about constructing a productive future'. Unconditional love is the one that bypasses anger altogether in a willful and free spirit because, among others, deeds may be denounced but people always deserve respect and sympathy (Nussbaum 2016: 76–77, 81, 86, 222). As appealing as her thesis might be, it seems to me that one should be mindful of three caveats about Nussbaum's analysis. First, it looks as if she is disjointing memory from unconditional forgiveness and unconditional love by virtue of the withering away of anger on the one hand and for the sake of an open-ended future on the other. Yet, as we said, forgiving is not coterminous with forgetting and, what is more, active trust and acting love should not and cannot be blind to past harms and transgressions. They exist and are to be valued despite bad and sad memories and that is exactly why forgiving is such a difficult and rare moral gesture to pursue. We cannot demand *in abstracto* that people who have suffered from the villainy of others disavow or sublimate their anger and bitterness; they are simply not saints. Arguably, though, we may expect that they work through traumatic experiences and find their own way to live in dignity and self-assurance. Second, although ideally it would be superior, unconditional generosity might be too much of a virtue for the right functioning of the political realm. Justice, fairness, empathy, and sympathy are just good enough. Even if victims remain resentful, institutional settings of retributive and reparative justice are adequate enough for public peace and social cooperation. Privately, people may harbor anger and spitefulness and may not want to forgive, let alone to love the perpetrators. Nobody can accuse them if that helps them maintain a coherent sense of themselves. This seems not to be an option for Nussbaum's moral rationality and Stoic moral maximalism.

Non-forgiveness

Derrida's unconditional forgiveness and Nussbaum's unconditional generosity follow a hyberbolic meta-ethical vision, similar to Ricoeur's vision of the heights of the spirit of forgiving as a flower of love. Yet Ricoeur admitted that it is within the discretion of a traumatized person not to forgive (Ricoeur 2004b: 483). Forgiveness must be given in a proper way and under the right circumstances, because there are unforgivable crimes. According to Jankélévitch's eloquent 1965 article in *Le Monde*, 'Pardoning died in the death camps'. In a much later piece of work he pointed out that 'it is pointless to search to reconcile the irrationality of evil with the omnipotence of love' and that since forgiveness is stronger than evil and evil is stronger than forgiveness there will be an infinite oscillation so than one may have good reasons not to forgive (Jankélévitch 1996). The ethical right not to forgive is a commonplace in the relevant literature, but it makes a difference whether non-forgiveness stems from a reflective recollection and interpretation of the traumatic past or from the victims' inability to overcome the alluring position of victimization.[23] The former stance defends dignity and self-respect, while the latter perpetuates powerlessness, self-pity and moral hypermnesia premised on the victim's resistance to dealing with the past in a healing way.

Most likely, non-forgiveness is well conceptualized as a case of *ressentiment*. Serving herein as a bridge, this emotional term will be scrutinized in the next chapter. For the time being it suffices to say that *ressentiment* is defined as an unpleasant moral sentiment that includes a chronic reliving of repressed and endless vindictiveness, hostility, envy, and indignation due to the powerlessness of the subject to express them, and resulting, at the level of values, in the negation of what is unconsciously desired. In these terms, I argue that the *ressentiment*-ful person cannot forgive, forgiveness is beyond his/her moral horizon; what actually counts is the sacralization of victimhood and traumatic restraint. Jankélévitch is thoroughly succinct on this point: 'Does the offended person who does not fully sacrifice his rancor ... truly forgive? For him the regimen of very vague and very remote *ressentiment* has supplanted aggressive rancor just as rancor had supplanted angry belligerence' (Jankélévitch 2005: 33).

Confronted with societal and political amnesia, the *ressentiment*-ful person exhibits a strong will to remember (Brudholm 2008: 110). Yet, the crucial point is that the kind of memory sustained by *ressentiment* is detrimental to societal reflexivity and working through of mourning. Of course, not every case of non-forgiveness is contingent upon *ressentiment*. Although Jean Améry, an intellectual who survived Auschwitz but committed suicide 40 years later, described his negative feelings for Nazi and post-war arrogant Germans who were conquering the world markets as *ressentiment*, he was careful to affirm that his special version of *ressentiment* was different from that of Nietzsche and Scheler[24] (Améry 1999: 66, 71). His internal struggle to deny forgiveness and redemption to the Nazis was not premised on a petty, resentful, and, in the last instance, narcissistic attitude of endless victimization, but on the moral duty of stating that the unforgivable ought not to be forgiven, that shameful amnesia and moral amnesty are never to be allowed. This did not lead him to a vicious rumination on evil or to small disguised acts of revenge and hostility. Struggling to preserve his own dignity and humanness, Améry was expressing a radical and deconstructive bitterness against the traumatic Nazi past and the postwar restitution.

By the same token, Alfred Polgar, a well-known Austria-born public intellectual, who had been raised in an assimilated Jewish family, fled to Prague when the Nazis came to power in 1933, and could not forgive his fellow patriots for their enthusiastic reception of Hitler in Vienna in 1938. In 1940 he moved to the USA. After the war, in 1949, Polgar returned, at the age of 76, to Europe and settled down in Zurich since he refused to go back to Vienna. He died there at the age of 82 on 24 April 1955 in a hotel room.[25]

Recently, several voices have been raised against the imperative to forgive arguing that refusal to forgive is a sort of moral dignity, a defense of the victim's integral subjectivity, and a moral protest against the unjustifiable evils and wrongdoings the victim has suffered.[26] Among the emotions the victim is left with after the traumatic experience and after the reluctance to forgive the perpetrators and get on with life are, of course, anger, hate, moral indignation, depression, humiliation, and shame. An additional and far more complex emotion that characterizes the post-traumatic experience is *ressentiment*, to be dealt with in the next chapter.

Moral acumen in forgiveness

Trauma, and cultural trauma for that matter, is a metaphor that we live by within our contemporary era. Its metaphorical meaning is constantly under negotiation, giving rise to different identity formation processes and concomitant moral-emotional reactions. The negotiation is part of the trauma drama wherein a number of different issue carriers and political power agents are involved. Because the trauma drama consists in memory claims and public representations of the wounds inflicted and about who the victims and the transgressors were, in many cases it amounts to an identity politics. At the same time, various therapeutic discourses on trauma in western countries claim that a successful working through of a traumatic experience amounts to forgiveness and the victim's reconciliation with the past. In this context, some stories are told and retold, while others stay untold and undisclosed. Whenever forgiveness and apology exceed the dyadic relationships within the trauma drama, they acquire sound political connotations because they become part of the public sphere and political agonism.

As arenas of the trauma drama, emotional public spheres (Richards 2010; Rosas and Serrano-Puche 2018) are, among others, intersected by push-and-pull factors for the initiation of remedial gestures and strategies as well as for the nurturing of narratives that keep the wounds open through endless melancholic repetition compulsions (Bellamy 1997: 8–9). These top-down, as well as bottom-up, discursive-narrative strategies perpetuate victimization and vindictiveness at personal, collective, political, and international levels and are symptomatic of some societies' inability to mourn. Trauma dramas everywhere evolve in different, unpredicted, and unpredictable ways: in all public negotiation of the meaning of the traumatic past, there are still silences, the 'violence of voicelessness' still haunts many victims (Anthonissen 2009: 100) through the intergenerational passing of 'chosen traumas' of benign victims. Restitution, reconciliation, and difficult forgiveness are alternative options for societal (re)organization, especially in post-conflict countries.

Apart from being a locus of debates, the trauma drama is a public generator of political emotions. As we have seen in Chapter 1, political emotions are deemed to be lasting affective predispositions, supported reciprocally by the political and social norms of a given society, playing a key role in the constitution of its political culture and the authoritative allocation of resources. In addition to political emotions, the trauma drama and the politics of identity generate vices and virtues at both the personal and the collective level. Hate speech and vengeance coexist with the demand for justice and reconciliation strategies. A crucial point regarding identity politics is whether its urge for recognition, apology, and forgiveness is rooted in positive and self-grounded praxis or in a hasty reflex reaction and disguised servility caused by a generalized powerlessness.

Arguably, as a secular virtue, forgiveness is a candidate for curing the 'spirit of revenge' (Griswold 2007: xx) and overcoming grunginess, self-pity, anger, and hate. Yet, forgiveness must be enacted in a principled manner related to whether, when, and how one is expected to forgive. For one thing, one should not forgive when punishment or even revenge is impossible; forgiveness out of powerlessness is the result of an effaced will. Furthermore, one should not hasten to forgive too soon or delay until too late. Being a virtue and in tandem with equity, forgiveness is allowed in its correct proportions, as delineated by Aristotle: in harmony with reason and in search of the mean state (*mesotes*), the 'liberal man', and the man of forgiveness for that matter, desires what he ought in the right manner and at right times. This is precisely the meaning of forgiveness given by Jankélévitch (2005) as *Acumen Veniae*. Choosing the auspicious and opportune time for an act of apology and an act of forgiveness at the interpersonal level is contingent upon the subject's skillfulness in the art of living, i.e., upon her/his emotional reflexivity and moral standards vis-à-vis the severity of the offense. As a virtuous responsible act, a gratuitous rare gift inspired by a cleansing spirit, the appropriate giving of forgiveness means that the victim is entitled and expected to give the right thing to the right persons at the right times. In this sense 'forgiveness would be miraculous specifically because God does not come to its aid' (Bloechl 2013: 108). To stress the point further, Tavuchis (1991: 88–89) rightly refers to the ancient Greek notion of καιρός, a time when conditions are right and propitious for the accomplishment of an important undertaking. Aristotle paid attention to forgiveness and pity in these terms (Sadler 2008; Sokolon 2006). Be it noted, however, that although they invented processes and schemes of reconciliation, the ancients (Greeks and Romans) did not hold an idea of forgiveness similar to the modern one influenced by Christianity (Konstan 2011). Their notion of forgiveness was more of a cognitive rather than moral or emotional nature and more cogent with the idea of asking for pardon rather than absolving or exonerating someone of his/her injustices (Konstan 2010: ix, 23, 57–58, 152). This is designated by the Greek word *syngnômê*, which derives from συγγιγνώσκω, which means to be cognizant (*cognoscere* in Latin). It is not accidental that in the *Nichomachean Ethics* Aristotle postulates that forgiveness is best applied to wrongdoers not aware of the actual circumstances under which they sinned (1109b30–111a2). In our own times, though, to the extent that imposed silence and commanded forgetting are not viable options, it is similarly unjustifiable to claim that living in an era of trauma and apology equals an easy, hasty, and unwarranted forgiving. This is expressed in many different languages by the oft-repeated plea, 'I want to forgive, but I must know whom to forgive and for what' (Boraine 2006: 309).

Another pressing factor pointing to the necessary differentiation between easy and difficult forgiveness is the widespread cynicism and moral relativism in most hypermodern western societies, referred to in Chapter 1. Many people do not believe in apology and the value of forgiveness (Shriver 1998: 219). In a society where 'anything goes', the existence of rational or even pragmatic criteria regulating public life is denied beforehand and, consequently, the very idea of the public good and good life is at stake. By and large, this tendency is pushed and pulled by the post-emotionality of contemporary western society (Meštrović 1997) referred to in the previous chapter. Social isolation, power-lessness, visual culture, and consumerist individualism drive many people to experience inauthentic emotions in the sense that they do not commit themselves to what they really feel. Oftentimes this ends in a kind of blasé affective expe-rience and a cynical attitude. As we mentioned in Chapter 1, swamped in bad faith and disdain, many cynics find every political argument pointless and they lack reflexivity and exhibit an almost paranoid distrust of political personnel and decision-making processes. Ultimately, they are indifferent to either good or evil (Lipowatz 2014: 309). Under these terms, there is no place in their hermeneutic horizon for the virtue of forgiveness and they do not believe in sin anyway.

Last but not least, one should take into account that the modern Greek desig-nation for 'forgiveness' is *sunchorese* (συγχώρεση), which comes directly from the ancient Greek word *sunchorein* (συγχωρείν), which means 'to be together with others at the same place'. This cannot be accomplished without some sort of societal reflexivity.

Synopsis

This chapter discussed forgiveness as an alternative moral-emotional response to trauma next to retaliatory revenge, justice, and non-forgiveness. Amidst a wide-spread emotional climate of cynicism on the one hand and a permeating therapeu-tic culture on the other, forgiveness stands as a moral virtue to be differentiated from ostensible apologies and pardons premised on political imperatives in post-conflict societies as well as from forgetting and social amnesia. With respect to trauma drama, the affectivity of forgiveness is related to mourning, emotional reflexivity and free will. Drawing, among others, from Jankélévitch, Arendt, Ricœur, Derrida, and Nussbaum, the chapter keeps a distance from the Abrahamic tradition, assessing forgiveness as a secular moral virtue, as an 'imperfect duty', to be performed toward the right persons, in the right manner and at right times.

Notes

1 www.rimini-protokoll.de/website/en/.
2 www.sgt.gr/eng/SPG1385/?.
3 https://fr.wikipedia.org/wiki/Rainer_H%C3%B6ss.
4 www.sgt.gr/eng/SPG1546/?.
5 www.tovima.gr/opinions/article/?aid=796163.
6 www.sanmetodio.eu/index.php.
7 These responses may be articulated in a secular, religious, or semi-religious spirit; argu-ably, they are not the only ones. There are other moral reactions too, that are worthy of in-depth commentary as they assume some form of theodicity. Among these are: (a) faith in a God who created and gave away the world out of love while at the same time letting humanity be really free to choose between good and evil; in this case, the

subject does not resign or withdraw but drastically makes decisions guided by memory, reason, and will; (b) active hope in the sense of the religious eschatology of the Final Judgment and the concomitant endurance of suffering and evil in the world, irrespective of evil harbored or evil caused by the subject (Lipowatz 2014: 426–433; Ricoeur 2004a).

8 https://quizlet.com/137807274/aristotle-rhetoric-the-emotions-flash-cards/.

9 Nussbaum's moral rationality is contiguous with Habermas' communicative rationality and what some sociologists of emotions call 'emotional reflexivity', e.g., Holmes (2010) and Burkitt (2012).

10 It should be noted however that, in spite of its immediacy, *jus talionis* curtails the possible arbitrariness of wild spontaneous revenge, on the one hand, while legally equalizing the powerful and the powerless, on the other; in this sense, it contributes to societal stabilization.

11 Conflating revenge and retribution somewhat, Elster (2006) spoke of five 'retributive emotions': anger, hatred, indignation, and contempt directed at the perpetrators, as well as pity for the victims.

12 Traumatogenic victimization as a derivative personality trait grounded on narcissistic wounds is somewhat relevant to the victimization Pascal Bruckner speaks about as a strategy to get away with duties and responsibilities (Bruckner 2000).

13 www.forgiveness-institute.org.

14 Wrongs fully excused are not blameworthy since there is nothing to forgive. As a lawful lawlessness derived from the monarchical tradition, official pardon or clemency is exercised by third parties as opposed to the victims; legal or political pardons reduce or even eliminate punishment, whereas forgiveness need not affect punishment whatsoever (Sarat and Hussain 2007: 6–7). In condonation one overlooks a wrong as if it did not exist or did not actually occur. Thus, condonation is a form of tolerating wrongdoing (Hughes 2016).

15 In cultural anthropology the emic perspective is seeing things from the actor's point of view whereas the etic perspective is the stance of the researcher trying to explain social realities using the theoretical concepts of social sciences (Geertz 1973: 14–15).

16 On this basis there is an ideal typical distinction in the relevant literature between intrapersonal decisional and emotional forgiveness. In emotional forgiveness there is a replacement of negative unforgiving emotions such as hatred and anger by positive ones like empathy, acceptance, etc. (Wade *et al.* 2005; Worthington and Cowden 2017).

17 Empathy is an emotional response arising from the apprehension of another's emotional state and it is very similar, if not identical, to what the other person is feeling (Eisenberg 2004: 677); therefore, it is someone's ability to imagine how the other feels in a certain situation (Ben-Ze'ev 2000: 108, 110). In contrast, in sympathy, the subject does not feel the same or almost the same feeling as the other person, but sorrow or concern for the other's misfortune (Eisenberg 2004: 678). Thus, sympathy is the counterpart in one person of another's sense of loss, sorrow, discomfort, abuse, and the like (Schmitt and Clark 2006: 469). See also Scheler (1954) and Smith (1976).

18 www.un.org/en/holocaustremembrance/2016/calendar2016.html.

19 Stepping on the same ground, in his *Lost Illusions* Honoré de Balzac thinks of repentance as a sort of indemnity for wrongdoing and as a means for absolution when gestured more than once. 'Repentance is virginity of the soul, which we must keep for God'; yet, periodical repentance is a great hypocrisy and 'a man who repents twice is a horrible sycophant' (de Balzac 2004: 496).

20 Counterpointing conditionality and unconditionality is a familiar philosophical stratagem found in Derrida's work. As with forgiveness, he discusses hospitality in exactly the same way: hospitality is certainly a moral gesture but it gains its ultimate value

when it is offered beyond any legal prerequisites as an unconditional welcome to the foreigner (Dufourmantelle and Derrida 2000).

21 Yet, for all his deconstructive demeanor and although he points out that only the victims have the right to forgive or not to forgive, he seems to endorse the idea that the Shoah is an instance of transgression that cannot be pardoned (Shoah Resource Center, 1998).

22 The Abrahamic tradition on the issue of forgiveness brings together Judaism, Christianity, and Islam.

23 It makes also a difference whether the issue of non-forgiveness is raised in low context individualist or high context collectivist cultural contexts (Ting-Toomey 1994; Worthington and Cowden 2017). In collectivist cultures it is reconciliation based on interpersonal and institutional trust which counts more than forgiveness. The latter is regarded as an atomistic gesture.

24 Namely, as the morality of the inferior. Chapter 6 details *ressentiment* in relation to resentment.

25 http://depts.washington.edu/vienna/literature/polgar/Biography.htm.

26 Thinking of resentment as moral anger linked to self-respect, Jeffry Murphy (1998: 17) contends that 'a too ready tendency to forgive may properly be regarded as a vice because it may be a sign that one lacks respect for oneself'.

Part II

The politics of
ressentiment

Chapter 6

On resentment, *ressentiment*, and political action

Introducing the twin emotional terms

At the conclusion of Chapter 5 it was mentioned that, among others, two possible emotional reactions to trauma might be moral indignation, coined mainly as resentment, and *ressentiment*. These are two powerful emotions which underscore political behavior and political predispositions in all countries in different periods of history. As will be argued later on, resentment is a less complex emotion than *ressentiment* in terms of its study and manifestation; yet either political emotion is of importance for the understanding of democratic and non-democratic, and anti-democratic politics of today, if for no other reason than because they may interact and not just be experienced independently of each other, or be elicited one after the other. For all their importance, there is dispute over their conceptualization, something that is however commonplace in the social sciences and humanities.

The conceptual relation between resentment and *ressentiment* has been theorized in more than one way. They may stand apart alongside the difference between the Nietzschean and non-Nietzschean lines of thought and moral stance (Demertzis 2006, 2017; Fassin 2013); as an encompassing negative emotion, *ressentiment* is said to prevail over resentment which is deemed a positive emotion conducive to civic engagement (Brighi 2016: 415; Ure 2015); resentment is considered as a generic moral emotion whose latest version, ever since the nineteenth century, is *ressentiment*, to be elicited in competitive societies (Moruno 2013); *ressentiment* is seen as a particular form of resentment fueling reactionary populism and 'anti-welfare' discourses (Hoggett, Wilkinson, Beedell 2013). Not infrequently, the two concepts are used interchangeably, especially when it comes to translations. For example, the translator of Marc Ferro's *Le ressentiment dans l'Histoire* translated *ressentiment* into resentment (Ferro, 2010). The same holds true in the translation of Tzvetan Todorov's *La peur des barbares. Au-delà du choc des civilizations*, where his 'countries of *ressentiment*' (primarily set in contrast with what he calls 'countries of fear') were converted into 'countries of resentment' (Todorov, 2010). What comes as a sort of surprise is that when discussing the very notion of Scheler's *ressentiment* the translator of Helmut Schoeck's book on envy (1969) deciphers it plainly and simply as 'resentment'. The two notions are also used alternately by scholars delving sociologically into the moral emotions discussion; for example, Stefano Tomelleri (2013) applies straightforward 'resentment' equally to all clear cases where the Nietzschean and Schelerian '*ressentiment*' is under consideration. Surprisingly enough, some moral philosophers mingle the two concepts as well (e.g., Annette Baier 1980) or deem the difference irrelevant (van Tuinen 2020). The same happens also with social philosophers who decode *ressentiment*' in terms of resentment even if they are aware of the relevant literature (e.g., Fritsche 1999: 109–114).

Despite their kinship, however, *ressentiment* and resentment are not identical. The former was introduced by Nietzsche's philosophy in 1887 (*Genealogy of Morals*) and since then has found its way into the works of many other social philosophers, sociologists and psychologists. No wonder then that there is no general consensus as to its meaning. The latter was elaborated much earlier, mainly by the Scottish Enlightenment philosophers Adam Smith and David Hume, and it has been sufficiently analyzed by moral philosophers and sociologists alike (Barbalet 1998: 63). In the remainder of this chapter I will attempt to follow and critically assess the main tenets of the theoretical discussion on the one hand, while trying to show how resentment and *ressentiment* are intertwined with the emotions-politics nexus on the other.

Of resentment

From a general point of view, resentment is understood as the emotion that arises when wrongs are done to us or to those we identify with. Seen somewhat differently, it is proposed that resentment is an emotional response to an intentional, unjust and harmful offence inflicted on a victim. As mentioned in the previous chapter, resentment is a kind of anger motivated by injustice and, in that sense, it can be assessed as a retributive, punitive, or moral emotion.

This is actually the way Hume wrote of resentment: 'when I receive any injury from another, I often feel a violent passion of resentment, which makes me desire his evil and punishment, independent of all considerations of pleasure and advantage to myself' (Hume 1739/1969: 465). First, it is to be noted in passing that, *more often than not*, 'passion' in Hume's writings is used alternately with 'emotion', with the latter sometimes denoting a less violent affective experience than the former. Second, it seems that, for Hume, resentment is a direct passion. In his scheme, a direct passion (a primary emotion in modern psychological terminology) arises immediately from good or evil, from pain or pleasure, whereas indirect passions (secondary emotions) proceed from the same principles by the conjunction of other qualities (Hume 1739/1969: 328). Thus, it is moral for a man who is robbed of a considerable sum to feel moral resentment against the crime because the distinction between vice and virtue is founded in the natural sentiments of the human mind (Hume 1748/2018: 80). To put it another way, resentment – like benevolence, the love of life, and kindness to children) – is an 'instinct' originally implanted in our nature (Hume 1739/1969: 464).[1] Third, as a violent emotion, resentment seems to have an inherent self-destructive dimension, since someone in the grip of resentment wishes to pay back the injury regardless of all considerations of pleasure and advantage to themself. In this respect, in Hume's account resentment is a violent and short-lived emotion like ordinary anger, cognate with what Butler understood as resentment: an emotion of mind against private or personal injury and injustice which may 'be called anger, indignation, resentment, or by whatever name anyone shall choose'.[2]

Greatly influenced by Hume, and Butler for that matter, Adam Smith offers a more nuanced account of resentment by irrevocably rendering it a moral emotion. Like its opposite, gratitude, resentment is premised on Smith's theory of sympathy and social approbation. For one thing, sympathy as a 'communication of passions' and as the ground wherefrom resentment, along with esteem, love, mirth, and melancholy, spring was first commented upon by Hume among the Scottish Enlightenment philosophers (Hume 1739/1969: 367, 445). Yet, it was Smith who actually delved into this faculty of human imagination influencing thereafter any other discussion on the subject.

Smith considers sympathy as a fellow-feeling with any emotion whatsoever, as the innate ability of everyone to identify with someone else's plight and affective situation. As long as humans are able to sympathize, they can make moral judgments; i.e., the notions of right and wrong, vice and virtue, are decided upon people's placing themselves in their fellows' situations by imagining what they would feel if they were to undergo the same detrimental or beneficial experience (Smith 1759/1976: 9–10, 22). This seems to be a 'natural sentiment' or an inherent faculty of human sociality which prompts persons to pursue social esteem. Actually, sympathy for Smith is the mechanism to 'acquire social approval, moral approbation and civic praise' which is ultimately the motivation for human interaction (Kalyvas and Katznelson 2008: 26–28). Psychoanalytically speaking, sympathy supports both the imaginary-narcissistic register of the ideal Ego and the symbolic Ego ideal, i.e., the fundamentals of self and personality. The ideal Ego function of sympathy in Smith's moral philosophy is carried out through the social recognition and applause conveyed by pride, self-estimation, selfishness, and self-love. On the other side, the Ego ideal function of sympathy is conveyed by propriety, self-commanded virtue, moderation (an Aristotelian notion), and mediocrity – all steered by the moral standpoint of the 'impartial spectator', a notion handed over from Hutcheson and Hume and elaborated in full by Smith. I would claim that the impartial spectator is not the functional equivalent of the ethical Freudian Superego but of the Lacanian 'name of the (symbolic) father', i.e., the disinterested moral law which consists in the subject's capacity to judge his/her own actions. This self-judgment occurs

> when he views himself in the light in which he is conscious that others will view him … he would act so as that the impartial spectator may enter into the

> principles of his conduct ... [whereas] he must humble the arrogance of his
> self-love, and bring it down to something which other men can go along with.
>
> (Smith 1759/1976: 83)

For all the importance of others' recognition and applause in everyday life sympathies, for Smith the ultimate source of the prudent person's self-esteem is the support and the reward offered 'by the entire approbation of the impartial spectator' (Smith 1759/1976: 215). Thus, the moral yardstick of any 'passion' and conduct is whether the impartial spectator can sympathize with it or not (Smith 1759/1976: 69).[3] It is in these terms that Smith ponders on resentment; by commenting on the qualities of conduct deserving punishment, he holds that the 'sentiment which most immediately and directly prompts us to punish, is resentment' provided that somebody has done us a great injury. To punish means to return evil for evil so that the offender is made 'to grieve for that particular wrong which we have suffered from him' (Smith 1759/1976: 68–69). Nevertheless, in its spontaneous 'undisciplined' immediacy, resentment is denounced as an 'unsocial passion' (Smith 1759/1976: 34); but as long as magnanimity and personal dignity motivate and steer the expression of this anger-like and revenge-seeking passion the likelihood is that it will find its way to become a moral sentiment. Which means that the impartial spectator will sympathize with resentment when it 'is guarded and qualified' in the sense that 'we should resent more from a sense of propriety of resentment, from a sense that mankind expects and requires it of us, than because we feel in ourselves the furies of that disagreeable passion' (Smith 1759/1976: 38). Then, it is only the violation of the most sacred laws of justice which qualifies resentment as a moral sentiment precisely because it raises the 'sympathetic indignation of the spectator as well as the sense of guilt in the agent' (Smith 1759/1976: 84).

To carry forward the issue to our times, the examination of resentment by the British philosopher of language Peter Frederick Strawson contributed to the broadening of the moral understanding of human sociability as long as, in their interactions, people make relationships invested with feeling, so that someone else's opinion about, and behavior towards an individual matters to the latter – the heart of social recognition. The importance of others for the construction of the self is expressed in emotionally laden 'reactive attitudes'. For Strawson, resentment is the negative reactive attitude that a person develops in the face of another person's intentional indifference, insult and injury towards him (Strawson 1974: 7, 14). By way of example: if someone accidentally steps on my hand as he helps me do something, the pain may be no less than if he did it on purpose in a gesture of contempt towards me. But while in the latter case I would feel deep resentment, in the former I might as well feel gratitude in the light of his good intent. As a negative reactive attitude, resentment implies a disapproval of the injurer who is considered responsible for his actions with good reason. Strawson thinks that if a small child, a mentally deficient person, a drug-addicted criminal or a sick man, causes us some sort of injury, small or big – it makes no difference – we cannot feel resentment towards them. This emotion presupposes moral responsibility of the wrongdoer, who is in principle a free, accountable and responsible agent susceptible to reparative gestures like apology, repentance, and forgiveness.

To stress the point further, for Strawson (1974: 14), resentment's moral nature rests not only on 'an expectation of, and demand for, the manifestation of a certain degree of goodwill or regard on the part of other human beings toward ourselves', but also on the moral bond between the resentable and accountable person and ourselves who are equally accountable. The resentable person is someone from whom we expect or demand a certain degree of goodwill, and to whom we ourselves owe a certain degree of goodwill (Ure 2015).[4] Although he admits a bit of confusion in his own terminology,[5] whenever a moral reactive attitude is provoked in virtue of another's injury or indifference by a wrongdoer, the likelihood is that a vicarious resentment will occur. This is the case of indignation which calls forth moral condemnation of the offender; and 'indignation ... like resentment, tend[s] to inhibit or at least to limit our goodwill towards the object of these attitudes, tend[s] to promote an a least partial and temporary withdrawal of goodwill' (Strawson 1974: 21). What ultimately counts for Strawson is the holding together of society as a moral community of free subjects; contrary to hatred, resentment and indignation presuppose that, for all the damage caused by the inflicted injury or indifference, the agents involved still see each other as bearers of moral responsibility and, as such, capable of restorative justice procedures.

In a similar way, John Rawls (1971/1991) conceptualizes resentment as a 'moral sentiment'. He incorporates it into his theory of justice, as 'moral sentiments' constitute the necessary condition for every 'rational' individual to realize, behind the supposed 'veil of ignorance', the two basic principles of justice as fairness: (1) every person is to have an equal right to the most extensive total system of equal basic liberties compatible with a similar system of liberty for all. (2) social and economic inequalities are to be arranged so that they are both (a) to the greatest benefit of the least advantaged and (b) attached to offices and positions open to all under conditions of fair equality of opportunity (1971/1991: 60, 83, 250). These two principles, ranked in lexical order, cannot be applied if individuals are not governed by a sense of justice[6] and moral sentiments. Moral sentiments are defined as families of dispositions and propensities regulated by a 'higher-order desire' (1971/1991: 192) which touch on the very sociability of humans (and here Rawls is not diverging from the Classical philosophical tradition): relations of love and trust between children and parents, trust and sympathy between friends, the love of humanity, and adherence to a common good, all of which 'presuppose an understanding and an acceptance of certain principles and an ability to judge in accordance with them' (1971/1991: 487).

According to the above criteria, Rawls, in almost the whole eighth chapter of his book, makes distinctions between moral and non-moral sentiments: anger, rancor, anxiety, envy, spite, jealousy, annoyance and grudging are *not* moral sentiments primarily because, in their manifestation and explanation, the individual does not presuppose a binding sense of justice and injustice. Together with guilt, shame, trust, indignation, obligation, infidelity, deceit and sympathy, resentment, for Rawls, is placed among the moral sentiments. He defines it (1971/1991: 484) as a sentiment, which arises when wrongs are done to us. As in Strawson and Smith, indignation is a moral sentiment caused when somebody else who concerns us is injured or insulted. Both emotions are founded on an 'acceptance of the principles of right and justice and in that sense 'egoists are incapable of feeling resentment and indignation' but only anger and annoyance, precisely because lack of a socially made sense of justice excludes someone from the 'notion of humanity' (1971/1991: 488). For the American philosopher, the resentful person is a moral agent.

By moving now from moral and social philosophy to sociological analysis, one should invariably refer to Jack Barbalet, the forerunner of the sociology of emotions in the European continent in the late 1990s. While linking resentment to a wide range of social and political phenomena, such as inter-class and intra-class antagonism, social inequality and citizenship, he draws heavily from T. H. Marshall and he points out that the antagonistic nature of class society generates a multiplicity of emotions and feelings, contrasting and/or complementing one another. Among them, resentment is of critical importance as long as class societies are characterized by horizontal and vertical mobility (1998: 68). It is precisely this emotion, he holds, that allows for the conversion of a structural-class contradiction into a class conflict, to real action in the public space. The combination of mobility chances and the concomitant social comparison is likely to give rise to resentment against social inequality.

Two points are worth noting herein: first is the 'relational basis of resentment' (Barbalet 1998: 66) which is understood as the emotional apprehension of undeserved advantages unequally distributed in class societies. In that sense, class awareness, and class consciousness for that matter, consists in the interplay of interest and emotion principally because persons do have emotions but belong to classes (Barbalet 1998: 64–65). Second, as long as, according to Barbalet, social-class phenomena are not reduced to individual-level interactions, one could claim that class resentment may be experienced as an individual and /or a collective emotion. This is because, on the one hand, high rates of social mobility can convert group-based conflict into individual competition (Giddens 1981a: 56),[7] while 'emotion connects different phases of social structure through time', on the other.

Viewed as indignation against inequality, resentment for Barbalet is the negative and unpleasant feeling that somebody is enjoying one or more privileges in an improper and unequal way. Resentment is directed not towards power but towards the normative content of the social order, in the sense that someone both: (1) judges unworthy the position that someone else has in the social hierarchy, and (2) thinks that someone else – a person or a collective agent – deprives him/her of chances or privileges that s/he her/himself could and should enjoy (Barbalet 1998: 68, 137). That is why relational resentment's motive is not only individual self-esteem, hurt honor, or personal ill-recognition but also 'the identification and protection of shared norms of justice that we believe do or ought to regulate social and political interaction' (Ure 2015: 600). Therefore, personal injury and restoration of individual loss are not sufficient and necessary conditions for resentment. Through comparison, frustration, and oppression, resentment and class resentment are deemed to arise as an action-prone emotion determined by the unbalanced social opportunities structure while, in turn, resentment shapes the level of class conflict (Barbalet 1998: 70–71). And most of the time societal conflict concerns rule of law rights, citizenship rights, and collective well-being demands. Under these terms then, Barbalet's resentment differs from Barrington Moore's notion of moral anger arising from systematic social injustices.[8] It might not be accidental that Moore does not use the term 'resentment' which is likely to have emanated from a generalized normative framework; instead, his 'moral anger' or 'moral outrage' is linked to vengeance against injustices caused by failures of authority regarding the distribution of the social product and the provision of security and fairness. In this sense, claims to distributive justice are rendered 'moral universals' (Moore 1978: 17, 45–48). Injustice is understood as a misfit of the mutual obligations that bind together the rulers and the ruled wherein a powerful person hurts plain men and women and this does no good to anyone except the perpetrator (Moore 1984: 106). For Barbalet, vengefulness is not always experienced together with resentment which, please note, is not by definition an emotion of subordination;[9] understood principally by the Australian scholar as moral indignation, resentment 'is an emotional apprehension of departure from acceptable, desired, proper, and rightful outcomes and procedures' (Barbalet 1998: 138).[10] This account makes resentment suitable for the analysis of societal phenomena and social movement action triggered by the dynamics of social, political, and discursive opportunity structures.[11]

A year after Barbalet, the sociologist, published his book, Jon Elster, one of the most prominent rational choice theorists took the initiative to systematically bring emotions in his analyses in a seminal book (Elster 1999).[12] He includes resentment in his list of social emotions, i.e., emotions triggered only by beliefs and appraisals that refer to other people and the social norms they abide by. He also ideal typically classifies social emotions into those aroused by comparison and those aroused by social interaction: emotions of comparison are triggered by favorable or unfavorable projective comparisons with individuals with whom we will never come to terms, e.g., celebrities; emotions of interaction arise when there is face-to-face or indirect contact with other persons, e.g., social media users (Elster 1999: 139–141). Yet, quite often, comparison and interaction work in tandem and this is the case with resentment. Therefore he defines resentment not simply as an emotion that comparatively stems from the perception that one's group is placed in an unjust subordinate position on a status hierarchy – an idea very close if not identical to that of Dennis Smith (2006); in a rather vague way, he re-defines it as an emotion where 'group A perceives itself as higher than B on a dimension of comparison while at the same time lower than B on the interactive dimension of power and subordination' (Elster 1999: 74, 142).

As a group-based social emotion, I think that, for Elster, resentment is shaped from the discrepancy between a favorable comparison *in abstracto* and an unfavorable interaction in real terms of power relations. As a way of a relative deprivation appraisal, resentful subjects feel that although their group deserves a higher position in the social hierarchy, it actually remains lower in everyday interactions with individual group members. Clearly, the emotional concern or the point in this appraisal is the group itself, not the individual as has been the case with other philosophical and sociological approaches to resentment.[13] Elster's dealing with resentment in his 1999 book is related to one of his earlier descriptions of what he called a 'feeling of being unfairly treated', premised on three appraisals: 'the situation ought to be otherwise', 'it is someone's fault that it is not otherwise', and 'it could be made to be otherwise' (Elster 1989: 64). He rightly claims that when one of these conditions is lacking, envy may arise instead of the feeling of being unfairly treated (Elster 1989: 64). An issue, however, is the label of the emotional experience of being treated in an unfair or unjust way; usually this is called resentment. Yet Elster seems to have confused envy with resentment (ibid) which he probably used later on interchangeably with *ressentiment*, theorized as an aspect of envy in social life (Elster 1999: 175). In this respect, I would like to note three points: first, resentment is not likely to occur unless the 'it could be made to be otherwise' assumption is intertwined with a strong sense of internal political efficacy, namely the belief that one is able to impact public affairs (Capelos and Demertzis 2018). Second, if the belief that an unfair situation could be made to be otherwise is lacking, envy is likely to arise. Third, if the belief that the situation could be made to be otherwise is lacking concurrently with the belief that it is someone's fault that it is not otherwise, it is most likely that *ressentiment* will be called forth, a reactive emotional response to injustice theorized by Dostoyevsky, Nietzsche, Scheler, and many others.

Of *ressentiment*

At any rate, resentment is a less complex emotion than *ressentiment*, a cluster emotion the analysis of which starts with Dostoyevsky's 'mouse-man' (1864/2000) and which is refined conceptually in the works of Nietzsche (1994: 20) and Scheler (1961). Ever since, the concept has been extensively defined and redefined by philosophers and social scientists from different disciplines and subdisciplines. At first glance, it is considered as a feeling of the weak who are likely to pursue the logic of La Fontaine's fox. It was Nietzsche himself who maintained that *ressentiment* is a morality of the weak creatures 'who have been forbidden the real reaction, of the act' (Nietzsche, 1970: 35); the *ressentiment*-ful person[14] is governed by a frightened baseness that appears as humility, her/his submission to those s/he hates becomes docility, and her/his weakness is supposedly transformed to patience or even virtue. The basic characteristic of Nietzsche's man of *ressentiment* is vindictiveness in disguise that leads to inaction (Nietzsche 1970: 133). Whether assertive activity and powerless passivity ultimately mark the distinction between resentment and *ressentiment* is an issue to cope with later. For the time being I will not stick so much to Nietzsche as to Scheler, whose phenomenological analysis of *ressentiment* is replete with sociological overtones to influence major social theorists like Werner Sombart and Robert Merton.

Max Ferdinand Scheler (1874–1928)

As the founder of the sociology of knowledge and a highly respected philosopher in the phenomenological tradition, Scheler, on the one hand, broadened the concept of knowledge beyond established and institutionalized forms of knowledge to include informal moral, metaphysical, and religious kinds of knowledge emanating from different cultural contexts. On the other hand, he surmounted the traditional distinction between cognition and feeling by relating them into a functioning unity where 'subjectivity' and 'objectivity' are fused. From his phenomenological point of view, emotion and reason are of the same epistemological status (Scheler 1980).

Scheler was aware of the golden age Russian literature, especially Leo Tolstoy, Fyodor Dostoevsky, and Nikolai Gogol, which was full of *ressentiment*-laden heroes as a result of the autocratic oppression of the Tsarist regime that left no valves to release the affective repression of the people. Scheler inherited Nietzsche's negative conception of *ressentiment*, which was in the first place greatly influenced by *Dostoevsky's Notes from the Underground.* Nevertheless, he differentiates himself in at least two aspects; first, he rejects the genealogical explanation of Nietzsche according to which it is the Christian worldview, and particularly the Christian notion of love, that fuels the servile *ressentiment*-ful attitude. For Nietzsche (1895/2014: Chapters 51 and 45) 'Christianity remains to this day the greatest misfortune of humanity' and forms a 'morality born of *ressentiment* and impotent vengefulness'. Scheler opposes this, arguing that 'Nietzsche ignores that Christian morality does not value poverty, chastity, and obedience as such, but only the autonomous *act of freely renouncing* property, marriage, and self-will ... which are considered as *positive goods*' (Scheler 1961: 184 n25). For Scheler, Christian morality is founded on the wealth of the open soul and has nothing to do with the ungenerous and repressed aspirations of *ressentiment*.[15] For him, the genealogy of *ressentiment* is to be found in the bourgeois morality that was gradually taking shape from the thirteenth century onwards and reached its peak in the French Revolution (1961: 81–82). Furthermore, he thought of the West as a victim of *ressentiment* and its civilization being in decline under the capitalist spirit.[16]

Another aspect where Scheler diverges from Nietzsche is his sociological understanding vis-à-vis a quasi-Social Darwinism advocated by Nietzsche himself (Ure 2015: 7), who first and foremost analyzed *ressentiment* is terms of *Schlecht* rather than *Böse*. In German, 'Schlecht' refers to a physiological evil or disorder whereas 'Böse' means evil from a moral point of view. For Nietzsche, Christian humbleness is but the morality of the sick who manage to infuse guilt into the way the powerful and healthy conceive of themselves: i.e., as if they do not deserve their heights and privileges. As long as 'the One God and the Only Son of God were products of *ressentiment*' (Nietzsche 1895/2014: Chapter 40), the brotherhood-like equality in front of God is seen as a collective degeneration of humankind, as the abnormal revenge of the unhealthy and powerless emanating from a disease he calls 'moraline' which ultimately grounds *ressentiment* (Nietzsche 1895/2014: Chapter 2). As he puts it, 'Christianity has the rancour of the sick at its very core – the instinct against the *healthy*, against *health*' (Nietzsche 1895/2014: Chapter 51). Scheler is at odds with this interpretation not only because he defended Christianity as such but also because he was interested in the interplay of the individual and the collective while analyzing *ressentiment*. As a philosopher of values, he did not see *ressentiment* abstractly, let alone as an individual state of mind; through his phenomenological sociology and social philosophy approach, he assessed *ressentiment* as a societal phenomenon, as a mentality, as it were, of modern societies over and above any axiological evaluations in terms of health and sickness. In this respect Scheler is not confined to micro-sociological analysis of roles played in modern social settings as Barbalet assumes (Barbalet 1998: 64). In this vein, Yankelovich (1975) asserted that *ressentiment* acquires socio-political significance when it is retrieved from the confines of private domains and is generated into the planes of public consciousness. It becomes a sociological issue when large numbers of people share the sting of frustrated hopes, thwarted plans, and helpless victimization and are prevented from overt expression.

Apart from these two points of difference, in one way or another the Schelerian analysis of *ressentiment* follows Nietzsche's track. To start with, there are two necessary conditions for the stirring of *ressentiment*, according to Scheler. The first is a not acted out vindictiveness, an unfulfilled and repressed demand for revenge. The second *sine qua non* of *ressentiment* is a chronic interiorized powerlessness, a sense of impotence and the inability to influence the order of things. So, while you want to take revenge you feel that you cannot do anything about it. In this sense, *ressentiment* is a retroactive and repeated emotional experience marked by non-acted out enmity. The subject's powerlessness is called forth by the prediction that at the next opportunity for action s/he will be defeated and, in this respect, *ressentiment* is an emotional passive reaction. In Dostoevsky's words, the *ressentiment-ful* person exists:

> … without believing either in its own right to vengeance, or in the success of its revenge, knowing that from all its efforts at revenge it will suffer a hundred times more than he on whom it revenges itself, while he, I daresay, will not even scratch himself.…

Dostoevsky further addressed this person as someone probably

> … grinding your teeth in silent impotence to sink into luxurious inertia, brooding on the fact that there is no one even for you to feel vindictive against, that you have not, and perhaps never will have, an object for your spite.…
>
> (Dostoevsky 2000: 19–20, 22)

Even though Scheler (1961: 46–48) refers to contiguous negative emotions or sentiments whose manifestations exhibit a climax (malice, annoyance, envy, grudgingness, rancor, jealousy, spite), he does not identify them with (or equate them to) *ressentiment*. They may be stages in its progressive formation, but in order to speak of *ressentiment* proper we have to exclude two possibilities: the genuine moral transgression through forgiveness or inner purification on the one side, and the active expression on the other. When both of these are absent, then *ressentiment* emerges. This is because it builds upon the intensity of the aforementioned emotions that demand real revenge and, at the very same time, upon the catalytic powerlessness of expressing them, due to fear and/or physical or mental inferiority. Scheler, echoing Nietzsche, states openly: '*ressentiment* is chiefly confined to those who *serve* and are *dominated* at the moment, who fruitlessly resent the sting of authority' (Scheler 1961: 48). Besides, it is not just repressed vindictiveness that leads to *ressentiment*; it is the repression of the imagination of vengeance too that contributes to this particular state of mind where at the end of the day the very emotion of revenge itself is evaporated (Scheler 1961: 49). This is another point where Scheler diverts from Nietzsche's conception, according to which *ressentiment* characterizes powerless natures who 'compensate themselves with an imaginary revenge'. Scheler's *ressentiment* is a far more repressed and thus a deeper idealized/rationalized displacement.[17]

Apart from the two aforementioned necessary conditions, his argumentation reveals some sufficient conditions for the nurturing of *ressentiment* in modern democratic societies, without however his naming them thus. The first sufficient condition is the gap between the perceived equality of social position and the rights that emanate from citizenship and the real power of the individual to enjoy them. This gap, says Scheler (1961: 50, 69), functions as psychological dynamite since a structural element of modern political democracies is the gulf between formal and substantive equality, much discussed ever since in social and political analysis. This condition would by itself lead to envy, class hatred[18] or moral outrage (Moore 1978), if it was not over-determined by chronic interiorized powerlessness (which is the second necessary condition) and the repressed wish for revenge.

The second sufficient condition is perpetual comparison, which is actually a constant of the human condition as we have already seen. If you do not compare yourself to others you cannot feel these hostile emotions which make up *ressentiment*. That is, you cannot feel vindictiveness, envy, jealousy, rancor or even hatred if you do not compare yourself to others. The 'other' could be an individual, or it could well be a reference group, especially so in open societies which are characterized by loose class differentiations, upward social mobility and perceived egalitarianism. The close coupling of envy and egalitarianism has been systematically analyzed ever since Aristotle (de la Mora 1987; Schoeck 1969). It is not just a painful emotion at the good fortune or prosperity of others but a disturbing and hostile emotional experience excited by the prosperity of people who are like us or equal with us after we have compared our own situation with theirs without necessarily getting any harm from them (Aristotle 1992: 1386b, 1387b; Nussbaum 2016: 51–52). True enough, envy is a 'democratic sentiment', as Tocqueville put it with regards to the eighteenth century United States (Tocqueville 1969: 130); in a perceived egalitarian society in which the burdens are supposed to be spread equally one suffers a damage to his/her self-esteem by admitting the superiority of another. In a way, this is a secularized violation of the Christian principle of brotherhood before God,[19] it is the unbearable difference that breaks the horizontal identification of the believers' ideal Egos before God (Freud 1949: 63–70, 85), or the semblance of people living in closed societies or small communities (Elster 1989: 69).[20] The seeking of dignity and social approbation of a working and lower middle class group of youngsters comparing themselves with the middle and upper middle classes in post-war Sicily occupies a good part of Helena Ferrante's *Neapolitan Quartet* which, among things, is about the emotional dynamic permeating the identity formation and class neuroses[21] of Lila and Lena, the two main characters of the novel (Ferrante 2013a, 2013b, 2014, 2015).

Finally, the third sufficient condition for eliciting *ressentiment* is the irrevocable nature of the injustice you feel subjected to. Each particular injustice, from which the revengeful attitude begins, has to be experienced as destiny, as something that cannot be changed in any way. The sense of injustice as unchangeable destiny is the quintessence of powerlessness, especially when hope for a way out is blocked. As a consequence, the *ressentiment*-ful person is likely to survive, i.e., to live on a minimum common denominator, instead of living her life in an open-ended way. By reconstructing Nietzsche, Raoul Vaneigem, who was aware of the difference between resentment and *ressentiment*, holds that as long as they live in an age of decomposition, our own age, the people of *ressentiment* are the 'perfect survivors', bereft of the consciousness of possible transcendence, people unable to grasp the necessity for a reversal of perspective, who, 'eaten up by jealousy, spite and despair' are 'trapped between total rejection and absolute acceptance of Power' (Vaneigem 2012: 143, 152).

So, it is through the combination of the above mentioned sufficient and necessary conditions that the Schelerian *ressentiment* unfolds. When these conditions are absent, we cannot speak of *ressentiment* in the proper, technical, meaning of the word. To take a helpful example: in her political autobiography Angela Davis recalls that while a juvenile she could not help feeling envy and jealousy against the white world. But precisely because she developed a strong sense of political competence, she 'never harbored or expressed the desire to be white' and her hard feeling of injustice for the 'untold numbers of unavenged murders' of her people and her hatred of jailers and the penal system did not end up in spurious opposition or political inaction. She never accepted racism as a destiny[22] (Davis 1988: 85, 253, 319).

Elaborating on the Nietzschean line of thought, a key, crucial point in Scheler's approach is the innermost intrapersonal process characterizing the *ressentiment*-ful person. This process is described as 'transvaluation', i.e., a reversal of values (*Umwertung*). Since the *ressentiment*-ful person does not possess the moral virtues and the proper psychological skills such as faith, high self-esteem, and sublimation capabilities, nor the social resources to manage the pressure the inferior social position exerts on him/her – but also is subject to the ever present existential anxiety (Scheler 1961: 52) – s/he proceeds to a chronic withdrawal into him/herself, thus avoiding acting out her revengeful attitude. In this way, Scheler says, s/he morally poisons herself. While at first s/he admires the values and privileges s/he does not possess (prestige, education, wealth, descent, beauty, youth, etc.), because s/he cannot acquire them s/he goes on to invalidate them, valuing the exact opposites. Since *ressentiment* is not rage or hatred, which have an 'expiry date' and a specific addressee, but is instead a chronic and complex emotional disposition with unclear recipients which is molded by the endless rumination of repressed negative affective reactions, it entails a compensating reversal of values, so that the person can stand and handle his or her frustrations. At first, I admire the prosperous, the gorgeous, the aristocrats, the educated, the renowned, etc. But since I am convinced I cannot become like them or compete with them, there is a silent hostility growing in me, a repressed vindictiveness for something that was unrightfully taken away from me. So, I start slowly to undervalue what I once admired. In Scheler's own words, 'the formal structure of *ressentiment* expression is always the same: A is affirmed, valued, and praised not for its own intrinsic quality, but with the unverbalized intention of denying, devaluating, and denigrating B. A is 'played off' against B' (Scheler 1961: 68). And value denigration emerges from the aching and unbearable difference between desire and impotence. By way of example, Ferrante's heroine Lena had a strong impulse to destroy their humiliating social setting and the insulting men who surrounded her;

> If nothing could save us, not the money, not a male body, nor even education, so much was it possible to destroy everything immediately. My anger grew in my chest ... I wanted the force to spread. But I realized that I was also scared. I understood only later that I know how to be unhappy in secrecy just because I am incapable of violent reactions, I fear them, I prefer to remain still cultivating ressentiment[23]
>
> (Ferrante 2013a: 20)

In psychoanalytic terms, we would say that it is a reaction formation, a defense mechanism against pressures exerted on the psyche. It consists in a systematically compensating imaginary devaluation of the object of comparison. In this respect, for all their kinship, one might unwittingly confuse *ressentiment* with envy or, as Simon Clarke (2004) does, think of envy experienced individually as the functional equivalent of *ressentiment* at the societal level, namely as a destructive emotion which damages future generations insofar as it represents pure negativity. Clarke refers to the Kleinean conception of envy, which is a projection of Thanatos and 'reminiscent of Nietzschean *ressentiment*' (Clarke 2004: 106). Where transvaluation is not central to Nietzsche's, let alone to Scheler's, argumentation I could agree with Clarke's insight. Yet, it seems to me that the Nietzschean version of *ressentiment* cannot be equated with or reduced to envy; rather, as a negative and complex emotion it contains and somehow submerges envy as long as perpetual powerlessness and relived inferiority block open destructive action or malicious expressions. Transvaluation, then, tames and modulates envy and, in that sense, although *ressentiment* is reactive it is not primarily destructive and aggressive. As Lewis Coser holds, as distinct from rebellion, *ressentiment* is not conducive to the ultimate affirmation of counter-values because '*ressentiment*-imbued persons secretly crave the values they publicly denounce' (Coser 1961: 24). With the question being what 'secretly' actually means in this context, it suffices to say that, as illusionary as it may be, on a purely individual level transvaluation could be seen as a sort of self-therapy. However, since Scheler is really interested in the apprehension of social and cultural phenomena, he argues that the *ressentiment*-ful mentality (or emotional climate in a more contemporary parlance) alters the entire cultural value landscape as well as the way in which one copes with power, knowledge, historical memory, social evolution, and social hierarchy and so on.

To stress the point further, reversal of values does not mean that in real time the *ressentiment*-ful person is conscious of the positivity of the values s/he contends; keeping a distance from Nietzsche, for Scheler the *ressentiment*-imbued person is not the rational, and yet impotent, actor who reacts according to the logic of the 'sour grapes' seething with bitterness. It is not about a '*rational self-interested*' attitude, but it has nothing to do with cynicism either. It's not as if the *ressentiment*-ful person knew and recognized the values but acted as if s/he didn't (that is, s/he knows and accepts education as an end in itself, but since s/he cannot be educated him/herself s/he devalues it, placing in its stead the spontaneity, let's say, of the common man). But nor is the hypocrite a model for *ressentiment*-ful people, since the latter do not reverse values in an ostensible way (Scheler 1961: 77). What Scheler means by 'transvaluation' is literally a substitution and depression: the old values stay backstage in the psyche, in a misty landscape of the soul, so that the *ressentiment*-ful man cannot see them as he acts at another level of values, which he has elevated to a positive normative context. The positive values are still felt as such, but they are overcast by the false ones (Scheler 1961: 59–60); it is a matter of an obscure awareness of true values which Scheler calls 'value blindness' or 'value delusion'.

Psychoanalytically, I would say that we are dealing with the result of a 'splitting' due to an intense narcissistic trauma, which displaces and/or negates the object of desire. In all likelihood, the subject of *ressentiment* is an ambivalent masochist who morally consummates through fantasies of the denigration and the canceling out of those ideals and values s/he cannot actualize in the first place due to powerlessness and lack of competence. So, the subject of *ressentiment* praises something because s/he has already repressed and negated something else.[24] Directed by a tyrannical Superego, repression not only stretches and changes the original object of desire, but it impinges on the emotion itself; for Scheler, 'since the affect cannot outwardly express itself, it becomes active within ... the man in question no longer feels at ease in his body' and the non-acted out other-directed negative emotions (hatred, wrath, envy etc.) turn against their own bearer resulting in self-hatred, self-pity, self-torment and revenge against oneself (Scheler 1961: 71–72). This process had been already grasped by Dostoevsky while describing the handling of anger and revenge by the 'cellar-rats' and 'maggot' men of *ressentiment*:

> anger in me is subject to chemical disintegration. You look into it, the object flies off into air, your reasons evaporate ... the wrong becomes not a wrong but a phantom, something like the toothache, for which no one is to blame.... So, you give it up with a wave of the hand because you have not found a fundamental cause.... The day after tomorrow, at the latest, you will begin despising yourself for having knowingly deceived yourself. Result: a soap-bubble and inertia.
>
> (Dostoevsky 2000: 27–28)

Nietzsche would have agreed completely with Dostoevsky since he thought of *ressentiment*-ful transvaluation as simply a means through which the sick momentarily alleviate their suffering. Yet Scheler wouldn't endorse this interpretation all the way; for him the *ressentiment*-ful man or woman is not someone who deceives themself – transvaluation 'should not be mistaken for conscious lying' for 'it is not that the positive value is felt as such and that it is merely declared to be "bad"'. Resulting from a deep-seated illusionary fantasy of transcending slights and inflicted transgressions – Scheler speaks herein of 'organic mendacity' – at the end of the day transvaluation renders the subject of *ressentiment* 'good', 'pure', 'human', 'true', and 'genuine', 'for the value it affirms is really felt to be positive' (Scheler 1961: 77–78). This is actually the case described by Theodor Adorno as half-education, half-formation or half-*Bildung* that permeates mass culture. In a media saturated society flawed by information the man in the street is likely to be content with being just updated rather than educated/cultivated. In this respect, half-education is not the mathematical absence of a percentage of *Bildung* as such, and the state of half-*Bildung* is worse than not-knowing at all. This is so because the half-educated admires masterpieces of art and science and at the very same time s/he repels them, substituting them with updated information and over-accreditation which confer a sense of self-esteem. For Adorno then, the people of half-*Bildung* are reified bearers of *ressentiment* as long as they replace the moral value of critical and self-enlightening consciousness with a whole-hearted self-affirming and compensating submission to the rule of production and consumption processes (Adorno 1997).

All in all, therefore, assessed upon the Schelerian premises, the person of *ressentiment* is *prima facie* sincere and honest but no less morally mutated. As mentioned before, at an ideal typical level of analysis, Scheler's *ressentiment*-ful person is not a hypocrite, nor is s/he a cynic; by virtue of transvaluation s/he becomes morally mutated. The issue then is whether transvaluation as a psychological mechanism rests on simple logical negation, or refers to something deeper. Retrieving a couple of psychoanalytic concepts, in my reading of Scheler the mechanism of transvaluation is certainly not a matter of foreclosure (*Verwerfung*), a defensive mechanism in which the repudiated element – the prized values in our case – is not just repressed or buried in the unconscious but expelled from the unconscious itself. Repression (*Verdrängung*) expels thoughts and feelings from consciousness and confines them to the unconscious; it does not destroy them. Springing from repression and as an act of denial, negation (*Verneinung*) is premised on the affirmation of the existence of an element which has been consciously registered in the first place (*Bejahung*). What then is cast in the unconscious through negation is the representation of the substance of the repressed object while the subject intellectually accepts its existence. Disavowal (*Verleugnung*) is another defense mechanism where the subject denies unbearable and/or traumatic but no less tangible elements of her external reality. For Freud and Lacan disavowal is always accompanied by the opposite stance, i.e., the recognition of traumatic reality (Evans 1996: 43–44, 65–66, 122, 165; Laplanche and Pontalis 1988: 118–121, 261–263, 390–394). Be it noted that the defense mechanisms comprising transvaluation are part and parcel and intertwined with the appraisal components of any emotion be it primary, secondary, or tertiary; namely, the subjective relevance of the triggering event (strength), its implications to the person's well-being (acceptability), the person's coping potential (availability), and the normative significance of the event (Scherer 2001; Frijda 2004b; van Troost, van Stekelenburg, and Klandermans 2013).

Gustav Le Bon, Robert Merton, Werner Sombart, and Alfred Von Martin

We know that the French word *ressentiment* was the ready-made linguistic label Nietzsche used to designate the slave morality. It was he who systematically upgraded that French word into a cultural-psychological concept, a *terminus technicus*. It has been argued that when he used that particular term, a French person would have understood it as the garden-variety connotation of 'resentment' in English. Although this was a word not spoken by the man in the street in France itself, in his aspiration to be a 'Good European' Nietzsche employed it because most educated Germans would use it to express *Groll* or *Verstimmung* since it was borrowed during the Enlightenment vogue for all things French (Birns 2005: 4–5).

If this is indeed the case, *ressentiment* must have returned to the French intellectual elite as a conceptual-linguistic counter-loan to interpret historical phenomena. The likelihood is that Gustav Le Bon (1841–1931), who was contemporaneous to Nietzsche (1844–1900), did not neglect the particular impact of *ressentiment* in the emotional dynamics of revolution. In 1912 Le Bon published his *La Révolution Française et la Psychologie des Révolutions* where he paid much attention to emotions such as hatred, fear, ambition, jealousy, envy, enthusiasm, and *ressentiment*. Among others, he observes that

> the effect of jealousy, always important in times of revolution, was especially so during the great French Revolution. Jealousy of the nobility constituted one of its most important factors. The middle classes had increased in capacity and wealth, to the point of surpassing the nobility. Although they mingled with the nobles more and more, they felt, none the less, that they were held at a distance, and this they keenly resented. This frame of mind had unconsciously made the bourgeoisie keen supporters of the philosophic doctrine of equality.
>
> Le Bon (1918: 54)

The reference to 'equality' clearly resonates with the influence of Nietzsche's orientation to the leveling of modern society and democracy.[25] In the original edition Le Bon did utilize *ressentiment* for the description of the emotional attitude of the bourgeoisie vis-a-vis the nobility.[26] With regards to the emotional dynamic of revolution Robert Merton deviates from Le Bon's account; in his notorious typology of individual adaptation to cultural values and institutional norms (conformity, innovation, ritualism, retreatism, and rebellion), he isolates the value and ideational substance of revolution or rebellion from *ressentiment*-laden societal orientations (Merton 1957/1968: 209–211). He points out that through the monopolization of social imagination by a new myth, revolutionary groups 'define the situation' in terms of a transformable social structure premised on genuine transvaluation. On the contrary, since *ressentiment* rests on the sour-grapes pattern the fox does not abandon the taste of sweet grapes but only refuses these particular grapes. For Merton, the *ressentiment*-ful attitude consists in the condemnation of 'what one secretly craves; in rebellion, one condemns the craving itself' (Merton 1957/1968: 210). Of course, he admits that organized rebellion may draw upon the reservoir of the discontents' *ressentiment* as structural dislocations deteriorate. Nevertheless, precisely because he refers to Scheler, a question is raised as to what Merton really means by the 'secret' craving of the socially prized values.

It seems to me that Merton didn't do justice to Scheler's transvaluation thesis, describing it in terms of logical negation. In my understanding, transvaluation is initiated by repression and stands in between negation and disavowal. These two defense mechanisms lean towards what Scheler calls value delusions and illusions; the *ressentiment*-laden person is mendacious but not actually a liar because even if 'deep down [in] his poisoned sense of life the true values may still shine through the illusory ones' (Scheler 1961: 77) the falsified reversal of values is experienced as moral achievement. By reducing the Schelerian transvaluation to the sour grape logic, Merton's account of *ressentiment* comes quite close to the fourth mode of adaptation, i.e., retreatism. People who want to escape from the requirements of social life are characterized by defeatism, quietism, internalized prohibitions, and resignation; they always remain humble, poor, hopeless, and isolated with no power to combat aggression against the weak and helpless. At the same time, however, these individuals are contemptuous of the incomprehensible social world and its values; it is only a step distance which differentiates Merton's retreatists from what he thinks of as the bearers of *ressentiment*, the substitution of mainstream values according to the logical negation conveyed by the sour grapes pattern (Merton 1957/1968: 207–209). However, one should credit Merton for taking in the affective dimension in his analyses during an emotion-proof period of academic sociology in the USA. The sociology of emotions dawned almost 20 years later.

Something like 40 years before Merton, a colleague and interlocutor of the author of *Ressentiment*, Werner Sombart (1863–1941), embraced Scheler's ideas and his analyses of various situations charged with this compound and painful sentiment (women, spinsters, older generations vis-à-vis younger cohorts, small anti-systemic parties against parties in power, apostates against their former beliefs, petty bourgeois strata and small officials against upper classes and big business, and so on). In his *Der Bourgeois. Zur Geistesgeschichte des modernen Wirtschaftsmenschen*, published in 1913,[27] Sombart referred systematically to the first edition of Scheler's book, while Scheler in turn commented at length on his friend's remarks in the second revised edition of his *Ressentiment* book (Scheler 1961: 63–64, 177, 188, 190, 194–195).

Sombart cast serious doubt on Weber's thesis about the inner link between the Protestant ethic and the capitalist spirit – composed of what he calls the entrepreneurial and the bourgeois spirit – by showing not only that there were diverse cultural and intellectual sources of the capitalist ethos, but also that some Protestant sects had nurtured a manifest antic-capitalist stance; he also documented that the Lutheran tradition embodied pre-capitalist and anti-capitalist value orientations (Sombart 1913/1998: 238–240, 258–267). About half of the book focuses on the moral and social sources of the capitalist spirit: a number of biological, philosophical, religious, and technological factors contributed to its emergence in multiple modalities in different European countries and the Americas. Throughout the book, Sombart mentions many early pioneers and later barons of capitalism classified either as 'heroes' or 'merchants' (e.g., Leon Battista, Benjamin Franklin, Carnegie, Siemens) in order to show the linkage between the capitalist sprit and capitalism as an economic system. More than anybody else, though, Benedetto Alberti – a well-off wool merchant in fourteenth century Florence – is centre-stage in the book, which is really dotted with the Italian's written thoughts, notes and diaries. Through this material Sombart documents an emotional climate in Florence of that period marked – among other things – by *ressentiment*. Bourgeois virtues were not valued for themselves but as a preferred reaction to the superior but no less envious value of the nobility, the 'Signori'. The bourgeois prudence, propriety, honesty, humility, and frugality were but the *ressentiment*-laden compensation for the lack of symbolic capital already occupied by the aristocracy; notably, Sombart declares that the most power-ful motive for Alberti to embrace the bourgeois world-view was *ressentiment* itself. But Sombart moves a step farther: he positively claims that in all historical ages *ressentiment* remained the most stable buttress of the bourgeois morality (Sombart 1913/1998: 345–346). It is an open question whether this is too bold a claim to be uttered in the early twentieth century, or if and to what extent it is valid nowadays. It is worthwhile, however, to note how he understands *ressentiment*; apart from a reference to the then freshly published book by Scheler, he makes no qualification as to how he understands this notion. Apparently, he adopts Scheler's theses, but did he really grasp the meaning of this multilayered sentiment? By the time he published his *Der Bourgeois* Sombart had already deserted his socialist political inclinations moving towards a 'heroic' reading of international relations and capitalism itself (Degli Esposti 2015). Sombart the economist was mostly influenced by Nietzsche and his notion of creative destruction (Reinert and Reinert 2006); it is unclear though how much he was influenced by Nietzsche otherwise. Yet, his statement 'even nowadays the pru-dent bourgeois denounces the sour grapes, thereby consoling himself' (Sombart 1913/1998: 346), discloses a rather shallow reading of the Schelerian transvalu-ation. Sombart lingers more or less on Nietzsche's version of *ressentiment*, while rather ostensibly embracing Scheler's approach.

No doubt, when Alfred von Martin published his *Soziologie der Renaissance* for the first time in 1932[28] the concept of *ressentiment* was commonly used by German sociologists and the academic public. That is why, although he uses the concept in four crucial points of his analysis, he did not bother to refer to Scheler's *Ressentiment* when scrutinizing the cultural dynamics of Renaissance Italy. Instead, he felt the need to cite Scheler's *Die Wissenformen und die Gesellschaft*, a book that carved out the sociology of knowledge as a discipline. Like Sombart, von Martin comments extensively on Alberti's writings and focuses upon the overt and covert antagonism between the rising bourgeois commercial social strata, land aristocracy, and upper clergy one the one hand, and the explicit and/or latent clash within the bourgeois class itself expressed as an antagonism between the proper-tied commercial class and the humanist 'intellectuals' on the other.

Thus, there were two vectors of *ressentiment* in Renaissance Italian society. The anticlerical and anti-feudal *ressentiment*-ful attitude was expressed by the idea of *humanitas* and the authority of Classical Antiquity (von Martin 2016: 36, 43). Work, erudition and personal achievement were viewed as the new *virtus* to replace the old *virtus* of the nobility; money and talent were forced together against the medieval tradition. Yet, at the same time humanists complained about the philistinism of the propertied classes who were accused of a sheer materialist outlook. Apart from the fact that contempt for money was chiefly an example of sour grapes (i.e., a negation of their insecure social position), the human-ists experienced an 'inner rebellion' against the propertied commercial class which was expressed by a feeling of strong *ressentiment* against an economic and political power that held them down (von Martin 2016: 50).[29] If that was not enough, the humanists found themselves caught between the old monks – the organic intellectuals of the medieval ruling class who pursued the ideal of *vita contemplativa* – and the new dominant commercial class who despised their intellectual gifts. They were only half at home in either social environment and were regarded as apostates by both sides. As a consequence, they emotionally defended cultivating a sort of 'humanistic *ressentiment*' characterized by an antipathy against scholasticism and ambivalence toward the well-off (von Martin 2016: 51–52).

Towards an appraisal

In our exposure of the Nietzschean accounts of *ressentiment* and the approaches to resentment it was argued, on the one hand, that they are either moral emotions or sentiments triggered by perceived injustices, slights, and wrongdoings. On the other hand, it was shown that *ressentiment* is a complex sentiment (Merton 1957/1968: 209), a cluster emotion that requires far more analysis than resentment. Nevertheless, it is proposed that, for all their moral similarities, they differ as to the intensity of the incurred injury and their duration: resentment is seen as more transient and episodic, and normally caused by minor insults while *ressentiment* persists and is temporally extended and elicited by major wrongdoings.[30] Otherwise, the argument goes, their difference is a difference of degree and in this respect one should ponder on either concept as designations of an ultimately unified emotional experience whereby *ressentiment* stands for the 'general sense' and resentment for the 'specific sense' (Meltzer and Musolf 2002: 242–243, 251; Aeschbach 2017: 48; TenHouten 2018: 10). In other approaches the argument is turned upside down; now it is resentment that stands as a general concept, as a generic moral emotion, with *ressentiment* deemed a nineteenth century and hitherto specific modality of the former, elicited in competitive societies (Moruno 2013). For others, the two concepts should be kept apart with the main difference being 'the sense of one's own powerlessness and the repression of other negative emotions that are involved in *ressentiment* but not in resentment' (Salmela and von Scheve 2017: 6).

To exhaust the literature and overhaul every single approach to *ressentiment* is not one of the purposes of this chapter.[31] What I intend to do in this section is to discuss my own understanding while commenting on others' contributions by giving due emphasis to the transvaluation process whose role in the emergence of the *ressentiment* phenomenon has not always been properly acknowledged.[32] I consider *ressentiment* to be an unpleasant complex moral sentiment with no specific addressees, experienced by inferior individuals including a chronic reliving of repressed and endless vengefulness, hostility, hatred, envy, and resentment due to the powerlessness of the subject in expressing them, and resulting, at the level of moral values, in the disavowal of what is unconsciously desired. This is a theoretical definition encompassing, among others, the criteria Aristotle (1992: 1378a 20–25) employs for the study of emotions.

Type, cause, and intrapersonal processes

Alike resentment, *ressentiment* is, on the one hand, a *social emotion* elicited from the comparison its bearer makes with other people's actions and fortunes, whereas valence-wise, on the other hand, it is primarily a *negative* and unpleasant emotion. The *cause* of *ressentiment* is the subject's incapacity to really act against perceived injuries, wrongdoings, undeserved slight, unfairness, or deprivation.

The repression of negative emotions and the concomitant transvaluation refer to the *intrapersonal processes* experienced by the bearer of *ressentiment*. The transvaluation process, already stressed in this chapter, has two functions; on the one hand, it diminishes the value of objects, symbols, traits, ideas, attributes, qualities, characters, and roles according to the sour grapes logic leading to the moral upgrading of one's shortcomings into virtues. On the other hand, it endorses a new value hierarchy based on the 'sweet lemons' mentality. The first is a kind of devaluation and the second an act of reevaluation. Although Aeschbach (2017: 260) rightly gives due emphasis to these two functions, he seems not to embed them into a unified mechanism as I do. He stresses the point so far as to separate a 'weak' and a 'strong' version of *ressentiment*, resting on devaluation and reevaluation respectively. In my opinion, this is too abstract a way of assessing *ressentiment*; instead, I think that the two versions are actually the two sides of the same coin, namely, transvaluation itself.

It has been argued that *ressentiment* is not actually an emotion, as complex as it might be, but a psychological mechanism (Salmela and von Scheve 2017: 16, 21)[33] through which repressed negative emotions like fear, insecurity and shame are transformed into anger, resentment and hatred towards perceived out group enemies. This transformation is carried out through defense mechanisms such as displacement, projection and attribution. Even if Scheler never refers to shame as a componential element in his *Ressentiment* book, one may accept that shame – a complex, high intensity, self-targeted emotion in itself – lies within the affective constellation of *ressentiment* (Demertzis and Lipowatz 2006: 106–107). By thinking, however, of *ressentiment* just as an intrapersonal defense mechanism, i.e., by substituting the intrapersonal affective and cognitive appraisals that make possible the very feeling of the emotion for the emotion itself, one is at pains to indicate the similarity and difference between *ressentiment* – thus understood – and other cognate defense mechanisms. As already indicated, the transvaluation process is in itself a configuration of defense mechanisms and therefore it seems to me that it confers no added value by dropping *ressentiment* out from the list of emotions.[34]

Moral standing

Its *moral* or *normative nature* is constituted in a two-stage mode; at the beginning, the *ressentiment*-ful person experiences a sort of immovable injustice and frustration, overwhelming him/her as an unbeatable destiny. At a second stage, after the transvaluation process is consummated, a new moral-hermeneutic horizon opens up for the person of *ressentiment*. While in the first stage s/he is morally injured by the damage to her self-esteem, in the second stage s/he is morally reconstituted, even if in an illusionary fashion. Despite the self-poisoning caused by the transvaluation process (Scheler 1961: 45), the *ressentiment*-ful person has ultimately convinced her/himself of her/his goodness, piety, piousness, self-righteousness, integrity, and honesty. Although he knows his moral deficiencies, in order to cope with them, Dostoyevsky's man from the underground describes himself like this: 'I was inwardly conscious with shame that I was not only not a spiteful but not even an embittered man, that I was simply scaring sparrows at random and amusing myself by it' (Dostoyevsky 1864/2000: 4).

From an *etic* perspective this person is morally disfigured, has espoused what Nietzsche calls value mendacity; from an *emic* perspective s/he is morally elevated, motivated by the desire to feel morally superior (Aeschbach (2017: 216). Feeling morally superior is a compensation transvaluation confers on the person of *ressentiment* for all her/his weaknesses and damaged self-esteem. In that sense, as Aeschbach holds (2017: 257), there is a pleasant aspect to the repressed and relived hostile, but no less negative, emotions that constitute *ressentiment*, a satisfaction that the wrecked minded are actually winners.[35] In other words, 'if this re-valuation works (i.e., reaches its goal) then inferiority and jealousy should disappear right along with the pain' (Posłuszna and Posłuszny 2015: 88).

Yet, the defense mechanisms of transvaluation cannot ultimately erase the positivity of the values which have been transmuted in the meantime; as disavowal is not foreclosure – namely the erasing of the repressed object, not only from consciousness but from the unconscious itself (Evans 1996: 64–66), the return of the repressed is always imminent, it remains *ante portas*. Illusions are not for ever precisely because the subjects are embedded into open-ended destabilizing social and societal contingencies. Therefore, as the *ressentiment*-ful person continues, deep inside, to be aware of the unreachable or unrealisable values s/he dismisses, '*ressentiment* is, in essence, unstable as it leads to a conflicted fox who judges the grapes to be sour while continuing to feel that they are sweet' (Aeschbach 2017: 259). In this vein, Nietzsche promulgated that *ressentiment* offers a temporary alleviation of victimhood and suffering.

Under these terms, a good many scholars adopt a negative stance towards *ressentiment* vis-à-vis resentment. The former is morally rejected due to its inherent value mendacity and/or illusionary valuation, while the latter is defended as the proper moral-emotional standpoint against culpable actions. Testing whether the American public's discomfort with political affairs shortly after the defeat in Vietnam was an expression of *ressentiment*, Daniel Yankelovich characterized it as a dangerous political emotion serving as a pool of psychic energy created by frustration and available for a politics of anger, retribution, destruction, and demagoguery (Yankelovich 1975).[36] From an international politics point of view, Elisabetta Brighi urges a defense of resentment as a dignified coping with failures of justice as against failures of recognition and status which are likely to elicit *ressentiment*. Hence, resentment is seen as the guardian of justice lying at the heart of democracy, concerned with the protection of shared norms that regulate social and political relationships (Ure 2015: 601), whereas *ressentiment* is understood as a corrosive emotional reaction to compensate for individual incapacities and lack of self-esteem. As she puts it, 'resentment starts from an affirmation of rights' while r*essentiment* 'starts from the perception of one's impotence or lack of rights' (Brighi 2016: 424). From a philosophical perspective, Aeschbach (2017: 261) denounces *ressentiment* because it cultivates lack of integrity, heteronomy and lack of courage on the one hand, and unwillingness 'to face the hard truth' on the other. Resonating with Nietzsche's biological idea that *ressentiment* is detrimental to individual health, from a clinical point of view it is seen as a psychological pathology characterized by obsessive repetition of past memories, anxiety, embitterment, and depression. Putting due emphasis on transvaluation, Elżbieta and Jacek Posłuszny identify an inherent problem with *ressentiment* as long as the ensuing alteration of the self lasts for more than just a short while. The compensating alteration of systems of values may involve tendencies toward fanaticism, fundamentalism and terrorism (Posłuszna and Posłuszny 2015). Because the defensive mechanism of transvaluation is never complete, the normatively reformed person needs a validation of her choice to denounce the prior value system. Even if the revaluation process compensates the weak and inferior for their powerlessness, it is nonetheless a claim to power in itself. In this respect, the substitution of values must be validated and the likelihood is that this validation or empowerment will acquire a fundamentalist twist (Posłuszna and Posłuszny 2015: 93). Relatedly, it was recently shown by qualitative research that extremist violent action is enhanced via the reinforcement of individual emotions (micro level) by group-based emotional dynamics employed in interaction ritual chains (meso level) so that people with less power shape their identity (Latif *et al.* 2018).

But is resentment such a bright and noble emotion, straightforwardly promoting dignity and retributive justice? Can one draw a sharp line separating it from *ressentiment*? If anything, for all their extant differences, resentment, *qua* moral anger, is one of the components of *ressentiment* even if in repressed form. Adam Smith's insights are of relevance on this point. Resentment's congruence with democratic politics and retributive justice is only possible if it is experienced and acted out in the proper way. As discussed earlier in this chapter, Smith maintains that resentment can become a detestable emotion if it is not agreeable to the conscience of the impartial spectator. This means that the ensuing punishment resulting from resentment should be reluctant and not directed 'from any savage disposition to revenge'. In effect,

> nothing is more graceful than the behavior of the man who appears to resent the greatest injuries, more from a sense that they deserve, and are the proper objects of resentment, than from feeling himself the furies of that disagreeable passion; who, like a judge, considers only the general rule,

which determines what vengeance is due for each particular offence; who, in executing that rule, feels less for what himself has suffered, than for what the offender is about to suffer; who, though in wrath, remembers mercy.

(Smith 1759/1976: 172)

Smith is aware that that there is an inherent proclivity for over-reaction in resentment directed by detestation, obstinate spite and disgust for the wrongdoer.

The man of furious resentment, if he was to listen to the dictates of that passion, would perhaps regard the death of his enemy as but a small compensation for the wrong, he imagines, he has received; which, however, may be no more than a very slight provocation.

(Smith 1759/1976: 160)

And if, despite the right punishment, apologies, and reparation one still resents the culprit one is likely to slip slowly into *ressentiment*. One might rationally claim that if the target of resentment has been punished, and s/he has apologized, there should be no reason to assume that resentment may lead to *ressentiment*. Alas, in social interaction where overt or covert antagonisms are at stake people do not always behave rationally; they usually over-react, urged by unconscious fantasies of omnipotence, sadism, or masochism. That is why, sometimes, repentance is not enough and the resentful subject might secretly wish the other person's destruction. It is precisely for this reason that it is argued that *ressentiment* may be seen as resentment's shadow (Ure 2015: 608).

The emotional object

As of its *object*, i.e., its focus of attention, during the transvaluation process *ressentiment* stands as an objectless emotion in the sense that it is not specifically *about* something or someone (Lamb 1987). Objectless emotions are not about specific objects; instead, they are 'generalized feelings about the world' (Goodwin, Jasper, and Polletta 2001b: 10–11). After the transvaluation process is consummated – as incomplete as it might be – it is directed, as Warren TenHouten proposes (2018: 10), towards or against larger social categories, the ruling class, the minorities, certain ethnic groups, or the whole cruel world.[37] Seen from the micro perspective, the objectlessness of *ressentiment* generates a mood-like 'lasting mental attitude' or 'a venomous mass' of negative affect (Scheler 1961: 45, 27); from the macro perspective it may be seen as an emotional climate or emotional habitus, a sort of collective-level *ressentiment*. To argue that, typically, *ressentiment* is first triggered by a particular object or person and may then progressively come to be prompted by more abstract entities (Aeschbach 2017: 50) is somehow misleading since what initially is felt against specific objects or persons is not *ressentiment* per se but anger, envy, hostility, hatred and/or resentment.[38] These emotions are incorporated and mutated into *ressentiment* insofar as the transvaluation process is put into motion initiated by the subject's incapacity to act out.

Complexity

It seems that there is a widespread consensus in considering *ressentiment* a complex or cluster sentiment; it also seems that there is no big discrepancy over the constitution of this sentiment, especially among those who differentiate it from resentment. Scheler himself described it as a 'unit of experience' caused by the systematic repression of certain emotions such as vindictiveness, envy, spite, the impulse to detract (*Scheelsucht*), malice, and *Schadenfreude*. These emotions are progressively felt and then are blocked due to incapacity and impotence (Scheler 1961: 45–48, 176). Merton asserts that *ressentiment* is the re-experiencing of diffuse hate, envy and hostility, coupled with powerlessness (Merton 1957/1968: 210). In a similar spirit, Aeschbach (2017: 44, 46, 215, 258) describes it as a broad and complex affective phenomenon and as a sentiment which is constituted by envy, resentment, indignation, revengefulness, hatred, *Schadenfreude*, malice, and anger.

Apparently, there is a nebulous alchemy in *ressentiment* emanating from the very protean nature of its constituent emotions. Not only is each one of them a secondary emotion in itself, but what seem to be of equal importance are the sequence and the order of their intensity. Although not absolutely necessary for my discussion, informed by TenHouten's typology one could preliminarily argue that *ressentiment* is a kind of double complex sentiment comprised by interlocked tertiary emotions. Their juxtaposition would look like this:[39] Envy = surprise & anger & sadness; Hatred = anger & disgust & fear; Vengefulness = sadness & anticipation & anger; Resentment = anger & disgust & surprise (TenHouten 2007: 148, 212, 236; TenHouten 2018: 7);[40] *Schadenfreude* = joy & anticipation & anger;[41] Powerlessness = anxiety & resignation & submissiveness. Herein a word of caution is needed: envy, vindictiveness, resentment, hatred, and *Schadenfreude* wouldn't add up to *ressentiment* without the catalytic role of powerlessness (designated also as impotence or incapacity in this text). For the purposes of my analysis powerlessness is construed as a combination of the three above mentioned secondary emotions which are defined as follows: anxiety = anticipation & fear; resignation = sadness & acceptance; submissiveness = acceptance & fear (TenHouten 2007: 111). Acceptance, anticipation, fear, and sadness involve the self in the sense that: (a) one accepts and absorbs one's social environment, (b) one explores and monitors one's personal and social territory, (c) one feels threatened by an external stimuli; (d) one experiences a loss of a temporary or permanent, imagined or real, valued object, trait or social relationship. It is through the combination of anxiety, resignation and submissiveness that the person of *ressentiment* feels pain as to his/her wounded self-esteem and experiences her/his inferior plight as a destiny.

But this is not the whole picture; secondary and tertiary, and primary emotions for that matter, are felt differently by different people in similar situations or by the same people in different situations which means that the appraisals of external and internal stimuli are contingent upon the social-cultural setting and the coping potential of the individuals. In our case, the mechanics of defense mechanisms involved in the transvaluation process are considerably linked to the dynamics that characterize the configuration of *ressentiment*'s constituent emotions. The likelihood is that the order of intensity of each primary and secondary emotion, as well as their sequence, makes a difference as to the actual feeling of *ressentiment* by particular individuals on the one hand, and its duration on the other. Unless one is prepared to make mental experiments, from this point onwards it is very risky to predict and describe the vicissitudes of this sentiment in real-life situations, even in an ideal typical fashion. Its psychodynamics are so complicated and precarious if only because, like any other emotion and sentiment, *ressentiment* is situated in a web of personality, normative, and socio-historical structural determinations triggering interpersonal antagonist and intrapersonal incongruent emotions which are likely to undermine its grasp and disintegrate its affective constituency. One might, for instance, assume that some of the constituent emotions of *ressentiment* have higher or lesser intensity than others. But it is only through empirical quantitative and qualitative, and hopefully triangulated, research designs that it is possible to explore the impact of *ressentiment* on public opinion and political action, something partly attempted in the following chapters of this book.

Action – inaction

So far in this section I referred to the cause, the valence, the object, the constituency, and the psychodynamics of *ressentiment*. An additional topic to be raised herein regarding the difference between resentment and *ressentiment* is their association with socio-political action. Distinguishing *ressentiment* from resentment is not merely academic but crucial in understanding and describing different aspects of seemingly uniform social and political phenomena. No doubt, the action readiness inscribed in resentment and indignation is intense, if not immediate, under certain circumstances, e.g., rebellions, demonstrations, etc. But what about *ressentiment*? Reading Dostoyevsky, Nietzsche, and Scheler in this matter suggests that the self-poisoned person of *ressentiment*, ruminating hostile emotions while grinding his/her teeth, remains in silent impotence. As these emotions are swarming in the *ressentiment*-imbued person, vainly craving some outlet, there is 'nothing left but to bottle up' his/her 'five senses and plunge into contemplation' (Dostoyevsky 1864/2000: 5, 54). It could be argued that, while resentment is normally associated with action, in *ressentiment* the link between affect, motivation, and action (Frijda 2004a, 2004b) is weakened, blocked or ruptured (Demertzis 2006; 2014). The built-in passivity of *ressentiment* was commented on in part by *Søren* Kierkegaard in his 1846 *The Present Age Book*. There, Kierkegaard – resonating with Tocqueville, speaks of the process of leveling which marks modernity itself and suppresses individuality to a point where the subject's uniqueness evaporates. The social-psychological leverage of this process is what he calls *misundelse*, an emotion which 'hinders and stifles all action' (Kierkegaard 1940: 49). Kierkegaard's translators translated 'misundelse' as *ressentiment* while its literally meaning is envy (Aeschbach 2017: 41). Nevertheless, to the extent that envy is a constituent element of *ressentiment*, what is of interest here is Kierkegaard's idea that *misundelse* hinders all action. In this vein, Brighi (2016: 425) is clear: 'far from being an active and positive mode of political action, *ressentiment* decomposes resistance and incapacitates contestation'.

But is activity–passivity such a crystal-clear distinction as to ultimately determine the difference between resentment and *ressentiment* at every point? And what exactly should we mean by passivity in this context? Albeit depressed, *ressentiment*-imbued persons do live their lives even if underground; they do not stand still. Do we refer, then, to a sort of passive aggressive emotional reaction which captures the morality of weak creatures 'who have been denied the proper response of action' (Nietzsche 1970: 35)? If that is so, how should we understand Scheler's assertion that the French Revolution is the greatest achievement of *ressentiment* in the modern era (Scheler 1961: 155, 177, 196)? And how are we to understand Marc Ferro's account that much of past and present religious conflicts, war atrocities, political revolutions, and anti-colonial movements are fueled by this sentiment (Ferro 2010)? Yet, if at the end of the day the persons of *ressentiment* are gripped by powerlessness and incapacity, if their negative feelings are bottled up, suppressed, and denied overt expression, how can they be or become agents of collective or even individual violent action?

Even if we assume that people feel only *ressentiment* while living their social and political realities and put to one side for a moment the fact that they are currently experiencing a constellation of primary or complex emotions, we could give a meaningful answer to all these seemingly odd questions. The crux of the matter lies in the transvaluation process; as long as this process is taking place, as long as the rumination and the repression of hostile and negative emotions are coupled by chronic impotence to act, the victimized and traumatized person of *ressentiment* remains passive, dormant, deactivated, ambivalent, quiescent, and sluggish. S/he cannot imagine taking revenge for the humiliation and the traumas s/he has suffered. But the moment the transvaluation is consummated with the reversal of the value system offering to this person a new identity based on a self-perceived moral superiority, s/he acquires every right to act accordingly in the name of God, Jesus, the nation, the people, or the working class.[42] The link between motivation and action is restored through the assertion – as illusory as it might be – of a new noble cause. The *ressentiment*-ful person may now maltreat, persecute, torture, and kill, not out of alleged spite or vindictiveness, but by putting her/himself in the service of a higher moral goal. The cunning of morality in all its magnificence! It is the moment when the 'psychological dynamite' of *ressentiment* explodes and impotent victimization ceases, replaced by other, more assertive and aggressive emotional stances. At any rate, however, the consummation of the transvaluation is an open-ended virtual process; it is not an individual's predicament, pointing to an inherent tendency for transformation, from being an impotent and despised and self-despising person to being an assertive force against 'oppression'.

The process of transvaluation may take time, several years or some decades for an individual, several decades or even some centuries for collectives. In either case, I regard it as an incubation period whereas the dynamics of the defense mechanisms implicated in *ressentiment* incapacitate its bearers regarding any effective initiative in the public space.[43] When the transvaluation process is over they feel entitled to act. On top of that, they may also adopt an active posture whenever the structural setting changes at such a pace as to sharply trigger action readiness and disintegrate *ressentiment* into its very own ingredients; in this case resentment, rage or hate prevails over powerlessness and therefore *ressentiment* is bypassed. And not only that: the likelihood is that the repressed resentment and rage have become more intense through their incubation, and the action motivated by them will be probably more violent and destructive than it would have been without *ressentiment*. In this respect, the notion 'organized political *ressentiment*', advocated by Yankelovich (1975), is a *contradictio in terminis.*

So, passivity and activity are not necessarily mutually exclusive attributes of *ressentiment* and resentment respectively, but also two versions of the action readiness immanent in *ressentiment* itself. How else then could one understand Scheler's or Sombart's explanation of the French Revolution premised on *ressentiment*?[44] The interpretation offered hereby will be employed later on for the analysis of populism and nationalism. For the time being there will be a brief digression to look at the role of *ressentiment* and resentment in modern and late modern politics.

Political bearings

For over a hundred years scholars have been writing about *ressentiment*, following in the footsteps of Nietzsche and Scheler. Thereafter, a host of nuances in the interpretation of this concept have come to the fore. Scheler, LeBon, Ferro, and others have connected *ressentiment* with the French and other Revolutions, rebellions, insurgencies,[45] religious wars, pogroms, and post-colonialism.[46] The force of the concept is such that prominent scholars argue that the politics of late modernity is a politics of generalized *ressentiment*, cropped by the uncertainties of late capitalism, and the individuals' diffuse sense of powerlessness, the public expression of which is not positive and self-grounded praxis but a hasty and dependent reaction which usually takes the form of 'identity politics' and ethicism (Brown 1995: 21–76; Connolly 1991: 22–23, 207). This argument goes in tandem with Bauman's early remark that class realignments in late capitalism and consumerist culture jointly produce a great potential for *ressentiment* via social comparison (Bauman 1982: 179–180). It also aligns with Pankaj Mishra's argument that, today, *ressentiment* is the defining feature of a world where the modernity's promise of equality is at odds with massive disparities of power, education, status and property ownership (Mishra 2017: 31).

Jean Baudrillard claimed that postmodern consciousness reflects a period of *ressentiment* and repentance generated by systematic efforts of governments and civil societies to memorize traumatic events of the past in a media saturated environment. Yet, the media are shaking the sense of reality through the proliferation of hyperreal simulations. The duplication of the real via all sorts of media brings about a hallucinatory semblance of what actually happened in earlier times so that acts of memorialization of the traumatic past become sorts of 'false consciousness exemplified in gestures of sanitization and revisionism' (Wyschogrod 1998: 94). In this respect, 'celebration and commemoration are merely the soft form of necrophagous cannibalism' forwarded by

> museums, jubilees, festivals, complete works, the publication of the tiniest of unpublished fragments ... [o]ur societies have all become revisionistic: they are quietly rethinking everything, laundering their political crimes, their scandals, licking their wounds, fuelling their ends...This is the work of the heirs, whose *ressentiment* towards the deceased is boundless.
>
> (Baudrillard 1994: 22)

And this boundless *ressentiment* is premised on the immanent incapacitation of postmodernism's simulacra harbouring the heirs of past traumatic realities.

In the same vein, other scholars argue that *ressentiment* is an essential trait of today's silent majorities, stirred by mediocrity and identity preservation (Pirc 2018: 112*)*. *Ressentiment* is often used to critically interpret political ideologies and doctrines such as Nazism, varieties of feminism, egalitarianism, nationalism, Communism, or fundamentalism. In the previous section we referred to Posłuszna and Posłuszny (2015) who explain extremism and fundamentalism as active validations of *ressentiment* which – despite the value reversal – cannot ultimately offer a permanent alleviation of the humiliation perceived by post-colonial populations and militant groups.

A strong case in this direction is put forward by Žižek (2008). Although Posłuszna and Posłuszny do not refer to his work on violence, it seems to vindicate their argument: terrorist fundamentalists are not true believers, they feel insecure about their cultural-religious identity and so break out against westerners' sinful lives. Should they be true believers, they would remain calm when encountering western hedonism or even the caricatures of Mohamed in European newspapers. Comparing themselves to others, they are not convinced about their own moral superiority; 'the problem with fundamentalists is not that we consider them inferior to us, but rather that *they themselves* secretly consider themselves inferior' (Žižek 2008: 86). Žižek maintains that terrorist violence results from *ressentiment* itself and not from the failure of the defense mechanisms involved in this complex emotion. This is so because he clearly equates or conflates *ressentiment* with envy, instead of seeing the latter as a component of the former. But even if he deviates from the typical understanding of *ressentiment*, whose core meaning rests on the transvaluation process, his interpretation corroborates my own and Posłuszna and Posłuszny's thesis that re-evaluation is always incomplete and thus the moral superiority the *ressentiment*-ful persons gained in the first place has to be sustained by every means at a later stage.[47]

In another chapter of his essay, Žižek speaks of resentment while critically commenting on Sloterdijk's arguments against modern totalitarianism. As in his *Critique of Cynical Reason* (1988/1983) a few decades ago, in his *Rage and Time* (2010/2006) Sloterdijk opts for a grand theory of western culture. First there was the great antagonism between cynicism and kynicism;[48] now there is the antithesis between Eros and *thymos* that permeates the history of the West. Under the designation *thymos* – a Homeric ancient Greek word – Sloterdijk accommodates envy, rage, completion, pursuit of pride and recognition.[49] Against Camus' sublimated notion of revolution, he holds (2010: 120) that thymotic irritability is what leverages leftist political movements, Communist terror (in either the Stalinist or Maoist version), and contemporary fundamentalisms. In the last chapter, Sloterdijk adopts Nietzsche's ethos of nobleness for building a 'detoxified worldly wisdom' premised on classic bourgeois virtues and values of the right to life, freedom, and property (2010: 227–228). In addition, he insists that 'under conditions of globalization no politics of balancing suffering on the large scale is possible that is built on holding past injustices against someone, no matter if codified by redemptive, social-messianic, or democratic-messianic ideologies' (2010: 228). What is needed then is *a rage free of resentment, a rage that successfully balances eros and thymos* to go thus 'beyond resentment' through matching entropic inclinations and pointing to meritocracy and relaxed antiauthoritarianism.

It is precisely here that Žižek criticizes Sloterdijk (whom once he admired for the cynicism vs kynicism schema);[50] for the Slovenian scholar, to go beyond resentment is not an option when heinous acts of the past cannot be denounced in the name of liberal progress. Alluding to Améry and Sebald, Žižek wants to 'rehabilitate the notion of resentment' (Žižek 2008: 189) towards an anti-Nietzschean direction. 'Authentic' resentment should not be related to the slave morality since it stands for a refusal to forget, to compromise, and normalize the crime and odd monstrosities. Against the moral praising of pure forgiveness I considered in the previous chapter, he sees resentment as an integral feature of contemporary politics of justice related to 'the triad of punishment (revenge), forgiveness, and forgetting' (Žižek 2008: 190).

So far, so good; there is, however, a conceptual-terminological vagueness in this debate. In his original book (2006) Sloterdijk uses *ressentiment* which is invariably translated into the English edition as 'resentment'; on his part, although he uses and refers to the German edition, Žižek adopts 'resentment' not only while criticizing Sloterdijk but also when he refers to Améry who always employed *ressentiment* to express his bitterness, detestation, and non-forgiveness for the Nazis. The negative thymotic meaning Sloterdijk gives to *ressentiment* is directly drawn from the Nietzschean tradition and only thus does his plea to move beyond it make sense. Therefore, I consider its translation into English as 'resentment' quite wrong. The question is whether Žižek wants to dismiss *ressentiment* altogether in favour of resentment adopting the translator's account. I tend to believe that this is actually the case on the basis of his neo-Leninist advocacy of politics (or one could say faith in them?) for the sake of which he put forward somewhat paradoxical positions based on over-extended and circumstantial interpretations of Lacan, Hegel, and Kierkegaard. How else can it make sense when someone claims that sometimes 'hatred is the only proof that I really love you' and that 'the domain of pure violence ... the domain of the violence which is neither law-founding nor law-sustaining, is the domain of love' (Žižek 2008: 204–205)? Resentment is kept apart as a valuable concept for all its affiliation with *ressentiment*.

From a quite different normative standpoint, in a systematic attempt to explain ethnic violence in the eastern European countries during the twentieth century, Roger Petersen develops an emotions-based theory of conflict and employs resentment as an 'instrumental' emotion that facilitates individual action to satisfy a particular desire or concern. Resentment prepares for the rectification of perceived imbalances in group status hierarchies and in this sense places someone against a group that stands in the way of achieving a blocked desire (Petersen 2002: 19, 24, 29, 40–41). Resentment functions either as a perpetuating factor of group animosities or as a catalyst of violence.[51]

Employing, as we saw above, a tailor-made notion of *ressentiment*/resentment which resembles envy, vengeance, and indignation, Scruton detests totalitarianism and fundamentalism emanating from a self-confessed conservatism. With his critique of Stalinism lagging well behind the western Marxist critical accounts of totalitarianism, Scruton sees both a utility and a danger in resentment; on the one hand, premised on a sense of justice, resentment helps us to keep a proper distance from each other with respect. It is 'to the body politic what pain is to the body: it is bad to feel it, but good to be capable of feeling it' as long as it contributes to mutual aid and co-operation. On the other hand, however, it becomes detrimental to social well-being when it 'loses the specificity of its target and ... becomes instead an existential posture: the posture of the one whom the world has betrayed' (Scruton 2006: 154–155). It is in the second version that resentment seeks revenge and supports totalitarianism in the name of groups of people or classes who have been hitherto unjustly subordinated, who are destined henceforth to be still excluded – but justly. Revolutionary justice is done through a generalized resentful enmity that closes the gap between accusation and guilt, depriving its victim not only of his/her worldly goods but of his/her humanity as well, something typical in political and religious terrorism (Scruton 2006: 157, 159–160). Scruton makes clear that it is the elites which gain advantage from the widespread resentment and that this emotion is released when totalitarianism takes power.

A big move toward discussing these emotional terms occurred just before and after Trump's electoral success, which is thought to be premised both on his own arriviste *ressentiment* and the Trumpists' deep-seated *ressentiment*-ful mentality (Hackett 2016; Salmela and von Scheve 2017, 2018). Although his white workers and less educated voters were unlikely to improve their class interests, his discourse seems to have vented their *ressentiment* against the professional classes, temporarily (Wimberly 2018: 189, 195). No doubt, however, Trumpists' *ressentiment* is stirred by grievances about extant inequalities and status anxieties (Hoggett 2018) and as there are a lot of people in the West who blame distant elites for unfairly allowing or accelerating their loss of economic and political power, Trumpism seems to be much bigger than Trump himself (Knauft 2018) pointing to what has been called 'the politics of resentment'. As important as they might be for international affairs, the politics of resentment are not confined to the polarized and fragmented political domain of the USA, as nicely described for instance by Katherine Cramer (2016). It draws much from global value change towards a 'silent revolution in reverse' and a sort of 'authoritarian reflex' (Inglehart 2018: 174ff.) and 'democratic recession' (Diamond 2015). The value authoritarian reflex 'is likely to fuel resentment both upwards toward elites and downwards toward out-groups of lower status' (Norris and Inglehart 2019: 123). Thus, resentment of the establishment among the 'left-behinds' gives rise to welfare populism driven by a sense of unfair treatment and discontented national identities within the wide-open space of globalization (Mann and Fenton 2017: 18, 22, 33).[52] Within this political cultural environment of backlash, people of different origins in different countries espouse a variety of responses while searching for meaning; infantile illusions, compliance, apathy, emotional self-destruction, anger, rage, and poignant postures, extreme radicalization, *ressentiment* and resentment are the responses discussed thoroughly in recent literature of all kinds.[53] At large, searching for meaning means that one is looking for recognition and dignity. In his Master-Slave dialectics, Hegel was among the first to realize that striving for recognition is a motor of history and, before him, it was Smith who founded human sociality on mutual approbation. As a constant of human conviviality, search for dignity in the public sphere is what the political organization of resentment is about, conveyed by today's postmodern identity politics because, if nothing else, Real-world liberal democracies never fully live up to their underlying ideals of freedom and equality (Fukuyama 2018: 46). In a way then, all revolutions and rebellions are revolutions of dignity. The same holds true regarding populist mobilization and populist political discourse, to be looked at in the next chapter.

Synopsis

Drawing from moral philosophy, historical, sociological and psychoanalytic literature, this chapter covered two kin yet different complex social emotions: resentment and *ressentiment*. The former is conceived as moral anger and the latter as an affective compensation for life failures. The weight of the chapter lies with *ressentiment*, the compensatory nature of which consists in value reversal experienced by inefficacious social agents through the disavowal of what is unconsciously desired. In a nutshell, the individual and societal roots of

ressentiment are: not acted out vindictiveness, chronic interiorized powerlessness, a gap between legal rights and real entitlement, social comparison, and
injustice experienced as destiny. Special emphasis is put on the political repercussions of *ressentiment* depicted from different historical contexts. Argumentation
is supported by recurrent references to novels and autobiographies showcasing
the composition of *ressentiment* and its roots in individual and societal dynamics.

Notes

1 Bishop Joseph Butler, to whom Hume owes a lot, preached that resentment has been
  implanted in man by God himself according to the Divine Providence. As an Enlightenment philosopher, Hume replaces God with Nature.
2 To be sure, however, Josef Butler distinguished between what he called 'hasty and
  sudden' and 'settled and deliberate' resentment (Butler, Sermon IIIV & IX). The first is
  sheer anger caused by harm; the second a remedial reaction against injuries, vice, and
  wickedness deemed as 'one of the common bonds, by which society is held together;
  a fellow feeling which each individual has in behalf of the whole species, as well as of
  himself'.
3 The issue of course is immense and cannot be given adequate treatment here, but suffice it to mention that Smith's impartial spectator shares much with Kant's Categorical
  Imperative. Just indicatively, see Fleischacker (1991).
4 It seems that this expectation is the hallmark of what Simone Weil (1962: 10) calls 'the
  sacred'. In her own words:

  > There is at the bottom of every human heart something that goes on expecting,
  > from infancy to the grave, that good and not evil will be done to us, despite the
  > experience of crimes committed, suffered, and observed. This above all else is
  > what is sacred in every human being.

  The good is the only source of the sacred. There is nothing sacred except the good and
  what pertains to it'.
5 'both my description of, and my name for, these attitudes are a little misleading' (ibid).
6 From a utilitarian moral psychological perspective, Joshua David Greene assumes
  that there is an inbuilt sense of justice that makes cooperation within groups possible. Together with other moral stances, it has been cultivated through natural selection
  (Greene 2013).
7 Collective emotions are experienced concurrently with others while group-based emotions are experienced on the basis of group identification, either alone or with other
  group members.
8 In the same vein, resentment was conceptualized as righteous indignation by Sennett
  and Cobb (1972: 117–118, 139, 148) in their study of the post-World War II American
  working-class consciousness.
9 On the contrary, Dennis Smith (2006) sees resentment as an exclusively emotional
  reaction to humiliation experienced as an undeserved social displacement that may
  occur by conquest, exclusion or relegation.
10 Nussbaum (2016: 262–263) actually dismisses resentment as a moral emotion as
  long as she holds that wrongdoings frequently call upon a payback wish inherent in
  anger.
11 In his comprehensive doctoral dissertation, holding that resentment differs from
  indignation insofar as the latter responds to impersonal wrongs and the former to
  personal inflictions, Sébastien Aeschbach diverges from the moral philosophical tradition which takes indignation as resentment against undeserved and unjust suffering

of sympathetic third persons. Confining resentment within the interpersonal cycles of interaction only, his definition goes as follows: (1) it is a kind of anger which is aroused when one is wronged; (2) it necessarily involves a desire for revenge; (3) it has the injustice of an unremedied wrong as its 'formal' object; (4) it has persons and their action as its 'proper' object (Aeschbach 2017: 13).

12 Ten years before, Elster (1989: 61) had initially welcomed the use of emotions in social analysis, thinking of them as the 'stuff of life' since 'creatures without emotions would have no reason for living'. For all his inspiring but no less brief argumentation over emotions, it was excessive on his part to claim that 'the importance of emotions in human life is matched only by the neglect they have suffered at the hands of philosophers and social scientists' (1989: 61).

13 Aaron Ben-Ze'ev relies on an individual level of analysis when it comes to resentment; he sees it as an emotional protest against perceived injustice and moral wrongdoing inflicted by other people who are held responsible. The perception of injustice occurs amidst downward interindividual comparisons. In this context, 'when we feel resentment the focus of attention is the other's improper activity, but the focus of concern may be its implications for our own status'; thus, unlike anger, hate, envy, and forgiveness, resentment is directed to the action rather than the actor herself. It lasts quite a lot longer than anger and it differs from envy because it focuses on the wrongfulness of our inferior situation and not on the inferiority as such; it conveys therefore a public and private accusation (Ben-Ze'ev 2000: 31, 283–284, 396, 465, 507).

14 Employing the term '*ressentiment*-ful person' or the 'man of *ressentiment*' might give the impression that I am referring to a role performance or a process of self-formation rather than to a composite sentiment. The truth is that I am using these terms colloquially, following some of my sources' terminology and certainly Scheler's own vocabulary. Scheler viewed *ressentiment* as fuelling different social roles like that of the spinster, the elder, the mother-in-law, the bourgeois vs the aristocrat, the retired official and so on. Jack Barbalet drew my attention to this issue.

15 It should be noted that when his book, originally entitled *Über Ressentiment und moralisches Werturteil*, appeared in 1912 (in 1915 it was republished under its final title *Das Ressentiment im Aufbau der Moralen*), Scheler was a deeply concerned and self-questioning Protestant who later on (1920) turned to Catholicism which he would soon abandon (1924) to espouse a vitalistic and no less pantheistic position. It is due to his philosophic Catholicism that he defended so ardently Christian love and morality.

16 However, insofar as Classical Enlightenment (Kant, Hegel, etc.) is not reduced to cheap and mundane versions of market-oriented liberalism and authoritarian socialism, Scheler seems to over-generalize his critique to encompass Enlightenment and Modernity *in toto*.

17 Compared with Aristotle's account of anger, the *ressentiment*-ful person in its Nietzschean version is an angry person, although not in an open way precisely because anger is 'attended by a certain pleasure – that which arises from the expectation of revenge' and 'by a certain pleasure because the thoughts dwell upon the act of vengeance, and the images then called up cause pleasure, like the images called up in dreams' (Rhetoric 1378b); on the contrary, the Schelerian version of the *ressentiment*-ful person is not and cannot be angry because the very imagination of revenge has already been repressed beforehand.

18 Class hatred could be viewed as a sort of compensatory conduct through which those who lack economic and cultural resources can claim some superiority (Sayer 2005: 183).

19 'The "equality of souls before God" – this fraud, this *pretext* for the *rancunes* of all the base-minded – this explosive concept, ending in revolution, the modern idea, and the notion of overthrowing the whole social order – this is *Christian* dynamite' (Nietzsche 1895/2014: Chapter 62).

20 It seems that Adam Smith's notion of envy deflects from the egalitarian assumption since he defines it as 'that passion which views with malignant dislike the superiority of those who are really entitled to all the superiority they possess' (Smith 1759/1976: 244).

21 For Vincent de Gaulejac (1987), class neurosis is the intersection of an individual's blocked psycho-sexual development with the blockage of the social self under conditions of downward and/or upward social mobility.

22 Nevertheless, she admits that her promise to herself never to desire to be white did nothing 'to drive away the wish dreams that filled my head whenever my desires collided with a taboo. So, in order that my daydreams [did] not contradict my principles, I constructed a fantasy in which I would slip on a white face and go unceremoniously into the theatre or amusement park or wherever I wanted to go. After thoroughly enjoying the activity, I would make a dramatic, grandstand appearance before the white racists and with a sweeping gesture, rip off the white face, laugh wildly and call them all fools' (Davis 1988: 85).

23 I use '*ressentiment*' as the translation of Ferrante's 'rancore' (rancor).

24 It might be argued that praising something is also part of repression and negation, not merely a consequence of these. Its function is convincing oneself more than others. I owe this insight to Mikko Salmela.

25 Together with other nineteenth-century thinkers such as Louis Blanqui, Abel Rey and J. M. Guyau, who advocated a cyclical cosmological view, Gustav Le Bon might have in turn influenced Nietzsche in formulating the idea of eternal recurrence (Williams 1952: 109; Brandes 1972: 49).

26 *Bien que s'y mélangeant de plus en plus, elle se sentait, néanmoins, tenue à distance et en éprouvait un vif ressentiment* (Le Bon 1912: 59).

27 The book was translated and edited in English very shortly after by Mortimer Epstein with the title The Quintessence of Capitalism: A Study of the History and Psychology of the Modern Business Man. New York: E. P. Dutton, 1915. In 1998 it reappeared as a Routledge Publication and was translated and edited in Greek. References are to the Greek edition.

28 Von Martin, *Soziologie der Renaissance. Zur Physiognomik und Rhythmik bürgerlicher Kultur*. Enke, Stuttgart 1932. For the needs of the analysis herein I used the 2016 German edition and the 1944 English translation. The *Sociology of the Renaissance* has long been a touch point for students of the Renaissance period. Be it noted that in the 1944 English translation *Ressentiment* was translated either as 'reaction' or as 'bitter resentment' (von Martin 1944: 36, 44, 45) meaning probably that the thematization of Nietzsche in British sociological vocabularies of that time was limited.

29 '*So kommt es zu dem inneren Aufstand der sich aus Einzelnen – und sich ihrer Einzelheit mit Stolz Bewusten – zusammensetzenden burgerlichen Intelligenz gegen die Masse des Burgertums, insbesondere gegen das Burgertum als Handelsstand, als kaufmannischen Berufsstand. Der Mann, der uber nur geistiges Kapital verfugt und von ihm leben will, wirdzwar erst auf burgerlichem Boden moglich; aber er fuhlt sich doch zumeist vom Burgertum, unten gehalten und reagiert mit Ressentiment gegen eine allzu burgerliche Geistverachtung bei der besitzenden, wirtschaftlich erwerbenden und politisch machtigen Schicht'* (Martin 2016: 50).

30 This, however, does not preclude the possibility that both resentment and *ressentiment* may arise from major wrongdoings.

31 Aeschbach (2017) has done terrific work to this end from a philosophical point of view, focusing mainly on the phenomenology of *ressentiment*.

32 See for example Roger Scruton's (mis)treatment of the Nietzschean *ressentiment* in his interpretation of totalitarianism. He merges resentment and *ressentiment* and states that Nietzsche himself 'would perhaps not have recognized' the way he uses the notion. Of course, as Meltzer and Musolf (2002) claim one should not reify concepts but the

crucial issue here is how much one is entitled to change them without losing their historicity, distinctiveness and clarity. Transvaluation is a *sine qua non* for the conceptual constitution of *ressentiment* and Scruton (2006) dismisses it altogether.

33 Aeschbach (2017: 10, 50, 258, 69) oscillates between thinking of *ressentiment* as a 'psychological mechanism that manipulates evaluations in a certain way with the goal of feeling better about oneself' or as the 'very mechanism that transmutes envy into moral emotions such as resentment and indignation', and as a long-lasting complex sentiment.

34 In their latest contribution, Salmela and von Scheve (2018) stress that social sharing of the transformed emotions which make up *ressentiment* reinforces the entire value transmutation process toward specific targets.

35 Thus confesses Dostoyevsky's mouse man: '… secretly, inwardly gnawing, gnawing at myself for it, tearing and consuming myself till at last the bitterness turned into a sort of shameful accursed sweetness, and at last – into positive real enjoyment! Yes, into enjoyment, into enjoyment! I insist upon that' (Dostoyevsky 1864/2000: 10).

36 He did not, however, set *ressentiment* apart from resentment; as a particular kind of political resentment, he viewed it as 'the second most powerful political emotion after instability' that 'really corrodes societies' (Yankelovich 2009: 26).

37 The same holds true for hatred, though not entirely, because one can hate a particular person in isolation and not just types of people, groups or categories; yet repressed hatred serves as a substrate from which *ressentiment* may spring forth. See, among others, Nussbaum (2016: 50) and Brudholm (2010). Hatred is likely to emerge from repressed anger that the person or group is not capable of expressing or acting out in a satisfying manner (Halperin *et al.* 2011).

38 In their qualitative study Hoggett, Wilkinson, and Beedell (2013) found resentment and grievance to be much more directed at one's 'neighbour' than at powerful but less visible groups and elites. They also found that their informant's *ressentiment* was objectless and expressed via a general complaining tone directed towards 'the authorities' and to categories of people perceived as being different (single parent, ex-offender, mentally ill, having special needs).

39 The symbol '=' means 'results from' and the symbol '&' means 'and' or 'joined with' (TenHouten 2007: 51).

40 Viewing *ressentiment* as a form of generalized resentment, TenHouten (2018: 10) maintains that it is comprised of the same three primary emotions (anger, surprise, disgust) and the same three secondary emotions (contempt, shock, and outrage) that resentment consists of. He also states that *ressentiment* is typically accompanied by other complex emotions and affective states of mind, which can vary greatly by situation and circumstance, such as vengefulness, hatred, bitterness, and malevolence. In view of Scheler's analysis, I do not think that TenHouten's position is valid as long as he disregards that the crux of *ressentiment* itself are those complex emotions which are allegedly circumstantial to its emergence.

41 *Schadenfreude* is a culture-specific emotion, meaning that, quite apart from the occurrence of punishment, the resenter takes pleasure in all bad things that rightfully happen to her/his wrongdoer. TenHouten has not thematized this emotion but one can safely use his equation and treat it as a tertiary emotion. For a sharp-witted analysis of *Schadenfreude* – which is also a *terminus technicus* – see Portman (2000) and Dijk and Ouwerkerk (2014). Emphasis should be placed here on 'rightfully', on the sense of justice that accompanies the vindictive joy one experiences facing the sufferings of a third person ('he asked for it', 'he should have watched his step'). If it wasn't for the dimension of justice we would be dealing simply with grudgingness or malice.

42 Adopting the position of the victim and endowed with a particular sense of moral superiority, the *ressentiment*-imbued persons imagine that they have all the virtue on their side, it is the other that is the villain, wrong, guilty, etc. (Hoggett, Wilkinson, Beedell 2013: 578).

43 Turning upside down the Nietzschean approach, Solomon (1994) claims that trans-valuation *qua* passivity is a strategy of the will to power that is specific to the weak. Consequently, powerlessness is not seen as a cause of *ressentiment* but as an outer manifestation of the weak persons' eagerness to take revenge in their own terms, as an expression of their own will to power. For Solomon *ressentiment* is the ever most clever and life-preserving emotion of the slave in his confrontation with the master. From a psycho-political perspective Tadej Pirc denounces this posture, claiming that Solomon has 'caught himself in the very web of the slaves' revolt in morality' which results from a double castration: the castrating lack of efficiency, and the castration of 'the master morality by the downtrodden' (Pirc 2018: 125).

44 At this particular point Meltzer and Musolf (2002: 244) discern a contradiction in Scheler's thesis without however really taking into account his account of transvaluation. Having done this, they would mix up resentment and *ressentiment*.

45 Although Albert Camus was well aware of Scheler's work, admitting that revolutionary rebellion is more than oftentimes contaminated with *ressentiment*, he nevertheless wished for an existential rebellious stance against the Absurd, cleansed of any trace of *ressentiment*. Revolution should be pure love and fecundity or it is nothing at all. Instead or rancor, malice and tyranny, Revolution should be an affirmation of life by men of flesh and blood, a passionate self-assertion and one's own creation of the Self premised on values that exceed the individual for the sake of a transcendental good; on this basis, Camus was at pains to 'understand why Scheler completely identifies the spirit of rebellion with *ressentiment*' (Camus 1991: 304–305, 18). Ultimately, Camus' *L'Homme révolté* and *Sisyphe* are one of a kind representing 'the "yes-saying" or loving attitude towards existence as opposed to the "nay-saying" attitude of the Nietzschean men of *ressentiment*' (Novello 2015: 12). Note that the translator of *The Rebel* substituted *ressentiment* in the original text for resentment, thus blurring Camus' argument for the English-speaking reader.

46 In a way then one is to wonder whether *ressentiment* is the father or the child of modernity. But I think this is more a kind of metaphor rather than an accurate interpretation of the complex emotional underpinnings of modernity (Becker 2009).

47 For a similar interpretation of Islamic terrorism alongside Scheler's account of *ressentiment* see Manfred Frings (2005).

48 See Chapter 1 on political cynicism as political emotion.

49 This is the way Ferro (2010) treats *ressentiment* and that is probably why his translator accounted for it as 'resentment' as indicated earlier in this chapter. He uses it as a catch-all or metaphoric notion conveying actual envy, resentment, jealousy, anger, and rage. For a similar critique see Aeschbach (2017: 253) who says that Ferro either has resentment in mind or fails to distinguish the phenomenon of *ressentiment* from resentment when discussing his cases.

50 In a more systematic way, Mario Wenning criticizes Sloterdijk's position as reformist because it suffocates productive rage rather than organizing it for the right causes. Sloterdijk's argument is based on a historical optimism that is difficult to reconcile with a political reality replete with blatant injustice which requires something more than an unconditional trust in civilizational learning processes (Wenning 2009: 97).

51 Relatedly, Greenfeld and Eastwood (2005: 263) holds that *ressentiment* serves as a decisive motive behind ethnic nationalist liberation movements.

52 Conflating the twin emotions, on the one hand and dismissing altogether the importance of the defense transvaluation process on the other, Mann and Fenton speak ultimately of envy rather than resentment or *ressentiment*. They attempt to relate class identities and resentfully based nationalism as they are reflected and mediated in contemporary British electoral politics.

53 For a psychoanalytic perspective, see David Levine (2018).

Chapter 7

Populism and the emotions

Introduction

Ever since late '60s almost every single publication on political populism[1] refers to its conceptual elusiveness. Scholars have been at pains to offer an adequate definition for synchronic and diachronic comparative analysis alongside a common minimal denominator balanced in terms of conceptual intention and extension (Rovira Kaltwasser *et al.* 2017). In spite of its being an 'elusive political animal' (Fieschi 2004: 237), populism ranks among the most fashionable and, concurrently, most essentially contested concepts in the social sciences, and in mainstream politicians' and media commentators' discourse (Glynos and Mondon 2019). It is indicative that in 2016, the number of searches for the word 'populism' documented on Google Trends increased by a factor of five, compared to its average in 2012–2015 (Guiso *et al.* 2017). Not infrequently, the conceptual over-extension of its meaning is such that competitive authoritarianism, nationalism and right extremism are squeezed under the unspecified categories 'populism', 'populists' and 'populist phenomenon' (e.g., Mounk 2018; Müller 2016). Throughout various waves of scholarship (Pappas 2016), whereas different theoretical approaches conjured (ideational, agentic, socio-economic, discursive/rhetoric, etc.) (Mudde and Rovira Kaltwasser 2017: 2–4), political populism has been analyzed mainly from the supply-side. It is scrutinized in terms of actors (people, elite, or leaders); organization (movement, party); style (moralistic, simplistic); consequences (polarization, homogenization); domain (old–new, left–right, democratic–nondemocratic, European–non-European); and normative implications (threat to or corrective of democracy) (Pappas 2016; Rooduijn 2014; Norris and Inglehart 2019: 22). Much research points to official political discourse (party manifestos, media content, party organization and electoral geographies, etc.) and as the lion's share of the research has focused on the definition, on the causes for its emergence and on the impact of political populism on public affairs (e.g., Norris and Inglehart 2019), less prominent in the literature are measurements of populist attitudes at the individual level. As the 'empirical studies of populist attitudes are still in their infancy' (Mudde and Rovira Kaltwasser 2017: 99), researchers from either side of the Atlantic (e.g., Hawkins *et al.* 2012;

Akerman *et al.* 2014; Andreadis *et al.* 2016; Castanho Silva *et al.* 2019) have recently tested different scales of populist outlooks in an effort to tap the essentials of the phenomenon at the micro level (people-centrism, anti-elitism, 'us and them' dichotomy). As a matter of fact, even less considered are the emotional underpinnings of political populism.

In this chapter I will discuss thematizations of the affective dimensions of political populism; the chapter will also focus on particular emotions that have been used or could be engaged in the understanding of the affective properties of populisms with a special reference to the Greek post-authoritarian populism of the 1980s and the 1990s.

Defining populism

A brief note is needed as to how populism is identified in this chapter. By putting myself at a distance from the mainstream ideational definition (Mudde 2017) and leaning towards the Essex School approach (Stavrakakis 2017a), I am upholding a 'schematic' conception of populism. I deviate from the mainstream approach because, even seen as a 'thin-centered ideology', populism does not cease to be considered as an 'ideology', i.e., a more or less organized set of ideas involved in the legitimation of political power relations. Yet populism appears in such a vast array of forms in so many different national contexts (Moffit 2016) that one can easily 'acknowledge the lack of coherence and continuity in terms of values, policies, programs' (Stavrakakis *et al.* 2016: 4). Its immense geographical dispersion and its multi-dimensionality are well depicted in a handbook of populism edited by Rovira Kaltwasser *et al.* (2017).

It is true that populism has an inherent incapacity to stand alone outside an ideological host vessel (Fieschi 2004: 237) and that its 'chameleonic', 'symbiotic', 'parasitic', and 'articulating' nature means that it always accompanies other political ideologies rather than the other way around, but this does not render it an ideology in itself – soft, thin-centered or otherwise. It attaches itself to mainstream ideologies not as a set of ideas arranged in a lesser, as it were, degree but as an orienting schema; i.e., as a mental structure by which individuals represent the political world and assimilate new information from their political environment. The populist schema has two principal slots through which individuals meaningfully orient themselves towards the political field: the 'people' and the division of society between two main power blocs. It is this schema which is likely to attach to different ideologies, contexts and conjunctures so that there are too many faces of populism to be manifested. Notwithstanding its roots in Kant and Bergson, the way I use 'schema' in this chapter draws not only from cognitive psychology, where it is registered as a perceptual, cognitive and emotive structure (Arbid 1985; Davou 2000: 288–295), but also from Bourdieu's (1977) *habitus* which consists of schemata of perception, cognition and practice that make possible the interplay of internalization and externalization of reality in social interaction. In a more or less similar vein, from a psychological anthropology perspective,

Friedman and Strauss (2018: 6–7) employ schemas to illustrate the multifaceted character of political subjectivities in a variety of political-cultural settings such as the politics of every-day life and social movements, concurrently taking into account cognitions, feelings and symbolic meanings in the way individuals make sense of politics. In this respect, they view schemas as affectively charged cultural models through which people understand how the world of politics works, who is 'us' and who is 'them'; they also use schemas to interpret their own experience 'imbuing perceptions with larger meanings and associated emotions'.

Although there is no unbridgeable gap between them, a difference between seeing populism as a schema and seeing it as a discourse composed of people-centrism and societal division (Stavrakakis 2017b) is that the latter is too formal and placed at the supply side, whereas the former, as an attribute of thinking, feeling and willful individuals, is located primarily at the demand side of the populist phenomenon. It is this two-dimensional schema that, given the contingency and variability of political antagonisms in terms of geographical dispersion and chronological depth, informs populist attitudes and sustains the symbolic and affective investments of populist demands.[2]

Emotionally analyzing populism

Populism can be studied, jointly or not, as a political discourse, as an ideology, as a movement, as a regime, as a party, as a code or style of rhetoric, as a mentality, as political cognitive schema, as a dimension of political culture. In one way or another, throughout all phases of its study ever since the late 1960s, many interpretations of populism have embraced the affective factor while pointing to charismatic leadership, to romantic folkish appeals, anti-elite stance, and the like. Analytically, however, its 'affective economy' has been mainly studied in a disguised or implicit way since well before the emotional turn in the social sciences and this continues; much more attention has been given to the ideological and discursive aspects of populism, to its organizational forms, to the structural drivers for its appearance, and of course to its very definition. Oftentimes, then, and similarly to the way they are treated in the media analyses discussed in Chapter 4, emotions and sentiments in the study of populism are used in a metonymic way. Namely, the analysis of populism in general and in each case study is carried out through the use of general affective categories and not through the interpretation and explanation of concrete emotions or sentiments. The concrete and particular emotions are hidden beneath generalities such as 'passion', 'affect', 'populist sentiment', 'affective economy' (e.g., Mudde and Rovira Kaltwasser 2017: 27, 35–36, 59, 102, 104, 108, 111; Hokka and Nelimarkka 2019). For example, it has been argued that the 'discontent' of the farmers and lower middle class caused by the enforcement of economic modernization, as well as the 'antipathy' and 'alienation' they felt towards the power elites (Hennessy 1969: 29, 46; Taggart 2000: 43) contributed to the appearance of North American and Latin American populism. Also, 'frustration' caused by neo-liberal fiscal policies is said to foster contemporary European populist demands (Mudde and Rovira Kaltwasser 2017: 38, 40, 102), the analysis of which should account for their 'affective investment' (Stavrakakis 2017b: 7).[3] But 'discontent', 'alienation', 'frustration', 'antipathy', and 'affective investment' are general affective categories which may cover a wide range of specific primary or secondary and tertiary emotions such as, for example, hatred, rage, envy, indignation, sorrow, etc. Likewise, in the interpretation of Russian populism (but not restricted to it) the romantic idealization of the peasant community by the intellectuals and the myth of 'the people' are referred to as being central analytic categories (e.g., Walicki 1969: 79; Taggart 2000: 46). However, idealization and mythologization are mechanisms producing imaginary constructions which only indirectly refer to 'real' sentiments, which actually are their 'raw material': joy, hope, nostalgia, admiration, pride, exultation, etc. The same holds with the logic of equivalence, advocated by Laclau and Mouffe (1985: 130), according to which the complexity of societal-class differences is simplified into two antagonistic camps with the signifier 'the people' as the nodal point which discursively organizes the populist space. Apparently, a lot of emotional energy is involved in this process but yet again the emphasis is placed on the description of the identification mechanisms and not on the concomitant emotions and sentiments supporting them.

Apart from the metonymic there are also incomplete uses of the emotions in the analysis of political populism. A case in point is the 'mythical heartland', which Taggart (2000: 95–8, 117) considers as a *sine qua non* of every populism. But apart from its imaginary substance, Taggart does not clarify the affective content of this notion. Another example is the notion of the 'populist mood' presented by Canovan (1999) as a fundamental ingredient of populist movements. She argues that populist politics cannot but be based on 'heightened emotions' for charismatic leaders, 'enthusiasm', and spontaneity. This is so not only for the historical cases of authoritarian populism (Nazism, Bonapartism, etc.), but also for the 'healthy' populisms which appear in western democracies and aim at the 'redemptive revival' of politics, beyond the managerial and pragmatic style of post-democratic governance. But apart from this general comment, Canovan does not elaborate on the populist emotional 'mood' any further.

At any rate, then, the emotive underpinnings of populism(s) cannot be ignored. Speaking in terms of movement activity, Minogue (1969: 197) was crystal clear: 'To understand the (populist) movement is to discover the feelings which moved people'. As a form of political thought, it has been recently argued that populism is 'no doubt based more on emotional inputs than on rational considerations' (Tarchi 2016: 12). Why then have emotions been under-labored for a long time in the analysis of populism? For one thing, it seems that this is due to the fact that many scholars, especially when cross national comparisons were at stake, focused themselves on the macro rather than on the micro level of analysis. Micro-analytical methods involve surveys, in depth interviews, ethnography, discourse analysis, and so on. In contrast, macro-level analysis involves comparative tools of historical sociology and structural-functional categories which are likely to overwhelm the allegedly individual-based emotions elicited in political populist phenomena. Nevertheless, the need to bridge the macro and the micro levels of analysis has been highlighted in post-Parsonian social theory and political sociology (Giddens 1984: 139–144) and thereafter the study of populism should follow the same research line, even more so because the sociological and the socio-psychological analysis of emotions inherently involves the linkage between social action and social structure (Barbalet 1998: 4). Another major reason for the misconception of emotions in the analysis of populism is the prevalence of an 'emotions-proof' style of political sociological research, a symptom, as it were, of the 'non-emotions period of sociology' (Barbalet 1998: 19). As already mentioned in Chapter 1, the 'rationalist' theorizing of political affairs was very much shaped by the western ideal of reason vis-á-vis passion.

Populism: an emotionally charged phenomenon

Conceptualizing emotions and political emotions is the necessary probe for ana-lyzing the emotionality of political populism within the intellectual *milieu* of the affective/emotional turn in social sciences and humanities. As already mentioned, the demand side of populism has started to be studied via attitude scales only recently. Quantitative and qualitative research on the emotional dimensions of different versions of populism is even newer and hence we know little about how emotions arise, at which objects (individuals, political issues, social insti-tutions, etc.) they are directed and how they are linked to social and attitudinal predictors of populist support. What, however, marks this type of research is the operational analysis of distinct emotions and/or clusters of emotions rather than metonymic or vague conceptualizations such as the ones referred to above. We know by now that in the context of particular populisms within particular national political cultures one can find a wide range of sentiments, which, among others, consists of nostalgia, anxiety, helplessness, hatred, vindictiveness, ecstasy, hope, gratitude, pride, dignity, melancholy, anger, respect, trust, confidence, sympathy, fear, indignation, shame, and envy. It is rightly argued that 'populism carries an emotional charge, which covers the spectrum from the negative *ressentiment* of the *laissés pour compte* to the positive extreme of the fusional love with the leader' (Ostiguy 2017: 92). However, we are just exploring ways to study these emotions and sentiments and trying to build up operative research agendas. Vari-ably, but crucially, the political cultural contexts determine 'which emotions are likely to be expressed when and where, on what grounds and for what reasons, by what modes of expression, by whom' (Kemper 2004: 46). For one thing, they do not emanate *within* the individual so much as *between* the individuals and 'the interaction between individuals and their social situations' (Barbalet 1998: 67); further, emotions are not actually felt one at a time but in flow and in relation to each other. In the reminder of this chapter I shall analytically refer to some of these emotions.

Nostalgia

Although it has been rarely researched and has only recently attracted systematic scholarly interest within memory studies, one cannot but think of nostalgia as an important driver of populism (Kenny 2017). Defining it as a backward-looking, anti-liberal, reactionary ideology, Betz and Johnson (2004) argue that radical right-wing populism reflects a deep sense of nostalgia for the good old days. Another case in point is the so-called 'Ostalgie', the German designation for the inclination of many voters of far right-wing populist parties and movements to long for the communist period and the way of living in the former GDR. Echoing Richard Hofstadter's liberalist understanding of American populism of the early twentieth century, Minogue (1969: 206) speaks of nostalgia as an integral element of the 'actual ideology of populism'.

At the micro-level, through an excessive attachment to the past, nostalgia is thought of as a compensation for the difficulties of current socio-economic settings and therefore as an emotional-mnemonic means to build a sense of continuity in one's own personal identity. In seventeenth century Switzerland and mid-nineteenth century France nostalgia was regarded as a mental disease, a *maladie de la mémoire* much like amnesia and hypermnesia. It was believed that without a treatment the disease was likely soon to affect major organs of the body and be fatal (Roth 2012: 27). As home-sickness and 'hypochondria of the heart', nostalgia was demedicalized by the late nineteenth century because, it was believed, technology, public education, hygiene, and progress in transportation 'were making people less vulnerable to the poisonous claims of the past' (Roth 2012: 37). Thereafter, doctors directed their attention to hysteria. Nowadays, the widespread discourse on the 'global epidemic of nostalgia' (Boym 2002: xiv) is meant as a cultural critique of late modernity rather than as a stern psychiatric problem. Therefore, nostalgia is not confined to individual biographies but has collective properties as well. At the macro-political level, nostalgia addresses the myth of the golden age which is very often evoked in populist and nationalist politics in tandem with other myths, like the myth of unity (of the people, the nation, the race) and the savior (charismatic leader) which are also related to nostalgia (Girardet 1997).

What is called 'historical nostalgia' (Stern 1992; Scherke 2018), differentiated from 'personal nostalgia', is a longing for times or situations that one is to experience vicariously, i.e., not first hand but through group identifications or mediatized historical accounts. That means that the past is selectively reconstructed, as in the case of the trauma drama. Hence, nostalgia is not necessarily a short affective episode but can be experienced as a long lasting mood formed in response to stressful situations. Understood in these terms, nostalgia is compatible with Taggart's notion of 'heartland' as an integral element of many populisms. Through selective memorization of the past, nostalgia is likely to intertwine with melancholy; i.e., one longs for something which actually never existed in the first place. In this sense, nostalgia's object is an undead past felt as stolen or abandoned. It is very likely linked to reactionary and conservative attitudes and right-wing populist outlooks (Capelos and Demertzis 2018; Mounk 2018: 236). Nevertheless, as Kenny (2017) shows with regards to British politics, nostalgia is not an inherently negative emotion normatively associated with traditionalism, pessimism, and restorationism; it can be associated with anti-establishment left-wing populism that raises strong objections to liberal elites and the post-democratic arrangements, blended with some sort of utopian sensibility. What is called 'radical' and 'reflexive' nostalgia (Boym 2002) may be engaged with progressivism and left-wing populism; radical nostalgia claims the past as a means for remaking the present, not as a refuge, and hence radical nostalgics do not wish to return to the past, but instead use it to cope with perceived historical injustices and traumas. As a case of political nostalgia, it expresses the refusal to familiarize oneself with an unjust political-emotional regime. The opposite, 'restorative nostalgia' – a notion similar to 'retrotopia' described by Bauman (2017) as the negation of utopia's negation – makes people to want to live within the past in terms of an ethnocratic exclusionary version of populist politics (Robinson 2016; Kenny 2017: 263). In a nutshell, contemporary populist currents can be understood in relation to these two sociohistorical modalities of (political) nostalgia which can be a mighty popular motivator, even more powerful than hope (Lilla 2016: xiv).

Anger, efficacy, and fear

The 'us and them' slot referred to above renders populist politics an inherently adversarial and polarizing enterprise and therefore the populist schema cannot endure unless it is affectively supported. As a primary and basic attack emotion, anger overwhelms the anti-elitist/anti-establishment posture by motivating individuals to participate in politics in conventional and unconventional ways, populist or otherwise. It is commonly accepted that the 2008–2009 economic crisis in the USA and many European countries has triggered angry reactions among disadvantaged citizens and the so-called losers of globalization and that, apart from structural and socio-demographic factors, anger is a strong predictor of populist attitudes and support for populist parties (Magni 2017; Davou and Demertzis 2013).

Anger is first and foremost a 'power emotion' in the sense that (a) it is triggered when individuals and/or collectives encounter physical or psychological restraints and blocks to need satisfaction or goal achievement (TenHouten 2007: 39–40) and (b) it helps mobilization and resistance against deprivation of resources or unfavorable conditions (Schieman 2006). As long as anger is elicited in the political realm it becomes a political emotion, the object of which is chiefly other persons rather than 'mankind' in general (Aristotle 1992: 1378a), while its evaluative concern is always one's own self-interest or one's own fellows' wellbeing. In political anger, then, agents are able to locate external accountability and blame others for the precarious, risky, threatening or even traumatic situations they face. Economic, political and intellectual elites and their representatives, as well as immigrants and refugees, are identifiable targets of grievances and blame attribution for current misfortunes. It is argued that 'angry citizens appear to be more receptive to populist discourse' (Rico *et al.* 2017: 12).

Yet political anger seems not to be enough for populist support; in tandem with appraisal theories of emotion, Magni (2017) has demonstrated that the effect of anger is conditional on (internal) political efficacy and that it is the angry and inefficacious citizens who are mostly prone to support 'anti-systemic' populist parties and align with populist discourses. Grown frustrated with mainstream actors, politically powerless citizens within the system are expected to engage in politics that offer a pathway to action outside the conventional party system. In contrast, angry and efficacious individuals are likely to support mainstream political parties and appropriate established political styles of discourse (Magni 2017). The same holds true of hopeful and politically powerless subjects (Davou and Demertzis 2013) and of fearful individuals who are likely to opt for political conservatism, since defending the status quo serves to reduce fear and uncertainty. In their analysis of current Spanish populism, Rico *et al.* (2017) found that it is anger rather than fear that is conducive to both right-wing and left-wing populist attitudes. The causes of angry populism have recently been traced in a number of triggering factors; one is a seeping 'silent counter-revolution' against the widespread post-modern and post-materialist cultural patterns and value priorities (cosmopolitanism, environmentalism, feminism, etc.). Pippa Norris and Ronald Inglehart (2019) argue that people from older cohorts and less affluent strata react wildly to liberal moral codes and socio-economic settings under the globalization processes, and to the immigration flows which they see as threatening their identity and status. Relatedly, by claiming that 'morality binds and blinds', the moral psychologist Jonathan Haidt attributes the angry populism of American Conservatists and the Tea Party voters to a set of divergent moral emotions which make them feel like a 'moral tribe' vis-à-vis the Liberals (Haidt 2013: 364–366). Similarly, Arlie Russell Hochschild (2016) grounds it on the 'empathy wall' that separates the American public into hostile 'red' and 'blue' political camps; she claims that it is not values themselves which invoke conservatives' anger but the way they feel these values.

Resentment and ressentiment

The action readiness of anger does not consist only in moving away impediments to one's aim in mid-task; more importantly, it encompasses intentional actions to remedy or rectify wrongdoings afflicting one's power, status and/or self-esteem (Kemper 1990) in which case the other actor is perceived as responsible for real or potential loss or harm. Due to blame attribution, anger is a 'highly social emotion' (Schieman 2006: 495); furthermore, whenever anger, and political anger for that matter, go beyond the removal of impediments for goal achievement and involve blame attribution, they acquire a moral pointer. As seen in Chapter 5, for Aristotle (1992: 1378a) anger is an impulse, accompanied by pain, for a conspicuous revenge for a conspicuous slight directed without justification towards what concerns oneself or towards what concerns one's friends. When there is absence of the 'right justification', the affront to fairness and justice renders anger more than self-targeted emotion and it is on this basis that populist discourse is so often moralistic in its anti-establishment polarization (Müller 2016: 21, 23).

As an extensive overview of resentment and *ressentiment* was given in the previous chapter there is no need to offer any further conceptual analysis. Given that overview, this section focuses on populisms' resentful and *ressentiment*-ful bases. To be sure, every now and then, populist outlooks have been ascribed to resentment sparked by grievances against globalization, immigration, and political corruption (Betz 2002: 198–200; Betz and Johnson 2004; Minkenberg 2000: 188; Mudde and Rovira Kaltwasser 2017: 103; Müller 2016: 11, 16–17, 57, 65; Fieschi and Heywood 2004: 292). The notion of resentment is so powerful in the analysis of populism that scholars frequently lose sight of its next of kin, *ressentiment*, which, as already documented, is an equally commanding and more complex emotional notion (Demertzis 2006; Müller 2016: 18; Meltzer and Musolf 2002; Brighi 2016).

Edward Shils was the first person who systematically used 'resentment' for the interpretation of the American populism of the 1950s; he thought of resentment in terms of moral rage and indignation and described populism as 'an ideology of resentment against the social establishment imposed by the long-term domination of a class, which is considered to have the monopoly of power, property and civilization' (Shils 1956: 100–101). Much later, without delving into the concept, Mudde (2004: 547) accounted for resentment as an affective element of the populist *Zeitgeist*.

In a somewhat more orderly way, while scrutinizing the emergence of the far-right European populist parties during the period 1990–2000, Hans-Georg Betz argues that in the early phase of their appearance they were greatly buttressed by the diffuse grievances of the working and lower middle-class electorate against globalization, immigrants, the fiscal crisis of the welfare state, and politicians' corruption, thus mobilizing grassroots *ressentiments* (Betz 2002: 198–200). Apart from Betz (2002), who endorses the ideational approach to populism, Pierre-André Taguieff, a prominent representative of the rhetoric approach, has also paid considerable attention to *ressentiment* as a constituting element of populism and racism in contemporary European societies. In his analysis, the moralistic character of populism is fueled by the sense of injustice ingrained in the resentful state of mind driven by the effects of global neo-liberal policies (Taguieff 2007: 75).

Yet, both authors oscillate between resentment and *ressentiment* and no substantive conceptualization of these terms is provided. As we have argued in the previous chapter, according to the Nietzschean and the Schelerian conceptualizations, *ressentiment* is not conducive to mobilization and explains political inaction rather than political action. It is 'resentment', *qua* moral anger or indignation, which may explain better the initial phase of populist mobilization. *Ressentiment* might be better used for the understanding of the emotional climate wherein resentful grievances have been incubated.

Building on previous work, Mikko Salmela and Christian von Scheve proposed an interesting interpretation of current right-wing populisms on the basis of the Schelerian version of *ressentiment*, which, as already mentioned in Chapter 6, they conceive as a 'psychological mechanism' rather than as a cluster-complex emotion. They carefully distinguish resentment and *ressentiment* and criticize Betz for mis-conceptualization of either term. Influenced by Scheff's theory of shame, they strongly argue that, besides its other affective ingredients, the crucial dimension in experienced *ressentiment* is the way shame is addressed by the individuals (Salmela and von Scheve 2017: 7–8). When shame remains repressed and unacknowledged due to chronic impotence, it is transformed into resentment, anger and hatred for alleged 'enemies' of the precarious self and associated groups, such as refugees, immigrants, sexual minorities, the long-term unemployed, political and cultural elites, and the 'mainstream' media. This is the case with right-wing populism. In left-wing populism, they argue, shame is acknowledged and thus it is morally transformed into indignation against neo-liberal policies and institutions. While the authors do not discuss transvaluation – a crucial part of Scheler' theory – it seems that they focus on the interplay between acknowledged/unacknowledged shame (Scheff 1994a) as the main trigger of the psychological mechanism of *ressentiment*.

Their approach to populisms' emotionality is innovative in more than one respect; I have already accepted – along the lines of their argumentation – that even if Scheler himself does not include it among the ingredients of *ressentiment* one could infer that shame could be thought to be such an ingredient. However, one should take heed of two issues: first, their statement that shame can be seen to be implicated in Scheler's account in feelings of inferiority when one compares oneself negatively with others has to be supported by empirical evidence, which, for the time being, is lacking in the relevant literature. Surely, quantitative (e.g., Ball 1964; León *et al.* 1988; Capelos and Demertzis 2018) and qualitative research (e.g., Demertzis 2004) on *ressentiment* is limited and their assumption remains to be supported by further evidence. Second, the fact that Scheler offers no discussion or reference to shame in his analysis of *ressentiment* may not be accidental; possibly, in his mind, impotence and inferiority are linked directly with envy, hatred, malice and rancor, rather than mediated by shame. Third, repressed and unacknowledged shame might not necessarily lead to resentment or an anti-social stance and behavior; it might be the case that during the transvaluation process unacknowledged shame transmutes into docility and submissiveness rather than backlash politics.

The case of Greek populism

In Chapter 3 I described the cultural politics of Andreas Papandreou's Socialist Party (PASOK) regarding the over-memorization of the Resistance during the

Axis occupation and the under-memorization of the civil war amidst the trauma drama of the 1974–1990 period. That politics had a strong populist imprint. Although Papandreou's period in office was undeniably characterized by the *par excellence* populist decade in European politics (Clogg 1993), it is probably the barrier of the Greek language that made it difficult for international scholars to include Greek populism of that period in comparative studies of populist phenomena (Pappas 2016). However, with the world-wide eruption of the 2008 economic crisis and the rise of academic and media interest in neo-populism, scholars are willing to focus on contemporary Greek left-wing populism promoted primarily by SYRIZA (Mudde 2015). Greece was among the first to be hit by the 'the perfect storm' of the sovereign debt crisis, the refugee crisis, climate change, and the crisis of public knowledge, as described by Brubaker (2017).

The political cultural and historical setting

Elsewhere I attempted to interpret that case of populism – which nevertheless is not confined to Papandreou's rhetoric – adopting a sociology of emotions approach (Demertzis 2006). In the reminder of this section I come back to that interpretation taking into account the impact of nostalgia, political inefficacy, resentment, and *ressentiment* as conceptualized hitherto. Methodologically, in the past I have selected evidence from a variety of prominent Greek novels, disclosing the emotional dynamics of the country's political life (Demertzis 2004). For the needs of the present analysis I will limit myself mostly to the emblematic novel of Kostas Mourselas (1996) which is accessible to the wider English-speaking public. Historically, my analysis will mostly refer to the period that starts after the restoration of democracy (*Metapolitefsi*) and lasts until the early 1990s, a period I described in Chapter 3 as the selective construction phase of the Greek civil war trauma drama.

To put the case in context, one should stress at the outset that PASOK was heavily supported by the new middle strata that emerged chiefly from the defeated of the civil war. An immediate aftermath of the defeat was the deportation of almost 30,000 fighters of the Democratic Army in countries of the Eastern bloc, just to escape custody. Another consequence was a massive internal migration from the countryside to big cities and primarily to Athens and Salonica; for many sympathizers of EAM – already politically and socially marginalized – the urban environment was a refuge from social isolation and the arbitrary persecutions of right-wing monarchical paramilitary groups dominating the countryside, small cities and villages. By the end of the 1950s and all the way throughout the 1960s, when the Greek economy started taking off, a very large number of them entered the labor market and, to some extent, the restored clientelistic networks. For the supporters of the victorious Right, moving to the cities was an opportunity to find a better job, preferably in the public administration sector. Subsequently, a mass of unprecedented chances for upward social mobility was carved out for both left and right. This has been described by George Ioannou, a significant Greek novelist who almost exclusively wrote about Salonica – the second biggest city of Northern Greece:

> The profession of doctor was highly estimated by the Salonica people in the 1950s. Next in the hierarchy were the lawyers. Being a secondary school teacher was much too financially degraded, a teacher in primary schools even worse. And of course, the adults were pushing us to become doctors or lawyers. The craze with the Polytechnic appeared later when the rebuilding of the country began.
>
> (Ioannou 1984: 30)

Apart from the manual workers in the construction industry, the former poor farmers and peasants and their children were turning into public servants, middle class professionals, and small-scale entrepreneurs (Lyrintzis 1993).

Nevertheless, for all their socio-economic incorporation and class mobility, the petit bourgeoisified left-minded groups were experiencing an unbridgeable antinomy; i.e., while their economic integration was becoming all too real, creating thus the conditions for social consensus and their gradual de-EAMification, the structure of the post-civil war state (palace, army, national-mindedness, etc.) did not allow for the lifting of their political exclusion (Charalambis 1989: 196). That antinomy, inflicting humiliation, anxiety, and fear[4] – already experienced after the war – as well as some hope and resentment, was begging for a political solution to bypass the inertia of the post-civil war state apparatus in view of the fact that Greece at that time was typically an open parliamentary democracy. The 1958 elections offered a way out since the United Democratic Left (EDA) entered parliament as the major opposition, taking 24.4 percent of the ballot. Since 1936, this was the first time that a leftist party accommodating Communists, open minded liberals, left-centrists and anti-monarchists had entered parliament in such a sound way.

Notwithstanding the legal outcast status of the Communist Party and the existence of political prisoners, the electoral success of EDA signalled a process of normalization of post-civil political affairs conferring even more optimism on its supporters. At the same time, however, it triggered a multifaceted reaction by the regime that pressed EDA to such an extent that it gradually lost much of its parliamentary stronghold. In the 1961, 1963 and 1964 general elections it received 14.6 percent, 14.3 percent and 11.8 percent of the vote respectively with a large part of its constituency moving to the Centre Union (EK). That party was formed in 1961 by George Papandreou – the father of Andreas – and was able to accommodate the EDA sympathizers' aspirations for upward social mobility and political inclusion. The 1961 elections were denounced by both EK and EDA who repudiated the result based on numerous cases of voter intimidation and irregularities, such as sudden massive increases in support for the National Radical Union (ERE), the dominant party of the Right, contrary to historical patterns, and voting by deceased persons. As a consequence, George Papandreou initiated an 'unrelenting struggle' until new and fair elections were held in 1963. In that struggle left-minded citizens of all kinds mobilized much emotional energy, primarily expressed as resentment and indignation against the blockage of democratic resolutions. In the 1963 election EK won the majority (42 percent) and took office having an agenda certainly more liberal compared to that of ERE. But still the rigidity of the regime was not meant to bend.

For all his liberalism, Papandreou, who was the prime minister during the December 1944 events against EAM (see Chapter 3), was suspicious of the Left; provided that ERE received only 39 percent of the vote, he could not accept the fact that the balance of power in the new parliament was held by EDA. He soon resigned, forcing a new election three months later, in February 1964. His tactics paid off and the EK was rewarded with a 53 percent share of the vote. Despite this unassailable parliamentary majority, Papandreou's reforms concerning the curtailing of the army's powers and the releasing of a number of political prisoners were strongly opposed by the Crown and the Right. On the occasion of the King's refusal to accept Papandreou's appointment as Defence Minister – as both Prime Minister and Defence Minister, Papandreou wanted to control the army, which was actually under the effective control of the crown – the latter resigned in July 1965 and a prolonged period of political turmoil was to follow (Clogg 1997: 145–168).

Once more, during that period open resentment was expressed by masses of democratic people, a large part of which continued after the end of the essentially anti-liberal – not just anti-communist – state which was blatantly backed by US foreign policy in accordance with Cold War priorities. Under popular pressure, the stillborn government which was formed in August 1965 under the auspices of the Crown collapsed within 16 days. A new coalition government, formed in September 1965, was to last until the end of 1966. Throughout that period there were many hundreds of violent demonstrations and peaceful petitions demanding the restoration of constitutional order and the abolition of a monarchy whose untimely interference in political life after 1946 had been more than obvious. The new election announced for May 1967 never occurred. The colonels' dictatorship put an end to the liberalization of the political system and halted the aspirations of the marginalized left-minded citizens and their offspring for political recognition and inclusion.

The seeds of ressentiment

To the extent that the seven years military dictatorship (1967–1974) overturned the democratizing potential of the mid-1960s, it can be argued that a significant portion of pre-dictatorial resentment and moral indignation during the dictatorship, expressed in public grievances and demands, was transformed into *ressentiment*. In the decade of the '60s and until the 1973 Oil Crisis there was a period of rapid economic development where large parts of the politically marginalized benefited in terms of socio-economic integration, and the dominance of the dictatorship was lived out as an inexorable and inescapable destiny which generated an experience of hopeless impotence and inferiority. The intergenerational trauma of the civil war and the rule of dictatorship overlapped and, in spite of their being rather well off by now, the politically marginalized did not feel at home. Their emotions and behavior were incongruent with their current economic position; even if they had been integrated into the national consumerist way of life, they did not stop feeling like victims who were not at peace with their own selves.

Apart maybe from the first-person narrator, all heroes and heroines in Kostas Mourselas' *Red Dyed Hair*[5] are immersed in just such an emotional climate: during the dictatorship, a group of friends with left-wing political convictions dating back to the 1940s in one way or another incorporated themselves into the system while retaining their old political identity as an empty shell.[6] Here is the ironic description of Liakopoulos, son of an executed Communist:

> And here we have the self-made man of the group. The perfect example of the Greek who starts out with nothing and reaches the moon.... He's become a great success, a shipowner, and plenty else besides. Nothing can stop him.... But the most important thing is that in spite of his money, he never abandoned his ideas, or his party. Rock steady, a man of principle.
>
> (Mourselas 1996: 333)

In the hero's own words:

> OK, nobody's going to go drinking the blood of the workingman ... Fine, don't get me wrong; we are all lefties, still are, in fact, but – you tell me – ... workers today, they're not the ideologues we used to know. You can't imagine what a shit-ass bunch they turned out to be. Sit on their ass and eat the grass.
>
> (Ibid.: 277)

When another member of the group, himself the son of an executed man, University professor Lazaris, invites his friends together with some who were financially and socially supported by their links with the colonels' regime, Liakopoulos addresses the narrator like this: 'Manolopoulos, we don't have a shred of honor left to our names, but when you come right down to it, we did what we could for the country. Besides, ideology is one thing, making a living is something else' (Mourselas 1996: 291). At any rate, 'we're not going to change our ideas, are we, just because we happen to drink a glass of wine with them? ... It's not as if we're going for a stroll around Constitution Square with them, is it?' (ibid.: 292–293). When it is disclosed that their buddy Panagiotaros has hung the King's portrait in his store he replies thus: 'So I put up a portrait of the King in a moment of weakness, big deal.... We all need a life preserver, right?' (ibid.: 317). Apparently, they all long for higher social status, to be achieved by instrumentally using the 'system', while simultaneously invoking their leftist past. They are looking for social recognition while being incapable of achieving it due to their political marginalization. This is the beginning of the transvaluation process; as lower, middle and higher-middle class members they realize that under the dictatorial regime they can never really join the elite. To bear that realization they start praising their own attributes and traits in a self-affirmative way. Offspring of the innocent and betrayed people, *they* are from now on the evaluative object of themselves. Acknowledging deep inside the retreat of their ideological principles, they act as if that is not the case, striving to overcome their political inferiority and powerlessness.

This striving is in contrast with the liberal stance of the central hero of the novel, Emmanuel Retsinas, called 'Louis' by his (ex)friends. Louis is described by Mourselas as an 'almost man', once and forever lumpen and anarcho-communist, sometime circus daredevil, part-time yacht captain, subverting propriety and moral order. Destroyer, builder, erotic, and provocateur. He is the 'other' of his friends from the 1940s. As he is the incarnation of the purged ideals of the Left, they secretly admire him while overtly denouncing him. He is the object of their envy and *Schadenfreude* but above all else he is the trigger of their *ressentiment*. As long as they cannot imitate Louis, they devalue him: 'That unavoidable question, "How much longer can you go on like that?" was the only subject under discussion at our group's last evening get-together. Hey, how much longer can he get away with it anyway, the sonofabitch?' asks Lazaris, stating after a while 'What the hell's the matter with him, anyway? There's got to be a limit. Either he admits he's a nut case, or he's a write-off. We've all watered down our wine. So, what's his big problem?' (ibid.: 321). For the hard-core Stalinist of the group, Antoniadis, Louis is dead meat: 'End of the line. This is where I get off. The guy's nothing but a bum. Worse, he's dangerous. Never grew up' (ibid.: 304). Addressing the narrator, who is Louis' cousin, Lakopoulos has an outburst in which he expresses his admiration for the hero, saying that he is not sure whether the independent Louis is a real or an imaginary person invented by themselves: 'Give us a break Manolopoulos. He's the one who takes his imagination for real life, not us ... you can tell him to stop insulting us, stop calling us bourgeois sellouts' (ibid.: 293). Once in the entire novel, the narrator himself (or Mourselas himself) thinks something of Louis that he never expresses openly: 'Wait a minute Louis, old pal. Who the hell do you think you are anyway? Everybody has to make some concessions. You want to live your whole life on credit?' (ibid.: 113). However, at a crucial moment in the novel he realizes that

> every last one of them depended on how Louis would end his life in order to justify their own.... What bound them together was their shared hatred. Well, maybe I shouldn't exactly call it hatred. A mess of confused emotions, really. Admiration, hatred, envy, but most of all a hidden, unspoken sense of awe. No one could forgive him for going square against everything they had wasted their lives in vain pursuit of....
>
> (Ibid.: 321–322)

I think that this sentiment Mourselas speaks about is no other than *ressentiment*.

The consequences of ressentiment

However, as soon as the Socialists took office in 1981 and the lower middle strata stemming from the defeated in the civil war (the 'non privileged' in Andreas Papandreou's populist rhetoric) found themselves integrated into the political system, *ressentiment* gave place to vengeance precisely because it could be released and acted out publicly. Party mass clientelism (Lyrintzis, 1984) and the 'green-guards' (cadres who dominated in trade unions, the public sector and state mechanisms) compensated for the 'stone years' of political marginalization. That was the imaginary moment where the 'people' clearly dominated over the 'establishment'. I strongly argue therefore that the psychic mechanism of transvaluation was the soil on which the post-dictatorial populism as ideology, movement, and practice flourished. I do not think that 'the people' could ever have been invested with mythical value so rapidly had they not previously been subjected to resentful transvaluation. Since the radicalized petit bourgeois strata could not become an 'establishment' they elevated 'the people' to the supreme legitimating and moral reference point. In other words, without being the only one, *ressentiment* functioned as a condition of possibility, paving the way for the formation of populism after the political changeover in 1974.

Apart from essentialist interpretations, I endorse the probes of discourse analysis approach which highlight the articulatory character of populist rhetoric. Namely, seen either top-down from the supply side, or bottom-up from the demand side, the 'people' and the 'us *vs* them' division provide the condition of possibility for the emergence of a variety of 'populisms' contingent upon different path dependencies, political cultural patterns, and political and discursive opportunity structures, as well as the situational drivers of each populist political scene. Historically and organizationally speaking, the Greek populism of the '80s and the '90s was a semi-peripheral populism where the mass of lower classes incorporated into the parliamentary political system in a vertical-patrimonial rather than horizontal-republican way (Mouzelis 1985). This was mainly expressed by the plebiscitarian relationship between Andreas Papandreou, the *par excellence* Greek populist leader, and the led, most of whom were previously excluded from the political arena. In his pro-people rhetoric, nostalgia was center-stage. Screened out by the widespread cultural nationalist patterns of citizenship, the Greek populist politics of that period faced new and multiple problems raised by the modernization and globalization processes.

Although, economic-wise, it was a country of the semi-periphery, Greece's entry into the European Union – by then named European Economic Community – in 1981 was mostly achieved on political premises since, economically and institutionally, Greece lagged behind the other eight developed countries of the Union. Modernization, and Europeanization for that matter, was taking place under conditions of incremental individualism, rapid upward social mobility, an atrophic civil society,[7] widespread clientelism and the generation of concomitant crony capitalism (Holcombe 2012). No doubt, those conditions did not provide for a predictable, reassuring and stable structural setting; multi-valence economic activity, fluidity of social roles, and the lack of an efficient welfare service and juridical system were rather the rule. This was the highly informal institutional setting that carved out a rift between the 'objective' open life chances and the way individuals were emotionally receiving them. So, for all the prospects of upward social mobility, the middle-class travelers themselves felt more anxiety than enthusiasm, more fear and concern than hope and optimism. Open life chances were seen more as a coincidence, rather than as the expected result of systematic individual and/or collective effort. A generated institutional fluidity led to new social positions being experienced more as ascribed than as achieved and thus society as a whole was largely seen as an already occupied and/or impenetrable field where very little institutional security was offered to anyone who took the initiative to cross over (Karapostolis 1985). The sole permanent safely net was the family and bonding social capital networks.

There has been, therefore, an incongruence within Greek 'structure of feeling' in the '80s and the '90s. Success and social approbation were experienced as being curtailed by envy; an invisible but no less effective malicious threat, emanating from a *ressentiment*-fully driven vertical and horizontal social comparison, was experienced either as a barrier to further development or as a force frustrating social recognition.[8] That is why the search for identity has been a perpetually insistent issue in the Greek public sphere. A pillar supporting the emotional regime of the post-authoritarian modernization process was the nostalgic appeal to Greek folk/popular tradition. As is repeatedly the case, the Greek populist notion of the 'people' was invented or even extrapolated through a highly selective evocation of the *Romeic* tradition, as opposed to the Hellenic tradition that draws from the Classical period. The *Romeic* tradition draws from the underdog folkish identities formed during the 400 years of Ottoman rule. The subjects of the Eastern Roman Empire were actually Romans (*Romaíoi* in Greek). Especially after the ecclesiastical schism in 1054, the designation *Romaíoi* or *Romioi* meant the subjects of the Empire and an Orthodox Eastern Christian; the same name was used by Ottoman Turks for their Greek subjects (Leigh-Fermor 1966: 98). In this vein, *Romiosyne* designated Greekness in its underdog and popular version; on the contrary, *Hellene* designated the Greek identity in its high-brow Hellenic version. Ever since its inception in the early nineteenth century, in its international relations and foreign affairs, the Greek nation-state permanently promotes an idealized Hellenic identity based on Classical heredity. When it comes to the interior, however, the state and other political organizations, as well as their organic intellectuals, generate the *Romeic* interpretation of Greekness as a sort of national cultural intimacy (Herzfeld 2005) which really speaks to the hearts of ordinary people.

Unlike the Hellenic inheritance endowed with universalistic claims, the *Romeic* tradition is conducive to particularistic identifications charged with nostalgia. Ultimately, Papandreou's populist movement did nothing but what other similar movements in the semi-periphery have been driven to do: to adhere to 'a belief in the possibility of controlling the modernization process' by searching for a synthesis between basic traditional values (in our case: folk religiosity, amoral familism, etc.) and modernizing forces and ideas (Stewart 1969: 182, 187). He was urging the need 'for a new national education of a mighty progressive nationalism for an independent Greece' premised on an idea that 'there is something called Greek personality ... expressing our common roots in the land of Greece ... our Greekness, this *Romiosyne* is the crossing point between our tradition and our goals' (Pantazopoulos 2001: 112). In a way then, while socio-economically the country was moving forward, much of its emotional energy was pointing backwards. From the mid-70s onwards a large part of cultural politics in Greece, promoted by Papandreou's populism, was driven by nostalgia for folkish customs and songs either adapted and revised into contemporaneous aesthetic codes or replicated only as a consequence of their initial selection. An exemplary case in point are the *Rembetica songs*, underground urban folk songs banned or censored during the Interwar period and turned into emblems of the popular psyche almost 40 years later. Through an astute 'invention of the tradition' process, the music of the poor, the dispossessed, the drug users, the refugees and the migrants who came to Greece from Asia Minor before and after the First World War (Petropoulos 2000), once despised, became part and parcel of the popular-national heritage and an aesthetic flagship of the Greek middle class radical populism of the '80s.

Among others, praising the *Romeic* identity and the re-evaluation of the *Rebetica songs* were two cultural politics repertories through which the 'people' were morally upgraded in the rhetoric of Greek populism during the last quarter of the twentieth century. I want to argue further that not only did either gesture result from *ressentiment* against local oligarchy but it also arose in response to the West as an economic, political, and cultural entity. It is at this point where populism and nationalism emotionally meet each other. If the 'Janus quality of the "populist situation" is reflected in the populist ideological synthesis of traditionalism and modernism' (Stewart 1969: 191), the same holds true with nationalism under the auspices of *ressentiment*. But this is the subject matter of the next chapter.

Perspectives for further research

Previous work on support for populist parties and populist discourse has often neglected the role of emotional factors and focused on socio-demographic characteristics (Cossarini and Vallespín 2019). In this chapter I offered an overview of different thematizations of affectivity in populisms by showcasing that, moving beyond diffuse and incomplete accounts, international research has started to focus not only on distinct emotions and sentiments but also on complex emotions such as *ressentiment*. To be sure, focusing on emotions for the understanding of populisms does not rule out structural, economic, and discursive approaches; on the contrary, it complements them, offering more nuanced empirical and theoretical analyses. And of course, pointing to emotions in populism research what is of real interest is not the psychosomatic etiology of individual affective dispositions – what William James once called 'medical materialism' – but their inter-subjective and moral significance. This makes for the *in-situ* interpretation of the meanings involved in the supply and the demand side of political populisms. Such an interpretation begs for the most appropriate theoretical approaches to be drawn from the theoretical and conceptual pool of the sociology of emotions on the one hand, and for sound and solid methodological triangulation on the other. What is more, one should take into account the concrete local, national or international political cultural landscapes where the populist phenomena occur and at the same time make sure that the level of analysis goes beyond the individual scale. This is what I have tried to do in this chapter referring to the Greek populism of the '80s and the '90s. Much recent research is being directed to the understanding of contemporary Greek populism conveyed mostly by the leftist SYRIZA party, espousing more nuanced theoretical and methodological approaches which start to recognize the need for the emotions to be taken into account (De Cleen and Stavrakakis 2017; Brubaker 2019). Since a good many scholars agree that there is a paucity of research on the demand side of populist phenomena, a lot remains to be done as to the specification of their emotive underpinnings.

Synopsis

Focusing on the demand-side of the populist phenomenon, and continuing from the previous chapter, this chapter has shed more light on the emotional underpinning of populist orientations and politics in relation to nostalgia, anger, fear, efficacy, resentment and *ressentiment*. I have kept a distance from the mainstream ideational approach by defining populism as a cognitive-emotional schema consisting of two principal slots through which individuals orient themselves towards the political field: the 'people', and the division of society between two main power blocs. Special emphasis was given to the post-authoritarian Greek populism of the 1980s, interpreted as a case of *ressentiment*-ful politics. The thrust of the chapter's argument is that there has been a tipping point in the analysis of populism(s) where structural and historical accounts are balanced by bringing emotions in so that more nuanced explanations are possible.

Notes

1 Although there are certain links between them, political populism is not to be conflated with cultural populism which is an intellectual assumption, promoted mostly by cultural studies scholars, that the symbolic experiences and practices of ordinary people are more important analytically and politically than high-brow culture. On this premise, advocates of cultural populism – post-modernists or otherwise – have become sentimental in their solidarity with ordinary people, losing sight of the critical stance against commercialization of popular culture products and practices. If anything, this drift is due to the misrecognition that 'ordinary people' is an intellectual category and not an unmediated

depiction of an authentic people's identity (McGuigan 1992). Pierre Ostiguy attempted an interesting link between cultural and political populism by introducing a 'high – low' socio-cultural division crossing the left-right dimension of political space. In this way, he takes into account the affectivity of populist narratives, defining populism as 'the antagonistic mobilizational flaunting of the "low"' (Ostiguy 2017: 84).

2 Influenced by the notion of populist *Zeitgeist* as described by Mudde (2004), Tarchi (2016) construes populism as a ubiquitous phenomenon in today's representation politics and defines it as a distinctive mentality, as a specific *forma mentis*, connected to a vision of the social order based on a belief in the innate virtues of the people whose primacy is viewed as the source of legitimacy of governmental activity and of any other political action whatsoever.

3 Rightly arguing against the ideational definition of populism, by promoting their discursive approach which, in their account, allows for more thoroughly taking into consideration populism's crucial strategic dimensions as well as its material, performative and affective dimensions and investments, De Cleen and Stavrakakis (2017) are actually calling for a linking of their approach to that of the (political) sociology of emotions. The likelihood is that the discursive analysis of particular populisms would be enhanced by the analysis of particular emotions.

4 Given this emotional climate, the narrator in Mourselas' novel comments on the political marginalization: *'If anyone squealed on me, if any stool pigeon noticed me, I was a goner. I'm on the blacklist ...'* (Mourselas 1996: 53).

5 Mourselas' novel was an editorial success with 200,000 sales during 1990–2007, something unusual for the Greek book market. As already mentioned, it was translated into English in 1996. In 1992–1993 it was transcribed into a 37-episode drama series by a private TV channel with unprecedented success. This also appeared on Australian television in 1995 and 1996 with English subtitles.

6 Critical historiography in Greece has repudiated public history's retroactive myth of an alleged massive resistance against the military regime, a myth congruent with the self-idealization of Greeks as a nation in perpetual resistance against all sorts of powers. The fact is that many offered their consent and many more tolerated the Junta, while the majority remained passive bystanders. Only a small minority of leftist, Centre-left and Centre-right groups and individuals took active initiatives to overthrow the military government.

7 In the sense of a non-lucrative third sector located inbetween the state and the market, pursuing public good and public service goals.

8 '(A)lthough they were different persons with different inclinations, the others were not supposed to move any further than we were standing or doing', Ioannou (1984: 102) recalls.

Chapter 8

The emotionality of the nation-state

Introduction

This chapter discusses the role of emotions and emotionality in the analysis of national identity and nation-building. From the outset, by virtue of the affective aspects of national and ethnic communities and the cultural inheritance of Romanticism, analysts never lose sight of the emotional dimension of nationalism; however, they couldn't scrutinize it further because until the 1970s the sociological or even the psychological discussion about the emotions was in its infancy. This chapter will briefly go through the emotional underpinnings of many theorists and proponents of different kinds of nationalism from the nineteenth century onwards and it will be shown that, apart from a few exceptions, most scholars didn't go far beyond the so called 'national sentiment'. In this respect, it will be argued that a political sociology of emotions can contribute to the specification of the catch-all and vague concept of 'national sentiment' by breaking it down into particular and distinct emotions in a more systematic way. Among others, drawing from Liah Greenfeld's work, the chapter comments on different cases of nationalism as they are built around particular emotions with *ressentiment* having a strong role to play therein. The chapter will end with a *ressentiment*-based interpretation of Greek nationalism.

The emotionally laden thought of nationalists

In this section I will pinpoint the emotionality inherited in the views of some prominent nineteenth century liberal advocates of nationalism, i.e., Fichte, Mazzini, and Lincoln. Also, I shall indicatively refer to two conservative proponents of integral nationalism. To start with, in his 1806 *Address to the German Nation* in a Berlin occupied by Napoleon, promoting the idea of the nation-state at a time when Germany consisted of many large and small states with no national unity, Johann Gottlieb Fichte urged that

> ... it must be love of fatherland that governs the State by placing before it a higher object than the usual one of maintaining internal peace, property,

personal freedom, and the life and well-being of all. For this higher object alone ... does the State assemble an armed force.

(Fichter 1995: 66)

This higher object is 'the promise of a life here on earth extending beyond the period of life here on earth – that alone it is which can inspire men even unto death for the fatherland'. Therefore, it is not the spirit of the peaceful citizen's love for the constitution and the laws which actually counts, but the 'devouring flame of higher patriotism which embraces the nation as the vesture of the eternal for which ... man joyfully sacrifices himself' (Fichte 1995: 66, 67). In his conception, the state is the means for the actualization of the nation:

> ... the State, merely as the government ... is not something which is primary and which exists for its own sake, but is merely the means to the higher purpose of the eternal, regular, and continuous development of what is purely human in this nation.

(Fichte 1995: 68–69)

It is the love of fatherland which 'must itself govern the State' (Fichte 1995: 69).

When it comes to the nation-state, the idealism of Fichte is a good match for the republicanism of Giuseppe Mazzini, flavored with Catholicism. In 1860 the Italian revolutionary was fervently claiming that 'your Country is the token of the mission which God has given you to fulfill in Humanity.... A Country is a fellowship of free and equal men bound together in a brotherly concord of labor towards a single end' (Mazzini 1995: 94). The advocated independence and equality were not so much a re-run of the 1789 Revolution as of a Christian egalitarian brotherhood: 'In the name of your love for your Country you must combat ... every privilege, every inequality.... Your Country should be your Temple. God at the summit, a People of equals at the base' (Mazzini 1995: 95). And again, the state, the country, is an instrument of this transcendental brotherhood:

> A Country is not a mere territory; the particular territory is only its foundation. The Country is the idea which rises upon that foundation; it is the sentiment of love, the sense of fellowship which binds together all the sons of that territory.

(Mazzini 1995: 96)

Just before what came to be called the American Civil War, being at pains to persuade his fellow Congressmen from southern states not to abandon the Union, Abraham Lincoln made a plea in the name of God: 'Intelligence, patriotism, Christianity, and a firm reliance on Him, who has never yet forsaken this favored land, are still competent to adjust ... all our present difficulty' (Lincoln 1995: 291).

If emotionality strongly marks the style of thought of liberal nationalists, it seems to carry even more weight in the conceptualizations of conservative and integral nationalists. Here I take two examples: the late eighteenth-century conservative philosopher and politician Edmund Burke heralded the idea of the nation as a self-referential and transcendental community of fate:

> Society is indeed a contract ... but the state ought not to be considered as nothing better than a partnership agreement in a trade.... It is a partnership ... not only between those who are living, but between those who dead, and those who are to be born. Each contract of each particular state is but a clause in the great primeval contract of eternal society ...
>
> (Burke 1995: 141)

In late nineteenth century Charles Maurras, obsessed with the idea of French decadence, developed his idea of integral nationalism according to which the life of the nation takes the lead over the life of its members: 'The social well-being of a nation, the material and moral interest each citizen has in its preservation, these are things which lift man up and sustain him in the highest spheres of his finest and proudest actions'. He also declared that 'the nationalism of my friends and myself bears witness to a passion and a doctrine. A holy passion, a doctrine motivated by ever-increasing human need' (Maurras 1995: 220, 217).

The emotions-driven studies of nationalism

Early theorists of nationalism

In the Anglo-Saxon world, nationalism started to be systematically studied at the beginning of the twentieth century. It was mostly historians rather than sociologists or political scientists who delved into its perplexities. The field was pre-empted by historians because, as Anthony Smith holds (1971: 3), sociologists were mostly interested in producing far-ranging generalizations on the one hand, and on the other hand, they were more attracted by stratification issues and social divisions rather than national solidarities. Apart from the work of Ernest Renan, Lord Acton, John Stuart Mill, and the so-called Austro-Marxists (Otto Bauer, Karl Renner, Karl Kautsky, and Rosa Luxemburg), until World War I just a scant amount of sociological study had been carried out on nationalism.

Among the historians of nationalism at that period (1882 onwards), the influence of Ernest Renan was unquestionable among those who explicitly included affectivity in their accounts. In his conceptualization, the nation is like a soul, a spiritual principle, the fruit of sacrifice and devotion; it is founded on 'a heroic past and glory ... a common will in the present, to have done great deeds together, and to desire to do more'. In this respect, a 'man is the slave neither of his race, nor his language, nor his religion, nor the windings of his rivers and mountain ranges'. S/he is a member of a nation to the extent that a collective moral consciousness is created by a great assemblage of people with warm hearts and the desire to live together: 'And as long as this moral consciousness can prove its strength by the sacrifices demanded from the individual for the benefit of the community, it is justifiable and has the right to exist'. In this sense then the existence of a nation is a daily plebiscite (Renan 1995: 153–154).

In this vein – sticking to just some cases for the sake of brevity – in 1905 Karl Lemprecht defined the nation as a kind of 'social soul', whereas in 1908 Friedrich Meinecke conceived it as a 'mental community'. A bit later, Otto Bauer (1924) viewed it as a group of people with different masses of mental images, and Sir Ernest Barker (1927) as a group of men [*sic*] inhabiting a definite territory who possess a common stock of thought and feelings acquired during the course of a common history.[1]

Interwar theorists of nationalism

The Interwar is a second period in the scholarly study of nationalism marked by more detailed analysis (Hobsbawm 1992: 2–5). Now it was not basically historians who took the lead; sociologists, psychologists and political theorists were involved as well. It is true to say that the 'twin founding fathers' of the academic study of nationalism after World War I, Carleton Hayes and Hans Kohn, were historians, but their analytical ability was far deeper and better in the sense that they took into account sociological and psychological interpretations. The same goes for Seton-Watson and Louis Snyder.

For Hayes (1933: 6), nationalism is a 'modern emotional effusion' aroused by the nation-state which is a locus of identification and the object of collective sentiments. For Hans Kohn, nationalism is inconceivable before the emergence of the modern state and the ideas of popular sovereignty preceding that in the period from the sixteenth to the eighteenth century. Well before the modernist and constructionist theorists of the nation and nationalism, he explicitly held, on the one hand, that the most important outward factor in the formation of nationalities is the sovereign territorial state. On the other hand, and in equal measure, precisely because nationalism is 'a state of mind', the abstract 'feeling of nationality' is what makes the territorial state an 'object of love' and what prompts 'our identification with the life and aspiration of uncounted millions whom we shall never know'. This is accomplished with the subjugation of most primitive and long existing feelings towards locality, home, family, language, and common descent into a broader emotional context that, out of a free volitional act, requires supreme loyalty to the nation as an abstract entity and as an idée-force which 'fills man's brain and heart with new thoughts and new sentiments'. Herein Kohn meets Renan (Kohn 1961: 3–6, 8–9, 12–19).

After World War II theorists

Going beyond those pioneers, and standing on their shoulders, prompted by the ferocious advent of decolonization after World War II and the traumatic reminiscences of the Great War and World War II, a third generation of nationalism scholars came forward, some great representatives of whom are Don Luigi Sturzo, David Apter, Clifford Geertz, Lucien Pye, Stein Rokkan, Louis L. Snyder, and Frederick Hertz. Now the emphasis was shifted away from historical questions as to the origins and nature of nations and was instead placed on practical and theoretical issues raised by political development and modernization processes, as well as the political culture and nation-building of the new states.

As it is impossible, even briefly, to comment on all of them, I will just limit myself to the contribution of only some of those thinkers. To start with, for Sturzo the nation is but an active collective consciousness and a moral bond contingent upon a stable geographical contiguity, economic interests and a cultural tradition.[2] In his seminal book, Frederick Hertz systematically uses the term 'national sentiment' which is occasionally replaced with 'national emotion' and 'national consciousness' (Hertz 1966: 16, 23, 82). As 'nationalism has proved more powerful than any other political creed', he places significant emphasis on 'national sentiment' as a feeling of love of the nation or the people, affection for the homeland, loyalty to the state, and zeal for its interests. Yet, he is careful enough to argue that this conceptualization is just the beginning for the understanding of nationalist politics. It might be true that 'the fundamental issue is that of national consciousness' because without this consciousness 'there is no nation', yet a host of other political, historical and cultural determinants need to converge in order to give rise to the nation as a modern type of political organization (Hertz 1966: 1, 8). Further, he also speaks systematically of 'national ideology' as a special set of ideas, sentiments, and aspirations which refer to and exert an influence on the 'character of the nation and that of other nations in the past, present, and future … on its mission in the world, on the tasks of the State and on the duties of the individuals towards the nation' (Hertz 1966: 45).

What is somewhat bizarre, however, is that – contrary to other contemporaneous scholars of nationalism, and later ones for that matter – he applies the notion of 'national sentiment' in both pre-modern and modern times. In his own words, 'national sentiment has undergone many ups and downs. It was much stronger in ancient Greece than in later times, and there were epochs when it was almost completely obliterated by other emotions' (Hertz 1966: 31, 33). Of course, Hertz was aware of the difference between traditional and modern society and the role of the nation in modernity; yet, by locating national sentiment and nationality itself in pre-modern historical settings it seems that he added unnecessary conceptual confusion to the discussion, preparing thus the ground for primordialist students of nationalism.

Although Snyder's contribution is of an age, it is still valid for its thematic variety, bibliographic richness, and detailed structure. From his analysis we understand that most scholars of nationalism in the first half of the twentieth century, whether sociologists, political scientists, historians, psychologists or psychoanalysts, were taking as granted that any notion of nation rests on some sort of common, collective, or shared sentiment, (we-sentiment, in-group loyalty, feeling of oneness, consciousness of kind, and so on). From a psychological point of view, the concept of the nation is seen 'as an ideal center of reference for emotion' (Snyder 1954: 43). As a matter of fact, Snyder devotes a whole chapter to the 'sentiment of nationalism' on the basis that we live in an Age of Nationalism: 'nationalism is a powerful emotion that has dominated [the] political thought and actions of most peoples since the time of the French Revolution' (Snyder 1954: 74). As a sentiment, it is

> a condition of mind among members of a nationality in which loyalty to the ideal or to the fact of one's national state becomes superior to all other loyalties, and in which pride is exhibited in the intrinsic excellence and the mission of one's national state.
>
> (Snyder 1954: 77)

That is why, for Snyder (1954: 31) – taking a somewhat different approach from Hertz – the existence of a nation implies a common 'political' sentiment, not just any kind of collective sentiment. And this political sentiment ensues from a complex process of social-historical construction since it 'is re-created in each generation by acculturation and is transmitted from mind to mind by education' (Snyder 1954: 90).

Snyder places emphasis on processes of identification as well as on other psychological defense mechanisms and symptoms implicit in various types of nationalism. For example, he refers to the frequently built-in 'sense of inferiority' hidden under exaggerations over the superiority of one's nation's art, political virtue, literature, and traditions; exaggerations like these are actually overcompensations for deep-seated feelings of inferiority and, not infrequently, express themselves in terms of megalomaniac-paranoid trends. He also cogently refers to the nationalist sentiment as a case of: (a) carry-over of parent and family fixation called forth by authoritarian parental identifications; (b) deep-seated fear and hostility caused by unsolved Oedipal complexes; (c) an outlet of aggression suppressed by Superego's imperatives; (d) abatement of anxiety at the group level (Snyder 1954: 97–98, 104–108). It seems to me that the more subtle psychoanalytic interpretations of nationalism which appeared later in the mid 1990s did not move the agenda much further than Snyder did based on Freud and other psychoanalysts, psychologists, and psychiatrists of his time.

From the 1960s onwards

From the end of the 1960s a fourth period in the study of nationalism appears; during this period the mass of the literature on nationalism overwhelms the number of works produced during the previous periods (Hobsbawm 1992: 4). The debates extend well beyond Anglophone audiences, and most of the scholarly production is multidisciplinary. Indicatively, some of the prominent figures in this fourth period are Karl Deutsch, Benedict Anderson, Ernest Gellner, Anthony Smith, Eric Hobsbawm, John Breuilly, Etien Balibar, Elie Kedourie, Immanuel Wallerstein, Charles Tilly, John Armstrong, Anthony Giddens, and Michael Billing.

Just to mention a few of them, Gellner (1994: 64) and Hobsbawm (1992: 46, 77, 92, 143, 147, 151, 192) held that in real life politics nationalist ideology cannot be separated from 'nationalist sentiment'. Yet, Gellner warranted that 'nationalism is not a shapeless free-floating unspecific unfocused feeling' (Gellner 1994: 65), but a political ideology and a political emotion which by and large is addressed to modern and late modern state power via the inclusion-exclusion interplay. Ten years earlier, he had defined nationalism as a political principle according to which the political and the national unit should be congruent. Upon this principle, he defined national sentiment as the anger aroused by the violation of this principle, or the satisfaction aroused by its fulfillment (Gellner 1983: 1). By the same token, Kedourie (1993: 136, 68) casts doubt on the idea that nationalism is some inarticulate and mighty feeling to be found always and everywhere; nationalism is at one and the same time a particular sentiment and a doctrine about the state and the individual's relation to it, giving rise to a post-traditional style of politics. Presumably, this sentiment translates the abstractness of the modern bureaucratic state into a mental community.

Benedict Anderson did not conceptualized nationalism as a political ideology; from a social anthropological point of view, he treated nationalism 'as if it belonged with "kinship" and "religion", rather than with "liberalism" or "fascism"' (Anderson 1983: 5). In his much-discussed book, he mainly scrutinizes the cultural roots of nationalism and in doing that he rarely uses emotional terms. Once in a while he speaks of 'political love', 'solidarities', 'fraternity', and so on (Anderson 1983: 141, 143, 133, 7, 84). Yet, his idea of the nation as a limited imagined political community *qua* 'horizontal comradeship' is full of emotional energy. Otherwise the ensuing fraternity wouldn't make it possible for so many millions of people to be willing to die and kill, and in addition, the colossal enterprises of every single nation-state to implement their own politics of memory and forgetting through education and cultural policies of any kind (museums, commemoration rituals, monuments, dictionaries, etc.) – something which Anderson focuses on a great deal – wouldn't make any sense.

Anthony Giddens is also hesitant to define nationalism as an ideology, or as any *–ism* whatsoever. He rather sees it as a primarily psychological phenomenon, an 'affiliation of individuals to a set of symbols and beliefs emphasizing communality among the members of a political order' (Giddens 1985: 116; 1981b: 193). In his approach, from the web of nationalism emanate different 'national sentiments' which are not 'ideologically neutral' since 'nationalism is the cultural sensibility of sovereignty, the concomitant of the co-ordination of administrative power within the bounded nation-state' (Giddens 1985: 218–219).

Nation-state: emotionally driven nation-ness – nationally laden emotionality

Evidently, nationalist discourse wouldn't have been articulated independently of affective investments and the use of emotional terms and metaphors. It is true that nationalism draws from primordial feelings of belongingness which are common to any kind of social bond. Yet, nationalism is a phenomenon of modernity and cannot be understood outside this cultural-historical context. It can be theorized in different ways, as a style of thought (Greenfeld), as a discourse (Laclau), as a political religion (Apter, O'Brien), as movement (Alter) or as a political ideology (Kendurie, Gellner).

From a weak social constructionist perspective, 20 years ago I defined nationalism as a modern political ideology through which intellectuals, social movements, and political associations seek to formulate collective and personal identities within a definite territory, and which is organized around and, at the same time, confers meaning on the empty signifier 'nation' (Demertzis 1996a).[3] Today, however, informed by the sociology of emotions, I would add that nationalist ideology employs structures of feeling and invents and imposes feeling rules.

As a rationalizing and secular system of enunciations, every single ideology offers ready-made justifications of extant political power relations, social inequalities and the ways individuals are engaged therein (Gouldner 1976).[4] This is a sociological conceptualization which adopts the 'double hermeneutic' perspective or, to put it another way, the theorizing angle of second order observation. It is one thing to describe what the nationalist discourse enunciates and quite another to theoretically understand why it does so (to observe the observation).[5]

Despite its variations, typical nationalism rests on three, at least, core assumptions: first, the nation is the natural unit of society known by certain distinguishing characteristics; second, the interests and values of the nation take priority over all other identities, interests and values; third, in order to exist each nation requires its own legitimate political sovereignty, i.e., its own state (see *inter alia* Kedourie 1993: 1; Breuilly 1985: 3; Tivey 1981: 5). As shown above, in the nationalist discourse the state is but a means in the service of the nation. However, from the vantage point of a second-order observation, i.e., from the point of view of the double hermeneutic, the nation confers legitimacy on the state and it is actually the state through nationalism which invents and socially constructs the nation and not the other way around (Gellner 1983: 48–49, 54–55; Anderson 1983: 6).

Of course, the nation is not arbitrarily constructed out of thin air; it is a product of on-going processes of selective formation of past ethnic features, cultural practices, political traditions, habituses, collective memories, identities and structure of feeling. These selective processes have been nicely described by many prominent scholars of nationalism like Eric Hobsbawm (1992: 46–79) who referred to 'proto-nationalist' historical-cultural elements which have been retroactively infused and articulated into different nationalist discourses, or Liah Greenfeld (1992) who has shown specifically how ethnic 'raw material' has been reorganized by the logic of nationalist identity. In a nutshell, it is raised out of the remembering–forgetting and the continuity–discontinuity interplay. In this sense, then, the nation is an empty signifier: its meaning is conveyed through the articulatory function of nationalist ideological discourse within time space coordinates. It is not a substance in itself and for itself. In this vein, Kedourie (1993: 141) maintains that 'it is very often truer to say that national identity is the creation of a nationalist doctrine than that nationalist doctrine is the emanation or expression of national identity'. In the same way, Gellner (1983: 7) attests that 'nations are the artefacts of men's convictions and loyalties and solidarities'.

Presenting itself retroactively, and in a no less mythical way, as an everexisting entity, the nation confers legitimacy on the modern sovereign territorial state; since the eighteenth century, the state as a post-traditional 'political community' derives its force and its legitimacy from its emotional foundations (Weber 1978b: 903). To this end, a common feature of the nationalist parlance, and banal nationalism for that matter, is the recurrent use of the emotional terms 'homeland', 'fatherland', and 'motherland' as family or kinship metaphors we live by, designating the nation-state as a community-like whole, an emotional container which functions as a quasi-parental protector that counterbalances anxieties (Richards 2018: 70–81) and the complexity of the modern state and the power asymmetries inherent in social stratification (Greenfeld 2018: 690). These metaphors as well as national rituals (Demertzis and Stratoudaki 2020) convey emotionally charged fantasies of common blood ties, an intimate bond of fraternity. They also convey security, pride, solidarity, gratitude, and dignity when the so-called national sentiment, the collective self-esteem, has to do with 'us' (Sullivan 2014a; 2014b); when the evaluative object is 'them' it is usually specified as hostility, distrust, antipathy, disgust, fear, hate, and envy for the national Other. These metaphors are expressions of two well-known unconscious defense mechanisms described mainly by Melanie Klein: splitting and projection. The first entails the division of the social world into good and bad; in the second, individuals and groups alike introject their good feelings and project their bad feelings onto an outside object (Vogler 2000; Richards 2019: 9–12). It is ultimately upon these mechanisms that the European adage quoted by Deutsch (1969: 3), King (1973: 5) and others, has come to the fore: 'a nation is a group of people united by a common error about their ancestry and a common dislike of their neighbors'.[6]

For Max Weber (1978b: 922) the concept of nation means that it is proper to expect from certain groups a specific sentiment of solidarity in the face of other groups. Nevertheless, Weber was cautious as to the coherence of this feeling precisely because these groups are far from being uniform; there is then

> an unbroken scale of quite varied and highly changeable attitudes toward the idea of the "nation" … which extends from emphatic affirmation to emphatic negation and finally complete indifference.
>
> (Weber 1978b: 924)

Therefore, despite its potency, national feeling and national habitus (Heaney 2013b) are contingent upon 'political meteorology' (Deutsch 1969); i.e., they are instituted alongside open-ended hegemonic and counter-hegemonic processes. Hence, legitimacy conferred by national sentiment is not stable and fixed once and for all. This is more true in postmodern societies, the flexible communications of which promote far more fluid and rapid mediations of national affectivity compared to the more or less stable mediatizations of early printed modernity.

In any case, of course, legitimacy is grounded not only on emotional but also on cognitive and rational foundations. Passions and interests are intermingled in the nation-building processes in many different and unpredictable analogies. Whether we speak of West (civic) or East (ethnic-cultural) nationalism, the nation-building process and nationalist movements generate legitimacy through: (a) the use of symbols and ritual performances (Breuilly 1985: 344–345); (b) the pacification and securitization of the body politic; (c) the establishing of civil equality. These are the minimal prerequisites which, emanating from the much broader Eliasean civilizing process, render 'modern nation-states vehicles of political emotion' (Berezin 2002: 41). People wouldn't be persuaded or enchanted by nationalist symbols unless they felt both secure and equal; security and equality are most clearly corroborated through internal and external surveillance and pacification, one the one hand, and citizenship, on the other (Giddens 1985: 4, 172–197, 181–192, 216–219). Provided that people recognize mutual rights and duties in virtue of their shared citizenship (Gellner 1983: 7), only under conditions of peace and security can social compassion and empathy, and civility, calm, friendship and confidence be planted and flourish (Berezin 2002: 38).

Before ending this section, two comments are needed. First, the emotional attachment to the nation-state ensures that citizens' loyalty to the nation is paramount and the national identity is the supreme identity that subsumes and orchestrates any other collective identity(ies). This superiority of nation-ness is proclaimed by both civic and ethnic versions of nationalism. Second, precisely because each nationalism invents the nation as an inherently limited and sovereign imagined community, citizenship rights within national territory are also limited: there is always someone who is excluded from the national body politic, be it slaves, aborigines, blacks, immigrants, refugees, the poor, and so on.

Beyond 'national sentiment'

I do not think that justice is ultimately done by Mabel Berezin's claim that

> discussions of nationalism have subsumed the emotional dimensions of political community and ... "revisionist" constructionist accounts of nationalism have "de-emotionalized" the nation or glossed over the emotionality of the nation-state.
>
> (Berezin 2002: 41–42)

Nor is justice done with Carolyn Vogler's account that the emotional under-pinnings of nationalistic conflicts and discourses tend to have been relatively unexplored by sociologists of nationalism, who lean towards cognitive biases (Vogler 2000). In this vein, Jonathan Heaney (2013a) has attributed too much cognitivism to Anderson's nation-ness as imagined community. Also, Thomas Scheff (1994a) criticizes Anderson and Smith for having underestimated the role of emotion in the analysis of nationalism.

To me these criticisms are a bit of exaggeration. As I have tried to show in the first two sections of this chapter, the emotional dimension has been ever present in both ideologists' and ideologues' nationalist discourse on the one hand, and in social and historical scientific analysis of nationalism on the other. It may be true that most scholars didn't go far beyond the so called 'national sentiment' but this does not mean that the emotional dimension has been neglected, underesti-mated, or has fallen by the wayside. It might be occasionally the case that emo-tions have been 'under-valued' in their theorizations but they were not overlooked altogether.

What a contemporary political sociology of emotions might confer on the 'emotional reading' of *homo nationalis* is the dismantling of the presumption that nationalism is a powerful but no less inarticulate and omnipresent feeling which invigorates the nation as a sort of 'horizontal comradeship'; a political sociology of emotions might be able to contribute to the specification of the catch-all and vague concept of 'national sentiment' by breaking it down into particular primary, secondary or even more complex emotions in a systematic way. Apparently, this is not an easy task but it has been modestly undertaken in a piecemeal manner for a good long time now.

The founding father of the sociology of emotions, Thomas Scheff, attempted to analyze 'fanatical ethnic nationalism', national conflict and violence by employ-ing shame and pride, as the master emotions of modernity on the one hand, and alienation on the other (Scheff 1994b, 1997a). In his words,

> most people find chronic shame and rejection an unbearable burden, and will do almost anything to relieve themselves of it. Gangdom, racism, nationalism and the crimes committed in their names provide for the shamed and rejected a haven of pride and acceptance.
>
> (Scheff 1997b)

Unacknowledged shame causes 'shame-anger cycles', i.e., spirals of increasing shame, followed by anger that the shame is being felt, which accumulates addi-tional anger, and so on. In addition, he claims that when individuals are deeply alienated from self and others, nationalism creates an illusion of imagined com-munity. The combination of alienation, unacknowledged shame, and broken pride provides the basis for offensive attitudes conducive to, or even driving con-flict. According to this scheme, which is nevertheless not void of psychological reductionism, Scheff offered an interpretation of the origins of both World War I and World War II. Based on Lasswell's *Psychopathology and Politics* (1960), he shows how Hitler's own psychopathology, his

> paranoia and continual humiliated fury produced a program responsive to the craving of his public for a sense of community and pride, rather than aliena-tion and shame. Since neither the alienation nor the shame were acknowl-edged, both Hitler and his public were trapped in a never-ending cycle of humiliation, rage, and vengeful aggression.
>
> (Scheff 1994b: 105)

From the point of view of historical sociology, economic sociology and anthro-pology, Liah Greenfeld (1992, 2001) has put sufficient emphasis on the emotional dimension of nationalism and national consciousness. Before and above anything else, like Kohn and Scheff, she makes it clear that '(n)ational identity is, fun-damentally, a matter of dignity. It gives people reasons to be proud' (Greenfeld 1992: 487); it is the 'sense of dignity that lies at the basis of national patriot-ism and commitment to national causes, which often strike outside observers as irrational' (Greenfeld 2001: 3). And its importance not only consists in the fact that dignity is linked with economic growth; it has also to do with the core of individuals' ideal Ego, i.e., it has vital significance for the facing of social and existential anxiety, social recognition and the effacing of class injuries (Richards 2019: 12–13).

Pushing this argument to the limit, she is taking the risk of overgeneralization by claiming that 'the world in which we live was brought into being by vanity. The role of vanity – or desire for status ... has been largely underestimated' (Greenfeld 1992: 488). This is actually not such a blatant statement compared to a couple of structural theories of emotions (Theodor Kemper, Robert Thamm, Joseph Berger, and Robert Shelly)[7] according to which the loss or gain of status (recognition) in connection with one's power, role expectation and anticipated sanctions elicits a host of primary, secondary and tertiary emotions. These emotions can be rather easily predicted when attribution of responsibility for one's loss or gain of status and power is also taken into account (and is attributed to self or others). In that case, these theories claim universal validity regarding the individual's emotional reactions and the ensuing styles of political action. Greenfeld is convinced that nationalism is very much explained by the elites' and powerful social classes' preoccupation with status and 'thirst for dignity' (Greenfeld 1992: 372).[8] Analyz-ing five prominent but no less dissimilar cases of nation-building, she concludes that, as far as status is concerned, 'the English aristocracy sought to justify it; the French and Russian nobility to protect it; the German intellectuals to achieve it, and Americans to defend it' (Greenfeld 1992: 488).

In addition, not only does Greenfeld acknowledge the importance of emo-tions but she very systematically brings to the fore the crucial role played by *ressentiment* in the emergence of French, German, and Russian nationalism. In order to really make sense of her quite complicated but elegantly expressed argument one has to bear in mind that, as a protean political-cultural phenomenon of modernity, nationalism has been called forth in the plural. It is nationalisms rather than a single nationalism which emerged as a post-traditional mode of domination. This means that from the very beginning various nationalisms have been inter-constituted, either through overt or covert antagonisms and comparisons. Concomitantly, students of nationalism have distinguished between 'primary' and 'secondary' nationalisms in the sense that in different countries and different time zones nationalist movements were formed either within the borders of already existing centralized and well administered states or in traditional quasi-patrimonial, and loosely organized multi-ethnic imperial states (Gellner 1994: 29–31; Kohn 1961: 329–330, 457; Trevor-Roper 1961). It is usually claimed that primary nationalisms acquire civic-libertarian characteristics whereas secondary nationalisms are of a more ethnic and collectivistic nature. Nevertheless, and despite the fact that 'the nation proved an invention on which it was impossible to secure a patent (since) it became available for pirating by widely different, and sometimes unexpected hands' (Anderson 1983: 67), what is probably more crucial in this respect is that newcomers cannot but compare themselves with the 'first-born' or the precedents from which they normally borrowed and engulfed nation-building ideas and ideals.

As for the genesis of nationalism, Greenfeld is adamant: the first nationalism appeared in early sixteenth century England where, due to a host of historical reasons, the ultimately empty signifier 'nation' started to apply not only to the elite but also to the population at large, thus becoming synonymous with the word 'people' which eventually came to be understood as the 'sovereign people'. Accordingly, 'by 1600 the existence in England of a national consciousness and identity, and as a result, of a new geo-political entity, a nation, was a fact' (Greenfeld 1992: 30). The nation was perceived as a community of free, rational, dignified, humane, and equal individuals participating in the political community. As the front runner of liberal nationalism, England utilized existing institutions while transforming itself: the development of science, the rapid upward mobility of the minor gentry and bureaucratic class which overtook the old nobility in tandem with the rise of the Tudor dynasty on the throne, and of course, steady and highly marketized economic growth, were the main 'objective' factors, the main conditions of possibility, that contributed to the imaginary institution of the 'nation' as a community of free and equal individuals entitled to political participation. This idea was ultimately engulfed by both the elites and the common people.

This was not the case with the latecomers France, Germany, and Russia, where, from the late eighteenth century onwards, the elites, in their relationship to central authority, had to confront and compare themselves with self-conscious established English nationalism. Confronted with the advent of modernity and fueled, among other things, by the experience of the Seven Years War (1756–1763) and the involvement of France in the American War of Independence, the disempowered intelligentsia and the French nobility appropriated the English originated idea of the nation in their own terms: they reinterpreted the values of liberty and equality by subsuming them into the grid of *volonte generale* (Greenfeld 1992: 182). After the death of the king, the Nation was assessed as the Supreme Being which had to be defended from foreign forces, especially the English. The traditional etatism of the French Christian Monarchy was reinterpreted into nationalist vocabulary(ies). Regarding the elites, this could only be done through the *ressentiment*-ful transvaluation of the English nationalist virtues and the cultivation of Anglophobia (Greenfeld 1992: 167, 177–180).

The *ressentiment* of the French nobility and intellectuals after Montesquieu and Voltaire, as well as the *ressentiment* conveyed by the German and Russian nationalism are called forth by the intellectuals' and nobility's lost status and power as a result of the civilizing and modernization processes:

> (t)he rapid disintegration of the traditional order threw the social system out of balance, and the strata composing the elite … found themselves in a situation of status-inconsistency. The divisions of the traditional order lost their meaning … the hierarchy of prestige no longer corresponded to the hierarchies of wealth, education, and power … all were affected and all alike were suffering from status-insecurity and anxiety which had led to a maddening itch of inconsistency, of the discrepancy between the possible and the existent, the frustrating apprehension of unfulfilled opportunity.
>
> (Greenfeld 1992: 152–153, 213)

In these terms, imported into France from England and into Germany and Russia from France, the nation as a notion was 'grafted on a body of indigenous traditions' (e.g., German romanticism versus *Aufklaerung*, *Gemeinschaft*-like understanding of society and nation-ness, the Russian feud against the West, Slavophilism, and so on); through *ressentiment* these traditions were rendered as defense mechanisms against felt inferiority vis-à-vis economically, militarily, and politically advanced nations. They have been elevated to become higher values so that the original civic and libertarian ideal of the nation was depreciated in favor of a collectivist, ethnic and authoritarian conception of nationality which permeated most of the national political culture (Greenfeld 1992: 156, 489).

The potency of *ressentiment*

Following the Weberian tradition of interpretative sociology, Greenfeld supports her arguments with a masterful analysis of innumerable primary and secondary documents in four different languages. However, her monumental scholarly accomplishment couldn't cover all aspects and details of the nationalist phenomenon. For instance, she didn't pay much attention to the proto-nationalisms Hobsbawm talks about in order to convincingly support the fact that the elites' national-ideological thinking was articulated with grass roots national sentiment(s). On top of that, her seeping top-down approach has no dialogue with the sociology of emotions whatsoever. In her later contribution (Greenfeld 2001), the sole reference to any sociological discussion on emotions is to Hirschman's *The Passions and the Interests*. What strikes one is that, although she rightly emphasizes the role of *ressentiment* in French, Russian and German nationalisms, she offers only a poor analysis of the concept.

Of course, she says that *ressentiment* cannot explain everything in nationalism as a style of thought and does not exhaust the national sentiment per se; rather, it is associated with what she calls an ethnic and civic-collectivist type of nationalism. But then again, *ressentiment* does not receive the treatment it deserves given the central role Greenfeld attributes to it; she devotes less than one and a half pages to its conceptual analysis. If Greenfeld had treated the concept more thoroughly she might perhaps have found that *ressentiment* does not stand still throughout national histories; precisely because emotions are relational constructs, not only can more than one be experienced at any one time (i.e., they are experienced 'in flow'), with different degrees of intention, but what is more, they may fade and change and, when the external environment permits it, may be elicited all over again. The likelihood then is that the so-called 'Tocqueville effect',[9] an 'iron law' enacted several times in the five national histories she explores, is not put into effect so much by *ressentiment* as by envy, one of the components of *ressentiment*, or even by resentment *qua* moral outrage (Greenfeld 1992: 16, 152–153, 213, 312). In another comparative analysis of nationalism, she pays a bit more attention to the conceptualization of *ressentiment* with a scant but correct note on transvaluation (Greenfeld and Eastwood 2005: 260–261). In the remainder of this chapter I will provide a general and no less brief outlook of Greek nationalism along the lines of Greenfeld's analysis.

A short emotional reading of Greek nationalism: the role of *ressentiment*

The historical context

As with all European countries, Greece's road to modernity had its own peculiarities and path dependencies. Founded in 1830, the Greek state developed from the 1821–1827 revolution against the centuries long Ottoman rule in the Balkans, which even today makes it not that easy to draw a neat boundary for 'Europe' along Greece's eastern boundary. That revolution, typical of the separatist eastern nationalist movements (Breuilly 1985: 107–111), was an historical case where an ethnic population under the domination of one of the great modern multiethnic empires undertook the project of independence by appealing to the ideal of the nation. There has been a long discussion among Greek and non-Greek scholars (Lekas 2005) as to the exact grid of drivers for the national war of independence which was much influenced by the French Revolution. The fact is that six years after Napoleon's downfall, which seemed to spell the end of nation-alism and liberalism.... The Greeks were the first successfully to raise the banner of nationalism and liberalism. With their 'war of independence' the age of nationalism in Eastern Europe was established.

(Kohn 1945: 537)

In Greek nation-building, the idea of the nation, and nationality for that matter, precedes the formation of the state, which, in turn, transforms nationality into nation as a modern political entity (Haddad 1977; Kitromilides 1990).

The making of the Greek state was feasible because of both the Ottoman weakness in the southern part of the Balkan Peninsula and the intervention by England, France, and Russia that aimed at keeping the 'Eastern Question' under control. In addition, it was affected by major competition and conflicts among the Greeks themselves, caused by great social inequalities, especially between peasants and local lords. The rationale of the Greek movement of independence was marked by the articulation of two different national discourses: a liberal vis-a-vis an indigenous discourse with localist overtones. Bearers of the latter, the traditional Greek elites of the landowners, military chieftains, and high clergy, ultimately prevailed in that inner struggle. Thus, a more conservative form of nationalism prevailed in the discursive practices of early modern Greek statehood. Yet Greek nationalism was, and still is, far from unequivocal (Koumandaraki 2002). On one hand, there existed those who defined the nation in terms of Classical herit-age and Enlightenment ideals such as freedom, rationality and secularization; prominent advocates of this outlook were the Greek merchant diaspora elites, mainly in Italy and France. On the other hand, one may find the proponents of the resurrection of the Byzantine Empire who understood the nation in terms of Orthodox Christianity.

Gradually then, two selective traditions have arisen; both have had pertinent effects on contemporary Greek political culture. The Hellenic and Enlightenment tradition has been less grounded in popular culture: for the most part, it has been cultivated from above by the westernized elites. In contrast, the Byzantine and Ottoman religious and political tradition has been much more ingrained in every-day *habitus*. This tradition is distinguished by a particular intersection of what Apter (1965: 83–87 and *passim*) calls instrumental and consummatory orienta-tions; the former relates to empirical ends and the latter to more transcendental and integrative values resistant to change. In a few pages I will come back to this distinction.

For the sake of my argument, some extra comments are needed concerning the second tradition: on the one hand, the 'patrimonial' political domination of the Byzantine Empire and the 'sultanist' type of the Ottoman rule, as well as the familial and communal kind of social organization and economic pro-duction, diminished the possibility of a 'contractual' idea of citizenship, which was to become so important in much of Western Europe. That was due to the personified, though not entirely arbitrary, style of politics of both empires.

As formally institutionalized legal and political procedures guaranteeing natural and civil rights and duties were absent, a set of instrumental and defensive outlooks towards authorities arose. This was reinforced by the realities of everyday life, which resulted in the mistrust of any sort of secular power and the belief that much of public life was based on nepotistic practices (Campbell 1983; Charalambis and Demertzis 1993). On the other hand, the Orthodox conception of the self, with its mystic and communitarian overtones, was the basic theodicy context within which individuals sought long-term justification and meaning-making. Unlike the Catholic and Protestant conceptions, where the relationship between person and God is mediated by individual reasoning and the law, the Orthodox Christian feels united with God in a non-mediated personal relationship. Consequently, a deep-seated fatalism has grown out of this conception, but this has not led to a single and all-embracing consummatory *Weltanschauung* that could harness the entire social and emotional energy. Partly, this was due to the elastic and instrumental character of Orthodox practical religion which, contrary to other Christian dogmas, has been flexibly adjusted to the day-to-day practical demands of life (Makrides 1995; Demertzis 1996b). The personified element made Orthodoxy more of a popular-religious practice than a prescriptively codified set of religious ethics, especially ones which required strict adherence. As a non-bibliocratic religious confession, Christian Orthodoxy was rather easily adjusted to the national imperatives through its politicization and nationalization. This was facilitated by the Byzantine tradition of a 'lesser degree of caesaro-papism'.[10]

As a matter of fact, the bifurcation between these two cultures caused numerous intense debates, social upheavals, policy disorientations, and feelings of insecurity, angst and ambiguity with the advent of the Greek nation-state in 1830. For instance, there was the great dispute concerning the name of the modern Greeks: should they call themselves 'Hellenes' or 'Romii? The latter term comes from the 'Romans', which is the way the Byzantines referred to themselves, i.e., habitants of the Eastern Orthodox Roman Empire; this debate was linked to the 'linguistic question' of the constructed and Classical-like *katharevousa* and the everyday *romeika*. The Hellenic designation referred to an outward-directed westernized image of the nation as the immediate heir of Classical culture, whereas the Romeic designation had strong connotations of the Orthodox religious origins of the Greeks (Herzfeld 1986: 18–21; Kohn 1945: 329–330, 539 and 457; Sugar 1969: 19–20, 34–35; Clogg 1997: 27–29). By the same token, in spite of their radical-liberal character, when compared with their western European counterparts, the concepts of citizenship and Greekness/Hellenicity in the first revolutionary constitutions were – and still are for the most part – conterminous with the idea of Christian believer.

One could pick up plenty of similar instances from Greek political history; however, the point at issue is that Greek nationalism has not only been more 'cultural' than it was 'political' but it created highly ambivalent collective identities as well (Triandafyllidou, Gropas and Kouki 2013). Moreover, grounded on the continuity of the Greek language, the idea of 'Hellenic-Christian civilization', introduced in the late nineteenth century and invariably deployed ever since in public education and political socialization, has never managed to effectively reconcile this ambivalence (Özkirimli and Sofos 2008: 45–55). While responding to the ideological needs of their emergent polity (Herzfeld 1986: 9), this 'Helleno-Romaic Dilemma' (Leigh-Fermor 1966) urged organic intellectuals of the nation-state to combine the 'two Greeces' into a single national narrative by inventing continuities and similarities between past and present. This was so because, ever since its inception, the Greek nation-state has been gripped in a double-mirroring of its own nationalism and the nationalisms of its western allies and its main adversary alike, i.e., Turkish nationalism. Greece has had to present itself as an integrated nation whose Classical Antiquity inheritance grounded western civilization and, in that sense, it had to comply to the standards of western modernity, adopting an Occidentalist self-image (Bozatzis 2014; Herzfeld 1995). In this respect, the notion that ancient Greece is the 'cradle of Western civilization' is a profoundly Orientalist idea (Gourgouris 1996). On the other side, however, soon after the establishment of the new state, with the decline of European Romanticism and the concomitant philhellenic exoticism (Sotiropoulos 2019), the great powers of England and France realized that the first Greek kingdom under their protection was actually a backward society which had almost nothing to do with Classical Hellas. The westernization of the new state, economy and society was set as a supreme goal to be achieved, hand in hand with the interest of local modernizing elites in consolidating a new nation amidst a declining oriental multiethnic empire. No doubt, those interests were premised on an orientalist gaze. Therefore, the Occidentalist self-image of the indigenous official national narrative has been mirrored onto the orientalist projections of western modernity's proponents.

The structure of feeling towards modernization

Provided, on the one hand, that nationalisms are ultimately inter-constituted and, on the other hand, that Greece was from the outset dependent on the Great Powers established by the Congress of Vienna in 1815, the orientalist-occidentalist double mirroring in the formation of Greek nation-building seems to be unavoidable. At any rate, idealized self- and hetero-mirroring has not been able to gloss over the fact that Greece did not fully comply with the alleged canon of western modernity, nor could it disguise the painful awareness of its inferiority when compared to its western counterparts with regards to technological and economic development. As a country of the semi-periphery, Greece has always been lagged behind in the hierarchy of the international division of labor and symbolic capital, experiencing the pressure of, as it were, a crypto-colonial dynamic (Herzfeld 2005: 56, 67–68, 214).[11]

This experience has been wittily expressed in the famous verses of George Seferis, one of the most important Greek poets and diplomats of the twentieth century and a Nobel laureate, written in 1936: 'Wherever I travel Greece wounds me, curtains of mountains, archipelago, naked granite, They call the one ship that sails AGONIA 937'. Be it noted, first, that the verse 'Wherever I travel Greece wounds me' has been a self-defining adage for many Greeks for several decades; second, the Greek name of the sailing ship is agony (meaning painful anxiety). In his wanderings Seferis the diplomat, the poet, the sublimated object of Hellenism itself (Gourgouris 1996: 201–226) feels anxious, traumatized and ashamed whenever the sheer reality of the nation is compared to its potentialities, and when comparison between Greece and more advanced countries is at issue.

Crypto-colonialism ensures a definite structure of feeling, a perpetually unbearable sense of deficiency begging for a rectification that almost never comes. For all the progress made throughout 200 years of independence, the carrying capacity of the socio-economic system has its limits. Although there is no place here for detail, suffice it to refer to three widely held postulates about Greece's economy since the early twentieth century. First, Greek economic capital has been more commercial and, in general, small business-based than industrial and productive. Second, most economic development has been mediated by the state as the principal mechanism of surplus distribution regulating market competition. Third, as an institutionalized capitalist market per se has never been a major feature of Greek development, economic life has been regulated by crony capitalist arrangements. As a consequence, neither a deep-rooted labor class nor a productive capitalist class has ever developed historically in Greece. Instead, petty bourgeois patterns of life and structures of feeling have tended to prevail. It is not accidental that since its accession to the EU in 1981 the gap in most, if not all, socio-economic indicators between Greece and the mean European values has remained wide open. The current and deep-seated economic crisis has already left Greece by the wayside of the Eurozone countries and put an end to aspirations for upward social mobility. As it has been repeatedly documented, these trends have elicited or enhanced a variety of extant negative emotions.[12]

Taking the risk of oversimplification, I would claim that, like the civilizational distance experienced by the Russian elites vis-à-vis the West, the awareness of this perpetual gap has been operating as a conducive condition for nurturing national *ressentiment* in Greece.[13] Against this unpleasant experience an expansive and deepening transvaluation process was initiated, a powerful reaction formation that concluded with 'cultural intimacy', an anthropological notion forwarded by Michael Herzfeld (2005) and adjusted for the purpose of my own analysis. As a compensation for the orientalist humiliating gaze, cultural intimacy has offered a symbolic dream space of communal familiarity, sameness, and existential security. Communal familiarity offers consoling explanations of 'apparent deviations' from formal rules and official public interest, and also offers 'a sense of defiant pride in the face of a more formal or official morality' promoted by the geopolitics of western cultural hierarchies. Although these deviations may be considered sources of external embarrassment, cultural intimacy provides insiders with the assurance of common sociality, and a familiarity with the bases of power that may at one moment ensure the disenfranchised a degree of creative irreverence and at the next moment reinforce the effectiveness of intimidation (Herzfeld 2005: 3, 9). Thus, cultural intimacy is not a fixed condition but a fluid interplay of autonomy and heteronomy, a socio-emotional space in the making, consisting of a 'structural nostalgia' for traditional *habituses* and archaic patterns of sociality and endowing with meaning nationalist and populist narratives and state ideologies (Herzfeld 2005: 22).[14]

Under these macro-structural coordinates, an eminent question, engaging a great deal of Greek historical and political sociologists and political analysts alike, concerns the cultural imprint of Greece's path dependency towards modernity. This concern draws much from the rather colloquial postulate of Greece's in-betweenness in relation to the West and the East. Oftentimes, analysts argue that Greece's culture is a transitory one or, what is more, that a sort of 'cultural dualism' permeates the entire social fabric (Diamandouros 1983, 1994; Mouzelis 1995; Papacosma 1988). The 'cultural dualism' argument proposes that two distinct cultural camps have been at war since the foundation of the modern Greek state: an introverted 'underdog culture' adhering to the 'tradition' of the Byzantine and Ottoman past and an extrovert cosmopolitan culture that 'draws its intellectual origins from the Enlightenment' and expresses the secular demands for western modernization.

The mechanics and the dynamics of national emotionality

From a different perspective, and in line with Shmuel Eisenstadt's hypothesis that transitional society as a concept is ambiguous and overloaded with western-centricity (Eisenstadt 1973: 100–101), I have argued that the 'transition' thesis is by definition obsolete and that the 'cultural dualism' claim is too schematic to encompass the multifaceted cultural reality of the country (Demertzis 1997). To this end, I have introduced the concept of 'inverted syncretism' designated to thematize the articulation rather than separation of tradition and modernity in Greece. Drawn from the sociology of religion, where it means the blending of diverse and heterogeneous religious practices, syncretism was used in comparative political analysis of the late 1960s to deal with the so-called political development of the new emerging nation-states after the decolonization process which took place amid West–East tension over the global world order. The normative interest of American comparatists was the introduction of liberal democracy in those countries where colonial administration had left powerful traces in political life, with the point being whether exogenous new democratic political structures, introduced or adopted in circumstances of acute pressure, were congruent with existing political cultural patterns emanating from time immemorial traditionalism. The borders of most new nation-states were more or less arbitrarily shaped by the colonial powers themselves in an attempt to gain an imperialistic stronghold in those territories. Consequently, ethnic communities were torn apart and traditional patterns regulating every-day life were shattered within the 'newly bordered power-container', as Giddens calls the nation-state (Giddens 1985: 120).

Third World countries entered modernity first through long periods of colonization and soon after through decolonialization settlements. Thus, their political culture was a mixture of anti-colonial nationalist orientations, archaic patterns, and modernizing political outlooks.

For comparatists like David Apter and Samuel Huntington, the future of liberal democracy in these countries was contingent upon the functional blending of tradition and innovative modernity, or, to put in their own parlance, between the local 'recipient' and the 'donor' political culture. And here comes 'syncretism'; it is an acculturation process where the patterns of one culture are domesticated to the schemes of another without losing their original function:

> Syncretism occurs when an overt form of one culture is not perceived in the same fashion by members of another but is perceived in such a way that it can be interpreted to conform to the borrowing culture's own patterns of meaning and yet retain essentially its original function.
>
> (Huntington and Dominguez 1975: 20)

That is, syncretism is a process of cultural reinterpretation which eventually facilitates modernization. Yet not every fusion of tradition and modernity leads to syncretism; there have been many cases of cultural resistance and radical ambivalence toward westernization which have led to terrorism, regional nationalist movements, and civil unrest. If anything, as European colonial powers originated in different political traditions (monarchical regimes, republics, centralized state authority, federal democracies, parliamentary or presidential democracies, multi-party or two-party systems, etc.), so too indigenous cultures were far from uniform. From an ideal typical point of view, syncretism is likely to occur when some traits of the 'recipient' and the 'donor' culture coincide or correspond with each other. In this respect, holding that values play a pivotal role in every single culture, Apter (1965: 81–95) and Huntington and Dominguez (1975: 18–27) analytically differentiate recipient traditional cultures as to whether they are mostly permeated by 'consummatory' or 'instrumental' values when confronted with structural changes.[15] These two value types correspond to the sacred and secular orientations towards world and social order. In consummatory traditional political cultures most societal relationships, everyday life practices, and gratifications are inexorably linked with highly elaborated, integrative, transcendental religious cosmologies which ascribe meaning to the entirety of social conduct. These cultures drastically resist change. On the other side, life courses and gratifications in instrumental traditional political cultures come from immediate empirical ends, from practical and ordinary acts of social life not absorbed by the religious sphere. In instrumental traditional systems there are tendencies toward functional differentiation between spheres of societal action which permit segmental changes that are not perceived as threatening or detrimental to the social fabric. In this respect, 'innovation is made to serve tradition' (Apter 1965: 85).[16] Comparative political analysis holds that syncretism is most likely to occur in instrumental rather than in consummatory cultures due to the former's adaptability to acculturation processes and socio-political changes (Huntington and Dominguez 1975: 19).

All in all, then, syncretism relates old with new through the selective reinterpretation of the former for the sake of the latter. In the modernization processes of countries like Japan, Taiwan, and South Korea syncretism occurred because the new modernizing cultural patterns were assimilated to the traditional ones, retaining though their original function and making change possible. Greece's path dependence towards modernity acquires much from the syncretic acculturation process. However, this is not the place to delve into historical details. Suffice it to say that, by and large, Greek traditional culture was instrumental enough for syncretic patterns to arise after the War of Independence. Most notable has been the elasticity of the Orthodox practical religion, referred to above, due to the fact that under the *millet* system of the Ottoman rule the Orthodox Grecophone Ecumenical Patriarchate of Constantinople was officially recognized as an institution of the Empire responsible for the fiscal, administrative, and judicial affairs of the Empire's Orthodox subjects, having thereby not only duties but also certain privileges (Clogg 1997: 10–11; Smith 1986: 114–115; Xydis 1980: 217). The instrumental elasticity of Orthodoxy was also pursued by the 'nationalization' of the Church in 1833, a few years after the establishment of the new state through a government decision which declared the church of Greece 'autocephalous'; since then 'the church could stand to gain a great deal more by assuring those in authority of its loyalty' (Stavrou 1995: 47).

Three other important pre-modern instrumental drivers for Greek syncretism were: (a) clientage networking, (b) customary every-day regulations within the Ottoman multi-ethnic state, and (c) communal familism. Unlike Stathis Gourgouris (1996: 53–70), who somehow dismisses the importance of clientelism as an archaic means of interest intermediation, it seems to me that clientage networks and customary regulations complement each other. Clientage/patronage networks offered a common political cultural code shared by rulers and ruled alike, so that a critical social space of predicted regularities and protection could open up vis-à-vis official autocratic sultanism (Petropoulos 1968: 24–37; Diamandouros 1983: 44–45; Campbell 1964). On the other side, as long as the patrimonialist style of Ottoman authority could not, and would not, guarantee the continuance of the economic privileges allocated to various subjects in different *millets*, societal entropy was evaded by the positing of customary regulations serving as functional equivalents of customary law. Overt antagonisms over the distribution of the social product were bypassed through precarious arrangements which envisaged *maximum* privileges for the sovereigns and *minimum* disadvantages for the ruled (Gourgouris 1996: 59–60). This unstable equilibrium demanded much emotional energy, negotiation skills and constant coping with contingency and uncertainty. Hence, it cultivated instrumental cultural patterns. Most of the time, in turn, these patterns were not followed in isolation, on an individualistic basis; it has been repeatedly documented in the relevant historical research that agrarian economic activity in the South Balkan peninsula between the fifteenth and late nineteenth centuries was carried out via local communities and the extended family system, already spiritually under the auspices of the Orthodox church. On this basis, I would claim that this instrumental *Gemeinschaft*-like tradition not only paved the way for the emergence of cultural Greek nationalism, but has also contributed to the forging of the cultural intimacy referred to above, as well as to the shaping of 'amoral familism', the unidimensional loyalty to (extended) family members in the face of public ethics compliance (Banfield 1958).

To recapitulate, I deem flexible religious practices, clientelism, informal socioeconomic arrangements, and communal social conduct to be crucial drivers for the syncretic path dependence of Greece towards modern socio-economic and political modernity. True enough, 'Greece appears balanced between a troubled tradition and a desired modernity' (Triandafyllidou, Gropas and Kouki 2013: 11); yet, as long as there exists no archetypical road to modernity since 'multiple modernity(ies)' are eventually the rule (Eisenstadt 2013; Mouzelis 2008: 145–159), to maintain that 'Greece is a *par excellence* case of failed modernization' (Gourgouris 1996: 105) actually makes no sense. It is true that Greek economy and society – placed in the semi-periphery of the world-system – differ considerably from the capitalist 'center'. Yet, the point is not that Greece deviates from an alleged canon but to comparatively clarify the substance of this difference, namely to understand the way(s) archaic and modern cultural patterns, the *romeic* and the *hellenic* identities, have been linked over the last 200 years or so, thereby affecting national *habitus*. To this end, I have argued that Greece's path to modernity has traversed a pattern of 'inverted syncretism' (Demertzis 1997). By inverted syncretism I mean that by retaining a formal status, modernizing cultural patterns and practices lost their original function while the instrumental traditional ones remained intact or even became rejuvenated. In this sense and in the main, Greece has been a country in which 'vocabularies imported from the West are used to conceal and /or legitimize institutional arrangements that are a far cry from the political modernity seen in western European parliamentary regimes' (Mouzelis 1995: 23). In a paradoxical way, Greece exited from traditionalism through the survival of tradition fused with modernity. A striking and almost self-perpetuating example is legal formalism. The adage among Greeks that 'we need a law to enforce the enactment of laws' is symptomatic, on the one hand, of a mass of complex and detailed legislation over a vast array of jurisdictions – often more advanced compared to the other western states – that is, however, interwoven to such a degree that it is frequently almost impossible to enact it in an effective way. On the other hand, the enactment of many well-crafted laws is contingent upon the issuing of ministerial and presidential decrees which either are just never issued or are issued with great delays, thereby rendering the relevant regulation obsolete. The gap between the legal-formal rule and the actual regulation is filled by traditional party clientelism, administrative corruption and free-rider economic behavior.

To make myself clearer on the inverted syncretism issue, let me give another example coming from cultural politics. In Autumn 2003, a period when the pro-European modernizer prime minister Costas Simitis was in office, a significant international Forum of Modern Art named *Outlook* was organized in Athens. Having already attracted 22,000 visitors, 45 days after its opening the Organization for the Promotion of Greek Culture decided, on the direction of the then Minister of Culture, to remove a painting by the Belgian artist Thierry de Cordier. The painting represented an ejaculating penis next to a Christian cross. A far-right politician was the first to complain, to be followed by the official Church and the leader of the main opposition party, who demanded the punishment of those 'responsible for that shame'; as soon as the issue became prominent on the media agenda, the ensuing wave of moral panic in the name of 'Orthodox religious faith' swept away any intention of cultural politics decision makers to keep the field of avant-garde artistic creation, even in its provocative persona, safe and free.

If Greece should stand alone outside and above the world-system and global power and status hierarchies, there would be no concern about inverted syncretism. From an *emic* point of view, however, ambivalence towards the West, self-victimizing discontent, and fluctuations between an inferiority syndrome and megalomania, as well as the self-characterization of Greece as a 'brotherless nation' (Gavriilidis 2008), which nevertheless is the 'cradle of western civilization', are in all likelihood created by the complexity of inverted syncretism itself. The more the traditional and archaic patterns are fostered and disguised by modern or even hyper-modern configurations, the more the sting of wounded pride is felt when comparisons with advanced capitalist societies are made. And these comparisons are taking place all the time as long as Greece is deeply embedded in the international division of labor, the global phantasmagorias of consumption, and global flows of information. Low technological expertise, poor administrative performance, and public debt and financial deficit are fields of relentless international comparison which elicits humiliation and embarrassment in most social echelons. But as was explained above, such negative feelings are dealt with through a compensating *ressentiment* which has become an emotional ingredient of the narcissistic national identity. Using a metaphor from physics, I would say that the mechanics of the emotionality of the Greek nation-state are the nexus of inverted syncretism, while its dynamics are, in tandem, cultural intimacy and *ressentiment* against the West.

The poetics of ressentiment

Such an emotional reaction is clearly depicted in Odysseus Elytis' thought – or at least in some part of it. Elytis (1911–1996) was another great Greek poet who also won the Nobel Prize for Literature in 1979. Alike Seferis, though younger than him, he belonged to the generation of the '30s, i.e., an interbellum modernist generation of artists concerned primarily with the quintessence of Greekness/Hellenicity. Influenced by surrealism, his poems were written in rich language, full of images from history and myths; his early work celebrated the mystery of the Greek light, the sea, and the air. Elytis was actually a public intellectual who greatly affected Greeks' self-awareness, especially through the setting to music by Mikis Theodorakis of his 1959 *Axion Esti* [*Dignum Est*]. That long poem is about holiness, innocence and persecution experienced by Greeks all through their history. The musical version, magnificent in itself, was presented in 1964 and has been performed innumerable times since then. The verses are sung by Greeks who love it for what it says about all injustice and resistance, and admittedly for its pure beauty and musicality of form. Describing 'the tribulations of the small and innocent Greek people, the efforts of evil foreign forces to destroy it, and its final aestheticized triumph and glorification' (Gavriilidis 2008: 147), it has almost become a second national anthem for Greece. Greeks themselves are poetically assumed to carry the attribute of pious victims, destined nevertheless to triumph over the powers of their adversaries. Overtly, the poem centers around trauma, suffering, inferiority, and sublime vindictiveness. In a nutshell, it contains most of the ingredients of *ressentiment*.

This emotional-moral stance became all too real a decade before Elytis came up with his *Axion Esti*. In 1948–1952 he lived in Paris where he audited philology and literature seminars and was well received by the pioneers of the world's avant-garde. That was a time where Paris was the heart of the international intellectual life with existentialism being its center of gravity. Although in the beginning he was thrilled with that intellectual environment and his acquaintance with great poets, novelists, philosophers and artists, he, the romantic modernist who was seeking the meaning of Hellenicity, the would-be national poet of Greece, gradually fall into despair. Ultimately, he could not feel at home there. In his voluminous *Open Papers*, where Elytis exposes his theoretical and philosophical ideas, he accurately describes his state of mind during that period:

> An entire literature, headed by Baudelaire, had sought for half a century to dominate the undeveloped and unexploited psychic regions ... by giving more weight to "human exception" than to the "human" and by combining its idolatry of Evil with other pathologies of beauty: pleasure in the forbidden and in guilt, remorse, pain as Virtue, the Artificial and Natural juxtaposed, and Hell's glory replacing and impoverished Paradise
>
> (Elytis 1994: 120–121)

Although he deemed existentialism counter-poetic he was striving to find a way out of the intellectual climate he could not feel familiar with. Thus, he admits,

> I literally "boiled in my juices" for three and a half years ... I went up and down the studios of poets or painter friends, listening a lot and speaking little, making huge but futile attempts to be reconciled with the new "universal'

spirit", always about to be born, whose only carriers were, as far as I could see, the barefoot, bearded throng on its nightly promenade through the spring-time sidewalks, holding an unread book underarm.

(Elytis 1995: 119–120)

Feeling alienated from the modern spirit, he turns to what he had always regarded as his own:

> The throng launched a costume I detested. A thousand times to be nude! I shut myself in my little room and began to read Plato, with a hedonism only the parched can feel when reaching clear water! It hardly made sense that Greek tradition would find a way to stretch a hand to nudge my shoulder among those Gothic churches, rococo salons and howling boites de nuit.
>
> (Elytis 1995: 120)

The transvaluation process was under way:

> Even at its best, The West ... was ... always and exclusively on **this**, not **that** side of "the Curse" ... I tried my best to isolate the exceptions, the handful of European heretics on whom I lavished my overfed admiration. In short, I had become a fanatic. I faulted the Gothic churches, the formal gardens of Versailles, the remnants of the Napoleon and Louis eras. I even faulted the language though – how strange – I had been raised in it more than I had in my own. Not to mention the Renaissance! It was the great enemy, the great counterfeiting of a Spirit which I felt, at that hour, entrusted its defense to my meager powers.
>
> (Elytis 1995: 123)

Thus, despising what he previously admired, he honestly affirms:

> I felt like an aristocrat, the only one privileged to call the sky "sky", and the sea "sea", indistinguishable from Sappho and Romanos thousands of years ago. Only so could I truly see the pale blue ether or hear the pelago roar.
>
> (Elytis 1995: 124)

Here we have a successful circle of a *ressentiment*-ful transvaluation: the initial embarrassment turned into pride in the name of a living tradition.

Explanatory range and boundaries

One could say that there is a family resemblance between inverted syncretism and Clifford Geertz's idea of the interplay between what he calls 'essential-ism' and 'epochalism' in Third-World nation-building (Geertz 1973). The entrance to political modernity pushes the new nations into the imperatives of contemporary history and society; it requires them to be aligned with the spirit of the new epoch of development, cosmopolitanism and the like. This is what he calls 'epochalism'. On the other hand, however, the new nations couldn't just start their journey into modernization out of thin air; their nationalisms were heavily based on indigenous ways of life and traditional norms and values which were supposed to offer an essential, a real ages-old, ground of collective and personal identity, a trend which he calls 'essentialism'. Faced with socio-economic change, the great bulk of the population is submerged 'into a vast confusion of outlooks' wherein 'essentialist and epochalist sentiments are scrambled' (Geertz 1973: 244). The two notions can be deemed functional equivalents of 'tradition' and 'modernity' or 'modernization' and thus the interplay between 'essentialism' and 'epochalism' resembles my inverted syncretism argument. Yet, what he misses is the mechanics of the interplay; as ingenious as his conceptualization may be, the 'vast confusion of outlooks' that spring from this interplay doesn't say much about its specification, something accomplished principally with the concept of inverted syncretism.

Certainly, inverted syncretism does not explain everything in Greek political-cultural life but it is a powerful *long-dureè* factor that molds much of the institutional setting and political behavior of elites and mass public alike. Most notably, in tandem with cultural intimacy, it has set the stage for *ressentiment*-ful transvaluation, both individually and collectively, for bearing the national loneliness Greece experiences when it is compared to advanced capitalist countries. It might not be accidental that while reviewing the merging of archaic and western cultural patterns, Herzfeld (1989: 110) pointed out that the reproduction of

> the stereotypical failings ... whereby interventionist Europe decried the Greeks as flawed and corrupt may also be part of the Greek catalogue of 'Romeic things', i.e., an attempt to articulate a non-reactive identity opposing Western hetero-determination.

Thus, as far as the emotional energy of national identity is concerned, inverted syncretism confers individual and collective pride and dignity, assuaging shame and humiliation. This is exemplified in the widespread anti-americanism which is actually a permanent feature of Greek political culture since late 1950s, manifested in media events and political rituals (Demertzis and Stratoudaki 2020).

For all its macro-perspectival angle, inverted syncretism may be concurrent with forceful hyper-modernizing cultural tendencies (Panagiotopoulos and Vamvakas 2014). That is so not only because the inverted syncretism itself may be eroded by culturally homogenous global flows of symbols and ideas, but also because, as Ernest Gellner has claimed, modernity, and post-modernity for that matter, have been spreading across the globe through 'tidal waves'; i.e., different countries at different time-points may remain, as it were, traditional with regards to their public administration or economy while exhibiting at the same time a non-traditional cultural and intellectual life (Gellner 1965).

A cognate apposite idea espoused by Eisenstadt (1973: 102) is that of 'partial modernization' of segregated parts of a still 'traditional' society which need not necessarily bring about an overarching change towards modernity. More than that, partial modernization 'might even reinforce traditional systems by the infusion of new forms of organization' (ibid.). Considering the Greek case, this a real but partial truth; inverted syncretism is but an instantiation of the 'multiple modernities' thesis, i.e., a variety of modernity. Macro-structurally, Greece paved its own path to modernity through a historically specific and multilayered constellation of socio-structural and agentic-cultural modalities that gave rise to an emotional regime where the political emotions of *ressentiment* have been center-staged.

In this path the constituent elements of Greek inverted syncretism have been not only intrinsically interdependent but have constituted a complimentary contradiction; namely, the relata of the syncretic relationship – tradition and modernity – are not just in 'tension', as is usually assumed (Triandafyllidou, Gropas and Kouki 2013: 2), but they contrast with each other in a symbiotic way. The question, however, in the long run is whether this is going to change and in what direction. A less likely historical scenario might be that tradition and modernity in Greece form a mutually exclusive and exhaustive contradiction; in a more conceivable scenario they are likely to form a mutually exclusive but not exhaustive contradiction. Whereas in mutually exclusive and exhaustive contradictions a third possibility does not exist, in the case of a mutually exclusive but not exhaustive contradiction other possibilities do exist (Israel 1979: 112–113). In the latter case then the future is open for societal agents to refigure the relationship between tradition and modernity and postmodernity on a systemic or sectoral basis.

Synopsis

Hatred, envy, pride, dignity, love and fear are likely to become political emotions when deployed in the discourse and the emotional practices of national habitus. From an emotions sociology point of view, after briefly tracing the emotional underpinnings of the main theorists of nationalism ever since the late nineteenth century, this chapter has demonstrated the relational societal procedures involved in the affective bases of the nation state by deciphering the so-called 'national sentiment'. Directly connected to Chapter 6, where a theoretical scrutiny of *ressentiment* explored its close knitting with resentment, the present chapter has exemplified the dynamics involved in *ressentiment* as a key programmatic political emotion stipulating Greek national identity.

Notes

1 For precise references see Karl W. Deutsch's concise account of the academic literature on nationalism during the first half of the twentieth century (Deutsch 1953: 15–28).

2 (Sturzo 1946: 13), as referred to by Deutsch (1953: 23–24).

3 The same year, in an almost similar anti-essentialist logic, Brubaker argued that

> nationalism can and should be understood without invoking 'nations' as sub-stantial entities. Instead of focusing on nations as real groups, we should (see) 'nation' as practical category, institutionalized form and contingent event.... To understand nationalism, we have to understand the practical uses of the category 'nation', the ways in which it can come to structure perception, to inform thought and experience, to organize discourse and political action
>
> (Brubaker 1996: 7)

4 Even primordial ethno-nationalists, as well as moderate perennialists like Smith (1986: 18) wouldn't deny that nationalism is about the modern territorial state which gains legitimacy through the master – but no less empty – signifier 'nation'.

5 In equal measure, Weber stated that 'if the concept of "nation" can in any way be defined unambiguously, it certainly cannot be stated in terms of empirical qualities common to those who count as members of the nation' (1978b: 922).

6 It was mistakenly attributed to Renan's analysis of the nation.

7 See Turner and Stets (2005: 215–260).

8 Nevertheless, a top-down explanation is not enough. Even if nationalistic feelings have often been promoted by dominant elites and professional intellectuals, 'nationalism is not merely a set of symbols and beliefs force-fed to an unwilling or indifferent popula-tion' (Giddens 1981b: 192).

9 The 'Tocqueville effect' posits that radical change may not always come when things are bad, but when things are not as good as people expected them to be. Alexis de Tocqueville analyzed the causes of the French Revolution by arguing that the push for revolutionary change occurred, counterintuitively, when things seemed to be getting better. In other words, expectations of progress can lead to revolutionary desires if progress fails to come quickly enough. This issue was taken up by James Davies (1962) in his interpretation of revolutions, as well by theorists of 'relative deprivation' like Robert Merton and Walter Garrison Runciman.

10 For Max Weber, the caesaro-papist character of church-emperor relations was grounded on the patrimonialistic content of the Byzantine mode of political domina-tion, endowed periodically with hierocratic elements (Weber 1978b: 1010, 1401, 1162, 1978a: 231–232).

11 In May 1827 the National Assembly elected Ioannis Kapodistrias, a former foreign affairs deputy minister of Russia, as the first Governor of Greece. As a Greek of noble origin, Kapodistrias knew quite well the imperatives of the new state but he ignored the demo-cratic provisions of the constitution. His strong-handed style of governance collided with the priorities of local landowners and chieftains who organized his murder in 1831. By then the state territory occupied almost one third of the present sovereign geographical area. After his death – and this is an example of crypto-colonialism – it was not the almost dissolved Parliament that decided the future of the country but the Great Powers. In a treaty in 1832, which Greece was not involved in negotiating, Britain, France, Russia, and Bavaria 'settled the terms under which King Otto was to accept the throne and which placed Greece under the guarantee of the Protective Powers' (Clogg 1997: 47).

12 See *inter alia* Capelos and Demertzis (2018).

13 This was certainly not the case elsewhere, with the cosmopolitan Greek diaspora in big commercial cities like Alexandria, Odessa, Vienna, Smyrna, etc.

14 It was this sense of cultural intimacy, as an imaginary cradle of *ressentiment*-ful trans-valuation, that drove Nicos Athanassopoulos, a Deputy Minister of Finance in Andreas Papandreou's government and former Public Attorney, to blatantly declare to an EU official 'when we were building the Parthenon you were eating nuts on the trees'. That

was his reply to the accusation of having breached the law in 1986 by disguising the counterfeiting of 9,000 tons of Yugoslavian corn into Greek corn traded by a state company under his auspices and claiming EU subsidies. After economic audit, Greece was convicted at the European Court and consequently Athanassopoulos was condemned by the Special Court to three and a half years imprisonment. While the trial was held, a number of PASOK fans gathered to defend him and as soon as the verdict was announced they shouted against the jury 'Shameless, Shameless!' As the verdict was utterly disapproved of by the crowd, they chanted the National Anthem. Be it noted that despite his conviction, Athanassopoulos returned to political life and was elected as an MP in the 1993 elections.

15   This is of course an ideal typical distinction; as Eisenstadt (1973: 161) points out, beyond the most primitive tribes, no traditional society has been entirely organized along the criteria of ascription, diffusion, and particularism; specialization, achievement, and universalism 'tend to develop to some extent in different parts or spheres of their institutional structure;' so distinctiveness between different elites, the distinction between center and periphery, and some channels of occupational, political, and cultural mobility are brought forward.

16   Remaining in an ideal typical mode of analysis, I would not claim that consummatory and instrumental cultures correspond to what Weber called value rationality and instrumental rationality (Weber 1978a: 24–26, 85–86, 1978b: 1376–1377). Value and instrumental rationality (or, respectively, substantive and formal rationality) are two versions of rational social and/or economic action which Weber typically distinguishes from traditional action which is determined by ingrained habituation. Both consummatory and instrumental cultures are part of the traditional type of social action; however, traditional action is not devoid of a 'practical logic' able to organize the totality of an agent's thoughts, perception and conduct by means of a few generative, immediately mastered and manageable principles (Bourdieu 1977: 109–110). Permeated by affectivity, this practical logic is derived from a strong core of human rationality common to all cultures, either modern or traditional, which consists in the explanation, prediction and control of events, as well as in the use of analogical, deductive and inductive inference (Horton 1982: 256).

Postscript

Over the last 20 years or so, it has increasingly been realized that political sociologists no longer have to choose between the rational choice and the political cultural approach in order to analyze electoral behavior, political participation, social movements, and public preferences and interests, as well as the legitimacy of power relationships. This is so because in the meantime the 'emotional turn' which has engulfed almost all other major disciplines and sub-disciplines in humanities and social sciences has altered fundamentally the way the academic community conveys political analysis (Thompson and Hoggett 2012). Gone are the days where prominent scholars were hesitant to postulate that at all levels and in all its modalities, contentious or consensual, politics is deeply ingrained with emotionality; the question now is how exactly political sociologists and political psychologists are to theorize the taken-for-granted politics-emotions nexus (Demertzis 2013). To put it differently, what matters now are the nature of the linkages and the blend of cognitive rationality and emotionality in political action and thought in different time and space settings. In this respect, political structures (parties, governments, parliaments, trade unions, etc.) are interwoven with 'structures of feeling' which are procedural experiences of individuals and collectives understood as 'not feeling against thought, but thought as felt and feeling as thought' (Williams 1978: 132).

The seeds of the political sociology of emotions were planted many years ago by many political sociologists and sociologists of emotions alike. Viewed from the angle of the sociology of emotions, it is one of its sub-sectors; seen from the perspective of political analysis, as argued in Chapter 1, it is the emotive counterpart of political sociology itself. Its emergence is not due to merely disciplinary factors; the worldwide predominance of affective anti-politics (e.g., the securitization of immigration policies, reactionism, competitive authoritarianism, nationalism-populism, etc.) conveyed by analogue and digital means of total propaganda (Edelstein 1997) makes the – or at least a – political sociology of emotions increasingly necessary in making the prospects of democracy and republicanism in the twenty-first century more intelligible.

Of course, this cannot be done by a sentimentalist political analysis, namely, by a reductionist 'feel(o)logical' interpretation of political behavior. That would be analytically insufficient and would also lead to a one-sided subjectivisation and a concomitant anesthetization of political sociological theorizing. The inherent sociality of emotions – i.e., the fact that they do not emanate *within* the individual as much as *between* the embodied individuals[1] while living and cooperating in institutional settings and figurational power and status relations – infuses political sociological analysis with an inexorable relational interpretative perspective. As a matter of fact then, political entities and processes are denaturalized all the way through and the ontology of the Political acquires a less teleological and a more contingent and agentic dimension (Guzzini 2016).

As much as possible, the political sociology of emotions has to be conceptually clear, on the one hand, and methodologically plural, on the other. In this respect, a couple of interrelated caveats have to be considered.

The political sociology of emotions is by definition multi- and intradisciplinary; this however should not cancel out the issue of the boundary with cognate disciplines and sub-disciplines, most notably with political psychology, cultural studies, and social psychology. Admittedly, there can be no yardstick for drawing these boundaries because there is not a single political and social psychology carried out at different universities and research centres in various countries; nor is there a uniform style of conducting cultural studies. Yet, a differentiating criterion might be a reluctance of the political sociology of emotions to keep 'affect' among its chief concepts.

It is true that, most of the time, sociologists of emotions do not differentiate or use alternately the terms 'affect', 'emotion', 'sentiment', 'feeling', and 'passion' (e.g., Turner and Stets 2005: 2; Cossarini and Vallespín 2019: 3–4; Diefenbach, Kahl *et al.* 2019: 11). To compensate this looseness, sometimes they employ 'emotion' in the singular as a thought category which provides in perspective a common thread among the internal shadings of emotions (in the plural) which are actually experienced by people (Barbalet 1998: 26, 80; Demertzis 2013: 4–6) merging 'endlessly into each other' (James 1931/1890: 448). But since there is no definitional basket to hold the innumerable emotional differentiations and fluctuations, a generally accepted sociological definition of emotion and a universally accepted typology are not currently available.

Nevertheless, among a good many sociologists there is a consensus over the assumption that emotions are not of an autonomic and innate bio-physiological nature but mediate between physiological reactions and cultural norms. In this respect, and according to a 'mild' constructionist approach, it could be claimed that while emotions are not reducible to biology, not everything is a construction or is constructable with regards to emotionality. Beyond the biological substratum which simply cannot be denied, emotions themselves are extremely plastic, subject to historical variability (Thoits 1989: 319; Rosenwein 2001: 231). From this viewpoint, emotions are constructs which result from the collaboration between the body and society (Kemper 1991: 341).

It is more important, however, that many psychologists, political psychologists, and sociologists endorse a componential conceptualization according to which emotion is made of (1) an appraisal of an internal or external consequential stimulus, relational contexts and objects; (2) physiological changes and activation of key body systems leading to action readiness towards something; (3) overt, free or inhibited facial expression, voice and paralinguistic expressions; (4) a conscious subjective feeling; (5) an adaptation function to the environment; (6) culturally provided linguistic labels of one or more of the first three elements, and (7) socially constructed rules on what emotions should be experienced and expressed (Averill 1980; Thoits 1989: 318; Gordon 1990: 147, 151–152; Scherer 2009; Turner and Stets 2005: 9). Evidently, each of these components involves a huge variety of dimensions and disputed sub-issues such as the nature of the appraisal, the relation between emotion and motivation, motivation and action, the direct and indirect effects of emotion on political judgment, and so on. Be it noted that there is no need for all these seven elements to be present simultaneously for an emotion to exist or to be recognized by others. Nor is it necessary that all these elements are self-consciously experienced. In this respect, emotion can be viewed as a 'multi-component phenomenon' (Frijda 2004a: 60) and as an 'open system' (Gordon 1981).

For those who literally prefer to speak of an 'affective' rather than an 'emotional' turn in social sciences, affect is seen as an unmediated, non-representational, unconsciously, or at least less consciously, formed embodied dimension of human feeling (Hoggett and Thompson 2012). Anchored in Spinoza's and Tomkins insights, affect takes on an over-arching ontological status described as the 'central dynamic force of social connectedness, ranging from face-to-face encounters to various interactive dynamics between individuals and collectives as well as inter- and intra-group relations' (Slaby and von Scheve 2019b: 5). It is eventually seen as the foundational dimension of human sociality in its more primordial, embodied sense. Consequently, emotions are construed as relationally embedded instantiations and episodic realizations of affect (von Scheve and Slaby 2019: 43–47; Wahl- Jorgensen 2019: 8). Even more, emotions, cognitions and behaviors are considered as effects of affect (Papacharissi 2015: 14–15). In Chapters 1 and 4 I commented on Brian Massumi's notion of affect as a pre-linguistic dimension of life and human communication; actually Massumi, a Deleuzean political theorist and philosopher, is the leading figure of cultural affect theory which is laying the ground for the articulation of notions like 'affective societies', 'affective publics', 'affective economies', and 'affective politics', founded on the idea that affect is an all-encompassing, pre-social, and non-consciously experienced bodily energy and intensity enacted by the human attribute 'to affect and be affected', the quintessential of Spinoza's *conatus* (Massumi 2015: 3).

Despite the difference in parlance, there is not that big a distance between Massumi's idea that affect precedes cognition on the one hand, and neuroscientists' postulation that an affective reaction to external stimuli is much quicker than conscious awareness of it on the other. For Massumi, the 'missing half second' between the event and its conscious registering (Massumi 2015: 213) is the realm of virtuality and a space over-filled with sensual content, yet not assigned to any specific semantic register. Likewise, for George Marcus' 'affective intelligence theory' it is the pre-conscious 'disposition' and 'surveillance' affective appraisal systems which predate any acknowledged cognition and prepare for the agent's response (Marcus 2002). Similarly, due to the fact that the 'brain knows more than the conscious mind reveals', Damasio's 'core consciousness' vs 'extended consciousness' (Damasio 1999) corresponds to Massumi's 'affect' and 'emotion'. At any rate, however, when it comes to emotionality and affectivity, one has to be cautious as to what 'consciousness', 'non-consciousness', and 'unconsciousness' mean (Barbalet 2009). If anything, a non-consciously expressed emotion does not mean that the subject is not currently experiencing the emotion itself – even if s/he is not aware of it and s/he cannot linguistically describe it.

A good deal of the recent interest in affect comes from psychoanalysis and psychoanalytical social and political theory (e.g., Stavrakakis 2019; Richards 2018; Soler 2016). Here, in the splitting of the subject (i.e., the subject of the unconscious vs Ego), the body, the drives, and the concomitant defense mechanisms are taking the lead over the analysis of distinct emotions. This strand of thought focuses on the unconscious aspects of emotional experience and postulates that within the libidinal economy it is not emotions themselves that are repressed but their symbolic representations indexed by particular symptoms; it is a superegotic 'forbidden' word or idea (as in the desire to murder one's mother), which is repressed and not the experience of affect itself precisely because affect is not repressed but displaced (Yates 2019). In this sense, the psychic apparatus is organized affectively through the interplay between drives, language, desire, and the reality principle. What is more theoretically interesting, then, is affect as embodied psychic energy rather than its particular manifestations at the level of the Ego (or 'extended consciousness' in neuroscientific terminology).

Without actually differentiating affect from emotion, Margaret Wetherell (2012) gives special emphasis to what she eclectically elaborates on as 'affective practice'. Contrasting social psychoanalytic approaches and the cultural theory of affect, she sees affect as embodied meaning-making, as human emotion which is inextricably linked with the semiotic and the discursive. As forms of practical consciousness, affective practices are viewed as historically situated, on-going, open relational processes that permeate human sociality. Emanating from her discursive psychology background, she thus refrains from linking affect too much, if not exclusively, to the uncanny, the body as 'desire machine', and to the non-discursive qualia. That is why she ultimately makes no conceptual distinction between affect and emotion throughout her book. Be it noted that even scholars who endorse the somatic and visceral nature of affect are cautious about the 'risk of overlooking the importance of culture and sociality' (von Scheve and Slaby 2019: 49). Indeed, this overlooking is likely to make the mind-body dichotomy reappear, an epistemological stance which post-Parsonian sociological theory has supposedly overcome (Martin 2013).

Provided that a one-sided conceptualization of affect as a disruptive force emanating from 'fruitful darkness' is missing from psychoanalytic, ethnological, or cultural accounts, one is unlikely to find substantive difference between it and the componential definition of emotion delivered earlier in this postscript. Nevertheless, it is not merely a matter of terminological preference; it seems to me that a point of difference between affect and emotion is that the former is premised on a grant theory level; 'affect *is* the whole world', Massumi (1995: 105) emphatically claims. On the contrary, emotion(s) is more suitable for middle-range theoretical reasoning, as I attempted to show in the previous chapters of this book. Hence a strong case should be made that, although not sealed off from the ontology of the Political, the political sociology of emotions has a mundane orientation and is designed for the causal explanation and the interpretative understanding of the emotions-politics nexus in real societies and polities through a perspective of weak social constructionism.

At any rate, (a) the socio-psychological assumption that affect is at a higher level of reality in the sense that it is 'the central dynamic force of social connectedness, ranging from face-to-face encounters to various interactive dynamics between individuals and collectives as well as inter- and intra-group relations' (Slaby and von Scheve 2019b: 5, 14) with emotions being just derivatives and (b) the claim coming from political psychology that affect is one of the three mental functions alongside cognition and motivation so that emotions, feelings, sentiments and moods are all affect types (Capelos and Chrona 2018), provide adequate reason for the political sociology of emotions to become relatively separate and distinct from contiguous fields of study.

Due therefore to its lack of clarity, in its visceral designation and as unreflective practical consciousness, affect is of relevance but not an essential keyconcept for the political sociology of emotions. As we showed in Chapters 1 and 4, oftentimes it is used metonymically with regards to particular emotions of lesser or longer duration and intensity. So, it is not 'political affect' but political emotions or political sentiments proper that occupy the central role in the conceptual armor of the – or at least a – political sociology of emotions. If anything, the advent of the sociology of emotions – and the political sociology of emotions for that matter – has brought the precise analysis of particular emotions to the fore, steering away from abstractions like, 'affectivity', 'emotionality' or 'national sentiment'. That was a step forward for social science. Since, however, we humans feel in flow, not experiencing one emotion at a time, and since a particular emotion may trigger another emotion or more than one, the time has perhaps come to not only investigate handfuls of discrete emotions (as for example fear, anger or nostalgia in the analysis of populism) but also engage seriously with emotional clusters and elaborate on composite sentiments. Such a move was attempted in this book with respect, among others, to political cynicism and *ressentiment*. It is not only that each emotion is a compound of body, feeling and discourse, reducible however to none of them (Burkitt 2002: 152–153; Scherke 2015: 475; Bericat 2016: 505); it is also that there exist clustered, multilayered secondary and tertiary emotions or sentiments like envy, gratitude, trust, hope, faith, resentment, hatred, love, and shame whose impact in the constitution of the emotions-politics nexus is greater than that of basic and/or primary emotions (TenHouten 2007: 52) usually employed in quantitative analyses. Elicited and operating on the background and foreground of political praxis, these clustered sentiments are experienced individually but they can be group-based as well as collectively shared within groups, quasi-groups, or socio-political movements.

Given how widespread the emotionalization of everyday life appears to be in the public sphere and the political realm, due, among other things, to consumerism (Richards 2018), the informalization of manners (Wouters 2007), the concomitant 'tyranny of the intimacy' (Sennett 1977), and the inclination of contemporary politics to the display, the diffusion and the manipulation of emotions through 'emotional governance' and affective practices, there exists a wide open research agenda for the political sociology of emotions. In its attempt to complement the enquiry of the social bases of politics with the emotional bases of politics, its research scope may span individual political behavior and public policies and planning. Fortunately, there is a full-blown palette of qualitative and quantitative methodological tools to facilitate offline and online research endeavors (Flam and Kleres 2015). At any rate, the political sociology of emotions' future is contingent upon inter-disciplinarity and triangulating research methodologies.

From a weak social constructionist theoretical angle, employing phenomenological and comparative methodology and mostly supported by qualitative research material, this book has advocated for the political sociology of emotions as a clear subfield of the sociology of emotions. The book makes a case for trauma and *ressentiment* being broad stages to think about the emotions-politics nexus in the sense that the political sociology of emotion is one that understands our present society as traumatizing and traumatized and can expose the ways in which the mediatization of traumas may generate a politics of pity, in which the media not only makes a spectacle of suffering but can create space for sympathy and compassion as a basis for moral and political decision making premised, among other things, on processes of individual forgiveness and public apologies. Taking stock of grand theoretical considerations about the age of 'anger', the age of '*thymos*', and 'the politics of resentment' on the one hand, and middle-range theoretical accounts of 'backlash politics', 'identity politics' and 'anti-politics' on the other, the book theorizes how *ressentiment*, a complex moral-emotional response to inequality, injustice and individual inefficacy, shapes populist mobilizations and nationalist discourses in tandem with other political emotions like nostalgia, anger, and fear. Scrutinizing the subtle emotionality of political phenomena, the political sociology of emotions straddles the causal or structural sociological explanation and the interpretative understanding of their origin, figuration and duration. For instance, Trumpism and the rise of far-right political parties in Europe (e.g., AfD, Golden Dawn, UKIP, True Finns, etc.) cannot be explained away with reference only to socio-demographic drivers; what is far more at stake is status anxiety and the repercussions of identity formation processes interwoven with emotional energy and affective practices. The discourse of political agents is replete with emotionality with emotional governance being a crucial political technology in late modern societies; in this respect the political sociology of emotions is normatively concerned with the distinction between democratic emotionality and demagogic emotional manipulation, an issue raised, among others, by the dynamics and ambivalences of what is called 'monitory democracy' (Keane 2009) at national and supranational level. As was indicated in the previous chapters, the political sociology of emotions rests at the middle-range of analysis just to keep at bay either abstracted empiricism or micro-foundational reductions in its overall attempt to investigate the emotional bases of politics.

Note

1 Emotions are not founded in the bodies of oneself or other; they relationally emerge between interacting embodied, speaking and desiring subjects. As eloquently put by Ian Burkitt (2002: 159), it is as mistaken to look for our emotion in the body as it is to look for time inside a clock.

References

Aarelaid-Tart, A. (2006). *Cultural Trauma and Life Stories*. Kikimora Publications A 15. Vaajakoski: Gummerus Printing.

Abbott, A. D. (2001). *Chaos of Disciplines*. Chicago and London: The University of Chicago Press.

Adorno, T. W. (1997). Theorie der Halbbildung. In Adorno, T. W. *Gesammelte Schriften* 8: 93–121. Frankfurt/M.: Suhrkamp.

Aeschbach, S. (2017). *Ressentiment: An anatomy*. Universite Genève no. L. 909.

Aguilar, P. (1996). *Memoria y Olvido de la Guerra Civil Espanola*. Madrid: Alianza Editorial.

Ahmed, S. (2004). *The Cultural Politics of Emotion*. Edinburgh: Edinburgh University Press.

Akkerman, A., Mudde, C., and Zaslove, A. (2014). How Populist Are the People? Measuring Populist Attitudes in Voters. *Comparative Political Studies*, 47(9): 1324–1353. https://doi.org/10.1177/0010414013512600.

Alapuro, R. (2002). Coping with the Civil War of 1918 in Twenty-first Century Finland. In Christie, K. and Cribb, R. (Eds) *Historical Injustice and Democratic Transition in Eastern Asia and Northern Europe: Ghosts at the Table of Democracy*. (169–183). London: Curzon.

Alcorn, M. (2019). Trauma. In Stavrakakis, Y. (Ed.) *Routledge Handbook of Psychoanalytic Political Theory*. (174–186). London: Routledge.

Alexander, J. C. (2012). *Trauma. A Social Theory*. Cambridge: Polity Press.

Alexander, J. C. (2006). *The Civil Sphere*. Oxford: Oxford University Press.

Alexander, J. C. (2004a). Toward a Theory of Cultural Trauma. In. Alexander, C.J., Eyerman, R., Giesen, B., Smelser, J. N. and Sztompka, P. (Eds) *Cultural Trauma and Collective Identity*. (1–30). Berkeley: University of California Press.

Alexander, J. C. (2004b). On the social construction of moral universals: The 'Holocaust' from war crime to trauma drama. In J. C. Alexander *et al.* (Eds) *Cultural Trauma and Collective Identity*. (196–263). Berkeley: University of California Press.

Alexander, J. C. (2003). *The Meanings of Social Life: A Cultural Sociology*. New York: Oxford University Press.

Alexander, J. C. and Smith, P. (2002). The strong program in cultural theory. In J. H. Turner (Eds) *Handbook of Sociological Theory*. (135–150). New York: Kluwer Academic/Plenum Publishers.

Alexander, J. C. and Smith, P. (1998). Cultural Sociology or Sociology of Culture: Towards a Strong Program. *Sociologie et Sociétés*, 30(1): 107–116 (special edition).

Alexander, J. C. and Giesen, B. (1987). From Reduction to Linkage: The Long View of the Micro-Macro Link. In Alexander, J. C., Giesen, B., Münch, R., and Smelser, N. J. (Eds) *The Micro-Macro Link*. (1–44). Berkeley, Los Angeles, London: University of California Press.

Alexander, J. C., Ronald J., and Smith, P. (Eds) (2012). *The Oxford Handbook of Cultural Sociology*. Oxford: Oxford University Press.

Allardt, E. (2001). Political Sociology. *International Encyclopedia of the Social & Behavioral Sciences*. (11701–11706). Amsterdam: Elsevier Science Ltd.

Almond, G. and Verba, S. (1963). *The Civic Culture. Political Attitudes and Democracy in Five Nations*. Boston: Little Brown and Company.

Améry, J. (1999). *At the Mind's Limits: Contemplations by a Survivor on Auschwitz and its Realities*. London: Granta Books.

Andén-Papadopoulos, K. (2003). The Trauma of Representation. Visual Culture, Photojournalism and the September 11 terrorist Attack. *Nordicom Review*, 24(2): 89–104.

Anderson, B. (1983). *Imagined Communities. Reflections on the Origin and Spread of Nationalism*. London: Verso.

Andreadis, I., Stavrakakis, Y., and Demertzis, N. (2016). New Indices for Right Wing Populism. In *IPSA World Congress*. Poznań. http://paperroom.ipsa.org/papers/paper_64103.pdf.

Anselmi, W. and Gouliamos, C. (1998). *Elusive Margins – Consuming Media, Ethnicity and Culture*. Toronto, Buffalo, Lancaster: Guernica Publications.

Anthonissen, C. (2009). Considering the violence of voicelessness: Censorship and self-censorship related to the South African TRC process. In Wodak, R. and Auer-Borea, G. (Eds) *Justice and Memory. Confronting Traumatic Pasts. An International Comparison*. (97–122). Vienna: Passagen Verlag.

Antoniou, G. (2007). *The Memory and Historiography of the Greek Civil War*. PhD. European University Institute, Florence.

Apter, D. (1965). *The Politics of Modernization*. Chicago and London: University of Chicago Press.

Arbid, M. A. (1985). *In Search of the Person. Philosophical Explorations in Cognitive Science*. Amherst: The University of Massachusetts Press.

Arcel, L. T. (2014). *Pogrom in the Soul. The Asia Minor Disaster Trauma in Three Generations*. Athens: Kedros.

Arendt, H. (1973). *On Revolution*. Middlesex: Penguin Books.

Arendt, H. (1958). *The Human Condition*. Chicago: University of Chicago Press.

Aristotle (1992). *On Rhetoric: A Theory of Civic Discourse* (Transl. by George A. Kennedy). Oxford: Oxford University Press.

Assmann, A. (2009). From collective violence to a common future: Four models for dealing with a traumatic past. In Wodak, R. and Auer Borea, G. (Eds) *Justice and Memory. Confronting Traumatic Pasts. An International Comparison*. (31–48). Vienna: Passagen Verlag.

Averill, J. R. (1980). The emotions. In Staub, E. (Ed.) *Personality: Basic Aspects and Current Research*. (134–199). Englewood Cliffs: Prentice Hall.

Badiou, A. (2016). *Notre Mal Vient de plus loin, Penser les Tueries du 13 Novembre*. Paris: Librairie Arthème Fayard.

Baer, A. (2001). Consuming History and Memory Through Mass Media Products. *European Journal of Cultural Studies*, 4(4): 491–501.

Baier, A. (1980). Hume on Resentment. *Hume Studies*, 6(2): 133–149.

Ball, D. W. (1964). Covert Political rebellion as Ressentiment. *Social Forces*, 43(1): 93–101.

Banfield, C. E. (1958). *The Moral Basis of a Backward Society*. Glencoe, IL: The Free Press.

Barbalet, J. (2009). Consciousness, Emotions, and Science. In D. Hopkins *et al.* (Eds) *Theorizing Emotions: Sociological Explorations and Applications.* (39–71). Berlin: Campus.

Barbalet, J. (2006). Emotions in Politics: From the Ballot to Suicide Terrorism. In Clarke, S., Hoggett, P., and Thompson, S. (Eds) *Emotion, Politics and Society.* (31–55). London: Palgrave.

Barbalet, J. (1998). *Emotion, Social Theory, and Social Structure. A Macrosociological Approach.* Cambridge: Cambridge University Press.

Barbalet, J. and Demertzis, N. (2013). Collective Fear and Societal Change. In Demertzis, N. (Ed.) *Emotions in Politics. The Affect Dimension in Political Tension.* (167–185). London: MacMillan/Palgrave.

Barker, E. (1927). *National Character and the Factors in its Formation.* New York/ London: Harper.

Barnes, S., Kaase, M. *et al.* (1979). *Political Action. Mass Participation in Five Western Democracies.* Beverly Hills: Sage Publications.

Bartmanski, D. and Eyerman, R. (2011). The Worst was the Silence: The Unfinished Drama of the Katyn Massacre. In Eyerman, R., Alexander, J. C. and Breese, E. (Eds) *Narrating Trauma. On the Impact of Collective Suffering.* (237–266). Colorado: Paradigm Publishers.

Basaran, T., Bigo, D., Guittet, E.-P., and Walker R. B. J. (2017). *International Political Sociology. Transversal Lines.* London and New York: Routledge.

Baudrillard, J. (1994). *The Illusion of the End.* (Transl. Chris Turner). Stanford, CA: Stanford University Press.

Baudrillard, J. (1983). *In the Shadow of the Silent Majorities.* New York: Semiotext(e).

Bauer, O. (1924). *Die Nationalitätenfrage und die Sozialdemokratie* (2nd edition). Vienna: Brand.

Bauman, Z. (2017). *Retrotopia.* Cambridge: Polity Press.

Bauman, Z. (2006). *Liquid Fear.* Cambridge: Polity Press.

Bauman, Z. (1999). *In Search of Politics.* Oxford: Polity Press.

Bauman, Z. (1993). *Postmodern Ethics.* Oxford: Blackwell.

Bauman, Z. (1992). *Intimations of Postmodernity.* London: Routledge.

Bauman, Z. (1989). *Modernity and the Holocaust.* Ithaca, NY: Cornell University Press.

Bauman, Z. (1982). *Memories of Class. The Prehistory and After-life of Class.* London: Routledge.

Beck, Ulrich (1992). *Risk Society. Towards a New Modernity.* London: Sage Publications.

Becker, P. (2009). What makes us Modern(s)? The Place of Emotions in Contemporary Society. In D. Hopkins *et al.* (Eds) *Theorizing Emotions: Sociological Explorations and Applications.* (195–219). Berlin: Campus.

Bellamy, J. E. (1997). *Affective genealogies: Psychoanalysis, Postmodernism, and the "Jewish question" after Auschwitz.* Lincoln: University of Nebraska Press.

Ben-Ze'ev, A. (2000). *The Subtlety of Emotions.* Cambridge, MA: MIT Press.

Bennett, L. (2003). *News. The Politics of Illusion.* New York: Longman (5th edition).

Benski T. (2011). Emotion maps of participation in protest: The case of women in black against the occupation, in Israel. *Research in Social Movements, Conflict and Change,* 31(1): 3–34.

Benski, T. and Fisher, E. (2014). Introduction: Investigating Emotions and the Internet. In Benski, T. and Fisher, E. (Eds) *Internet and Emotions*. (1–14). New York: Routledge.

Benski, T. and Langman, L. (2013). The effects of affects: The place of emotions in the mobilizations of 2011. *Current Sociology*, 61(4): 525–540.

Berezin, M. (2002). Secure States: Towards a political sociology of emotions. In Barbalet, J. (Ed.) *Emotions and Society*. (33–52). Oxford: Blackwell Publishing/The Sociological Review.

Berezin, M. (2001). Emotions and Political Identity: Mobilizing Affection for the Polity. In Goodwin, J., Jasper, J. M. and Polletta, F. (Eds) *Passionate Politics. Emotions and Social Movements*. (83–97). Chicago: The University of Chicago Press.

Berger P. and Luckman T. (1967). *The Social Construction of Reality. A Treatise in the Sociology of Knowledge*. London: Penguin Press.

Bericat, E. (2016). The sociology of emotions: Four decades of progress. *Current Sociology*, 64(3): 491–513.

Betz, H.-G. (2002). Conditions Favoring the Success and Failure of Radical Right-Wing Populist Parties in Contemporary Democracies. In: Mény, Y. and Surel Y. (Eds) *Democracies and the Populist Challenge*. (197–213). Hampshire: Palgrave.

Betz, H-G. and Johnson, C. (2004). Against the current – stemming the tide: the nostalgic ideology of the contemporary radical populist right. *Journal of Political Ideologies*, 9(3): 311–327.

Bewes, T. (1997). *Cynicism and Postmodernity*. London: Verso.

Bhabha, H. (1991). Question of Survival: Nations and Psychic States. In Donald, J. (Ed.) *Psychoanalysis and Cultural Theory*. (89–103). London: Macmillan.

Birns, N. (2005). Ressentiment and Counter-Ressentiment: Nietzsche, Scheler, and the Reaction Against Equality. *Nietzsche Circle*. Available online at www.nietzschecircle.com/essayArchive1.html.

Bloechl, J. (2013). Forgiveness and its Limits. An Essay on Vladimir Jankélevitch. In Udoff, A. (Ed.) *Vladimir Jankélévitch and the Question of Forgiveness*. (97–110). New York: Lexington Books.

Bocock, R. (1993). *Consumption*. London: Routledge.

Boltanski, L. (1999). *Distant Suffering. Morality, Media and Politics*. Cambridge: Cambridge University Press.

Boraine, A. (2006). Truth and Reconciliation Commission in South Africa amnesty: The price of peace. In Elster, J. (Ed.) *Retribution and Reparation in the Transition to Democracy*. (299–316). Cambridge, U.K.: Cambridge University Press.

Borradori, G. (Ed.) (2004). *Philosophy in a Time of Terror. Dialogues with Jürgen Habermas and Jacques Derrida*. Chicago: University of Chicago Press.

Bottomore, T. (1979). *Political Sociology*. London: Hutchinson.

Bourdieu, P. (1977). *Outline of a Theory of Practice*. Cambridge: Cambridge University Press.

Boym, S. (2002). *Future of Nostalgia*. London: Basic.

Bozatzis, N. (2014). Banal Occidentalism. In Antaki, C. and Condor, S. (Eds) *Rhetoric, Ideology and Social Psychology. Essays in Honour of Michael Billig*. (157–176). London and New York: Routledge.

Brandes, G. M. C. (1972). *Friedrich Nietzsche*. Haskell House Publishers.

Breakwell, G. (2007). *The Psychology of Risk*. Cambridge: Cambridge University Press.

Breese, B. E. (2013). Claiming Trauma through Social Performance: The Case of *Waiting for Godot*. In Eyerman, R., Alexander, J. C. and Breese, E. (Eds) *Narrating Trauma. On the Impact of Collective* Suffering. (213–236). Colorado: Paradigm Publishers.

Breuilly, J. (1985). *Nationalism and the State*. Manchester: Manchester University Press.

Brighi, E. (2016). The Globalisation of Resentment: Failure, Denial, and Violence in World Politics. *Millennium: Journal of International Studies*, 44(3): 411–432.

Brooks, L. R. (Ed.) (1999). *When Sorry isn't Enough. The Controversy over Apologies and Reparations for Human Injustice*. New York & London: New York University Press.

Brown, W. (1995). *States of Injury: Power and Freedom in Late Modernity*. Princeton, NJ: Princeton University Press.

Brubaker, R. (2019). Populism and nationalism. *Nations and Nationalism*, 25(3): 1–23. DOI: 10.1111/nana.12522.

Brubaker, R. (2017). Why Populism? *Theory and Society*, 46: 357–385.

Brubaker, R. (1996). *Nationalism Reframed: Nationhood and the National Question in the New Europe*. Cambridge: Cambridge University Press.

Bruckner, P. (2010). *The Tyranny of Guilt. An Essay on Western Masochism*. Princeton: Princeton University Press.

Bruckner, P. (2000). *The Temptation of Innocence. Living in the Age of Entitlement*. New York: Algora Publishing.

Brudholm, Th. (2010). Hatred as an Attitude. *Philosophical Papers*, 39(3): 289–313. DOI: 10.1080/05568641.2010.538912.

Brudholm, T. (2008). *Resentment's Virtue: Jean Améry and the Refusal to Forgive*. Philadelphia: Temple University Press.

Burke, E. (1995). Reflections on the Revolution in France. In Dahbour, O. and Ishay, M. R. (Eds) *The Nationalism Reader*. (134–142). New Jersey: Humanities Press.

Burke, J. (2005). *Fear: A Cultural History*. London: Virago.

Burkitt, I. (2012). Emotional Reflexivity: Feeling, Emotion and Imagination in Reflexive Dialogues. *Sociology*, 46(3): 458–472.

Burkitt, I. (2002). Complex emotion: relations, feelings and images in emotional experience. In Barbalet, J. (Ed.) *Emotions and Sociology*. (151–167). Oxford: Blackwell.

Butler, J. (1827). *Fifteen Sermons Preached at the Rolls Chapel*. http://anglicanhistory. org/butler/rolls/.

Campbell, J. K. (1983). Traditional Values and Continuities in Greek Society. In Clogg, R. (Ed.) *Greece in the 1980s*. London: MacMillan.

Campbell, J. K. (1964). *Honour, Family, and Patronage*. New York and Oxford: Oxford University Press.

Campbell, K. (2004). The Trauma of Justice: Sexual Violence, Crimes Against Humanity and the International Criminal Tribunal for the Former Yugoslavia. *Social Legal Studies*, 13(3): 329–350. DOI: 10.1177/0964663904044998.

Camus, A. (1991). *The Rebel. An Essay on Man in Revolt* (transl. Anthony Bower). New York: Vintage Books.

Canovan, M. (1999). Trust the People! Populism and the Two Faces of Democracy. *Political Studies*, XLVII: 2–16.

Capelos, T. and Chrona, S. (2018). The Map to the Heart: An Analysis of Political Affectivity in Turkey. *Politics and Governance*, 6(4): 144–158.

Capelos, T. and Demertzis, N. (2018). Political Action and Resentful Affectivity in Critical Times. *Humanity & Society* 1–24. DOI: 10.1177/0160597618802517.

Capelos, T., Katsanidou, A. and Demertzis, N. (2017). Back to Black: Values, Ideology and the Black Box of Political Radicalization. *Science and Society* 35: 35–68.

Cappella, J. and Jamieson, H. K. (1997). *Spiral of Cynicism. The Press and the Public Good*. New York and Oxford: Oxford University Press.

Carpentier, N. (Ed.) (2007). *Culture, Trauma, and Conflict. Cultural Studies Perspectives on War.* Newcastle: Cambridge Scholars Publishing.

Caruth, C. (1995a). Introduction. In Caruth, C. (Ed.) *Trauma: Explorations in Memory.* (3–12). Baltimore: Johns Hopkins University Press.

Caruth, C. (1995b). Introduction. In Caruth, C. (Ed.) *Trauma: Explorations in Memory.* (151–156). Baltimore: Johns Hopkins University Press.

Castanho Silva, B., Andreadis, I., Anduiza, E., Blanuša, N., Corti, Y. M., Delfino, G., Rico, G., Ruth, S. Spruyt, B., Steenbergen, M., Littvay, L. (2019). Public Opinion Surveys: A New Scale. In Hawkins, K., Carlin, R., Littvay, L. and Rovira Kaltwasser, C. (Eds) *The Ideational Approach to Populism: Concept, Theory, and Analysis.* (150–177). Democracy and Extremism Series. London: Routledge.

Castoriadis, C. (1987). *The Imaginary Institution of Society.* Cambridge: Polity Press.

Chambers English Dictionary. (1990). Catherine Schwarz (Ed.). London: Chambers Harrap Publishers Ltd.

Charalambis, D. (1989). *Clientelism and Populism. The Extra-institutional Consensus in the Greek Political System.* Athens: Exandas Publications.

Charalambis, D. and Demertzis, N. (1993). Politics and Citizenship in Greece: Cultural and Structural Facets. *Journal of Modern Greek Studies*, 11(2): 219–40.

Charitopoulos, D. (2012). *Aris. The Lord of Mountains.* Athens: Topos Publications.

Chouliaraki, L. (2010). Journalism and the Visual Politics of War and Conflict. In Allan, S. (Ed.) *The Routledge Companion to News and Journalism.* (520–532). London and New York: Routledge.

Chouliaraki, L. (2006). *The Spectatorship of Suffering.* London: Sage Publications.

Chouliaraki, L. (2004). Watching 11 September: The Politics of Pity. *Discourse and Society*, 15(2–3): 185–98.

Clarke, S. (2004). The Concept of Envy: Primitive Drives, Social Encounters and *Ressentiment. Psychoanalysis, Culture & Society*, 9(1): 105–117.

Clarke, S., Hoggett, P., and Thompson, S. (Eds) (2006). *Emotion, Politics and Society.* London: Palgrave/Macmillan.

Clogg, R. (1997). *A Concise History of Greece.* Cambridge: Cambridge University Press.

Clogg, R. (Ed.) (1993). *Greece, 1981–89. The Populist Decade.* New York: St. Martin's Press.

Clogg, R. (1979). *A Short History of Modern Greece.* Cambridge: Cambridge University Press.

Close, D. (1995). *The Origins of the Greek Civil War.* Longman Publishing Group.

Clough, P. T. and Halley, J. (Eds.) (2007). *The Affective Turn. Theorizing the Social.* Durham and London: Duke University Press.

Cohen, S. (2001). *States of Denial: Knowing about Atrocities and Suffering,* New York: John Wiley & Sons.

Cooley, H. C. (1964). *Human Nature and the Social Order.* New York: Schocken Books.

Coonfield, G. (2007). News Images as Lived Images: Witness, Performance, and the U.S. Flag After 9/11. In Carpentier, N. (Ed.) *Culture, Trauma, and Conflict. Cultural Studies Perspectives on War.* (163–181). Newcastle: Cambridge Scholars Publishing.

Concise Oxford Dictionary of Current English. (1990). R. E. Allen (Ed.). Oxford: Clarendon Press.

Connerton, P. (1989). *How Societies Remember.* Cambridge: Cambridge University Press.

Connolly, W. (1991). *Identity/Difference: Democratic Negotiations of Political Paradox.* Ithaca, NY: Cornell University Press.

Coser, A. L. (1961). Max Scheler: An Introduction. In Scheler, M. *Ressentiment*. (5–32). Glencoe: Free Press.

Cossarini, P. and Vallespín, F. (2019). Introduction: Populism, Democracy, and the Logic of Passion. In Cossarini, P. and Vallespín, F. (Eds) *Populism and Passions. Democratic Legitimacy after Austerity*. (1–12). New York: Routledge.

Cowen, T. (2006). How far back should we go? Why restitution should be small. In J. Elster (Ed.) *Retribution and Reparation in the Transition to Democracy*. (17–32). Cambridge, U.K.: Cambridge University Press.

Cramer, K. J. (2016). *The Politics of Resentment: Rural Consciousness in Wisconsin and the Rise of Scott Walker*. Chicago and London: The University of Chicago Press.

Crook S., Pakulski J., and Waters M. (1992). *Postmodernization. Change in Advanced Society*. London: Sage Publications.

Cunningham, M. (2012). The apology in politics. In Thompson, S. and Hoggett, P. (Eds) *Politics and the Emotions: The Affective Turn in Contemporary Political Studies*. (139–155). New York: Continuum International Publishing Group.

Dalton, R. (1988). *Citizen Politics in Western Democracies*. Chatham/New Jersey: Chatham House Publishers, Inc.

Damasio, A. R. (1999). *The Feeling of What Happens. Body and Emotion in the Making of Consciousness*. New York: Harcourt Brace & Company.

Danto, A. C. (1985). *Narration and Knowledge: Including the Integral Text of Analytical Philosophy of History*. New York: Columbia University Press.

Davies, J. C. (1962). Toward a Theory of Revolution. *Source: American Sociological Review*, 27(1): 5–19.

Davis, A. (1988). *An Autobiography*. New York: International Publishers.

Davou, B. (2000). *Thought Processes in the Information Age. Issues of Cognitive Psychology and Communication*. Athens: Papazissis Publications.

Davou, B. and Demertzis, N. (2013). Feeling the Greek financial crisis. In Demertzis, N. (Ed.) *Emotions in Politics. The Affect Dimension in Political Tension*. (93–123). Palgrave/MacMillan, London.

Dayan, D. (2007). On Morality, Distance and the Other. Roger Silverstone's *Media and Morality. International Journal of Communication*, 1: 113–122.

de Balzac, H. (2004). *Lost Illusions*. Project Gutenberg EBook. [#13159].

De Cleen, B. and Stavrakakis, Y. (2017). Distinctions and Articulations: A Discourse Theoretical Framework for the Study of Populism and Nationalism. *Javnost – The Public*. DOI: 10.1080/13183222.2017.1330083.

De la Mora, F. G. (1987). *Egalitarian Envy: The Political Foundations of Social Justice*. New York: Paragon Publishers House.

de Rivera, J. (1992). Emotional Climate: Social Structure and Emotional Dynamics. In K. T. Strongman (Ed.) *International Review of Studies on Emotion*, vol. 2. (197–218). New York: John Wiley & Sons.

Dearing, J. W. and Rogers, E. M. (1996). *Agenda-Setting*. London: Sage Publications.

Deegan-Krause, K. (2007). New Dimensions on Political Cleavage. In Dalton, R. J. and Klingemann, H.-D. (Eds) *The Oxford Handbook of Political Behavior*. (538–556). Oxford and New York: Oxford University Press.

Degli, E. F. (2015). War as Social Regeneration: Sombart from The Quintessence of Capitalism to Merchants and Heroes. In Antonio L. Palmisano (Ed.) *Sombart's Thought Revisited. DADA. Rivista di Antropologia Post-globale*, 1: 41–53.

Dekker, P. (2005). *Political Cynicism: A Hard Feeling or an Easy Way to Maintain Distance?* Paper presented in the Political Psychology Section at the ECPR, Budapest 8–11 September.

Demertzis, N. (2017). Forgiveness and Ressentiment in the Age of Traumas. *Oxford Research Encyclopedia, Politics,* Oxford University Press, USA. DOI: 10.1093/acrefore/9780190228637.013.146.

Demertzis, N. (2014). Political Emotions. In Nesbitt-Larking, P., Kinvall, C. and Capelos, T. (Eds) *The Palgrave Handbook of Global Political Psychology.* (223–241). London: Palgrave/Macmillan.

Demertzis, N. (2013). Introduction: Theorizing the Emotions–Politics Nexus. Demertzis, N. (Ed.) *Emotions in Politics. The Affect Dimension in Political Tension.* (1–16). London: Palgrave/MacMillan.

Demertzis, N. (2009). Mediatizing traumas in the risk society: A sociology of emotions approach. In Hopkins, D., Kleres, J., Flam, H., and Kuzmics, H. (Eds) *Theorizing Emotions: Sociological Explorations and Applications.* (143–168). Frankfurt: Campus Verlag.

Demertzis, N. (2006). Emotions and Populism. In Clarke, S., Hoggett, P., and Thompson, S. (Eds) *Emotion, Politics and Society.* (103–122). London: Palgrave.

Demertzis, N. (2004). Populism and *Ressentiment.* A Contribution of the (political) Sociology of Emotions. *Science and Society,* 12: 75–114. DOI: http://dx.doi.org/10.12681/sas.768.

Demertzis, N. (1997). Greece. In Eatwell, R. (Ed.) *European Political Culture.* (107–121). London: Routledge.

Demertzis, N. (1996a). *The Nationalist Discourse. Ambivalent Semantic Field and Contemporary Tendencies.* Athens: Ant. N. Sakkoulas

Demertzis, N. (1996b). La place de la religion dans la culture politique grecque. In Mappa, S. (Ed.) *Puissance et Impuissance de l'État.* (223–244). Paris: Karthala.

Demertzis, N. (1985). *Cultural Theory and Political Culture. New Directions and Proposals.* Lund: Studentlitteratur.

Demertzis, N. and Lipowatz, T. (2006). *Envy and Ressentiment. Passions of the Soul and the Closed Society.* Athens: Polis (in Greek).

Demertzis, N., Papathanassopoulos, S., and Armenakis, A. (1999). Media and Nationalism. The Macedonian Question. *Harvard International Journal of Press/Politics,* 44(3): 26–50. https://doi.org/10.1177/1081180X99004003004.

Demertzis, N. and Stratoudaki, H. (2020). Greek Nationalism as a Case of Political Religion: Rituals and Sentimentality. *Historical Social Research,* 45(1): 103–128. DOI: 10.12759/hsr.45.2020.1.103-128.

Demertzis, N. and Tsekeris, C. (2018). *Multifaceted European Public Sphere. Socio-Cultural Dynamics.* London: Media@LSE Working Paper Series.

Derrida, J. (2003). *On Cosmopolitanism and Forgiveness.* London: Routledge.

Derrida, J. (1986). Foreword: *Fors:* The Anglish Words of Nicolas Abraham and Maria Torok. In Abraham, N. and Torok, M. *The Wolf Man's Magic Word. A Cryptonymy.* (xi–xlviii). Minneapolis: University of Minnesota Press.

Deutsch, K. W. (1969). *Nationalism and its Alternatives.* New York: Knopf/Random House.

Deutsch, K. W. (1953). *Nationalism and Social Communication: An Inquiry into the Foundations of Nationality.* Massachusetts: The M.I.T. Press (2nd edition 1966).

Diamandouros, N. (1994). *Cultural Dualism and Political Change in Postauthoritarian Greece.* Centro de Estudios Avanzados en Ciencias Sociales (CEACS).

Diamandouros, N. (1983). Greek political culture in transition: Historical origins, evolution, current trends. In Clogg, R. (Ed.) *Greece in the 1980s*. (43–69). London: The Macmillan Press.

Diamond, L. (2015). Facing Up to the Democratic Recession. *Journal of Democracy*, 26 (1): 141–55.

Diefenbach, A., Kahl, A. *et al.* (2019). *The Politics of Affective Societies: An Interdisciplinary Essay* (EmotionsKulturen/EmotionCultures). Bielefeld: Transcript-Verlag.

Dijk, van W. W. and Ouwerkerk, J. W. (Eds) (2014). *Schadenfreude. Understanding Pleasure at the Misfortune of Others*. Cambridge: Cambridge University Press.

Dostoyevsky, F. (1864/2000). *Notes from the Underground*. Athens: Govostis Publications.

Döveling, K. (2009). Mediated Parasocial Emotions and Community: How Media May Strengthen or Weaken Social Communities. In Hopkins, D., Kleres, J., Flam, H., and Kuzmics, H. (Eds) *Theorizing Emotions: Sociological Explorations and Applications*. (315–337). Berlin: Campus.

Döveling, K., von Scheve, Ch., and Konijn, E. A. (2011) (Eds) *The Routledge Handbook of Emotions and Mass Media*. London: Routledge.

Dufourmantelle, A. and Derrida, J. (2000). *Of Hospitality. Anne Dufourmantelle invites Jacques Derrida to respond* (translated by Rachel Bowlby). Stanford: Stanford University Press.

Edelstein, A. (1997). *Total Propaganda. From Mass Culture to Popular Culture*. London: Lawrence Erlbaum.

Edkins, J. (2003). *Trauma and the Memory Politics*. Cambridge: Cambridge University Press.

Eisenberg, N. (2004). Empathy and Sympathy. In Lewis, M. and Haviland-Jones, J. M. (Eds) *Handbook of Emotions* (2nd paperback edition). (677–691). New York/London: The Guilford Press.

Eisenstadt, S. N. (2013). *Comparative Civilizations and Multiple Modernities*. Leiden and Boston: Brill.

Eisenstadt, S. N. (1973). *Tradition, Change, and Modernity*. New York: John Wiley.

Ekman, P. (1993). Facial Expression and Emotion. *American Psychologist*, 48: 384–392.

Eliasoph, N. (1998). *Avoiding Politics. How Americans Produce Apathy in Everyday Life*. Cambridge: Cambridge University Press.

Eliasoph, N. (1990). Political Culture and the Presentation of a Political Self: A Study of the Public Sphere in the Spirit of Erving Goffman. *Theory and Society*, 19 (4): 465–494.

Ellis, D. (1999). Research on Social Interaction and the Micro-Macro Issue. *Research on Language and Social Interaction*, 32(1/2): 31–40.

Elsaesser, T. (2014). *German Cinema-Terror and Trauma: Cultural Memory since 1945*. London: Routledge.

Elster, J. (2006). Retribution. In J. Elster (Ed.) *Retribution and Reparation in the Transition to Democracy*. (33–56). Cambridge, U.K.: Cambridge University Press.

Elster, J. (1999). *Alchemies of the Mind. Rationality and the Emotions*. Cambridge: Cambridge University Press.

Elster, J. (1989). *Nuts and Bolts for the Social Sciences*. Cambridge: Cambridge University Press.

Elytis, O. (1994). *Open Papers – Selected Essays*. Washington: Copper Canyon Press.

Entman, R. M. (1993). Framing. Toward Clarification of a Fractured Paradigm. *Journal of Communication*, 43(4): 51–58.

Erikson, K. (1995). Notes on Trauma and Community. In Caruth, C. (Ed.) *Trauma: Explorations in Memory*. (183–199). Baltimore: Johns Hopkins.

Evans, D. (1996). *An Introductory Dictionary of Lacanian Psychoanalysis*. London: Routledge.

Eyerman, R. (2015). *Is this America? Katrina as Cultural Trauma*. The Katrina Bookshelf. Austin: University of Texas Press.

Eyerman, R. (2012/2017). Cultural Trauma: Emotion and Narration. In Alexander, J. C., Jacobs, R. N., and Smith, P. (Eds) *The Oxford Handbook of Cultural Sociology*. Oxford: Oxford University Press. Online Publication Date: Jun 2017. DOI: 10.1093/oxfordhb/9780195377767.013.21.

Eyerman, R. (2011a). *The Cultural Sociology of Political Assassination. From MLK to Fortuyn and van Gogh*. New York: Palgrave/Macmillan.

Eyerman, R. (2011b). Intellectuals and cultural trauma. *European Journal of Social Theory*, 14(4): 453–467.

Eyerman, R. (2008). *The Assassination of Theo Van Gogh: From Social Drama to Cultural Trauma*. Durham, NC: Duke University Press.

Eyerman, R. (2001). *Cultural Trauma. Slavery and the Formation of African American Identity*. Cambridge: Cambridge University Press.

Eyerman, R., Alexander, J. C., and Breese, E. (Eds) (2013). *Narrating Trauma: On the Impact of Collective Suffering*. Boulder, CO: Paradigm.

Eyerman, R., Madigan, T., and Ring, M. (2017). Cultural Trauma, Collective Memory and the Vietnam War. *Croatian Political Science Review*, 54(1–2): 11–31.

Fassin, D. and Rechtman, R. (2009). *The Empire of Trauma. An Inquiry into the Condition of Victimhood*. Oxford and Princeton: Princeton University Press.

Fearon, J. and Laitin, D. D. (2003). Ethnicity, Insurgency, and Civil War. *American Political Science Review*, 97(1): 75–90.

Ferrante, E. (2015). *The Story of the Lost Child*. New York: Europa Editions.

Ferrante, E. (2014). *Those Who Leave and Those Who Stay*. New York: Europa Editions.

Ferrante, E. (2013a). *My Brilliant Friend*. New York: Europa Editions.

Ferrante, E. (2013b). *The Story of a New Name*. New York: Europa Editions.

Ferro, M. (2010). *Resentment in History*. Cambridge: Polity Press.

Fichte, J.-G. (1995). Address to the German Nation. In Dahbour, O. and Ishay, M. R. (Eds). *The Nationalism Reader*. (62–70). New Jersey: Humanities Press.

Fieschi, C. (2004). Introduction. *Journal of Political Ideologies*, 9(3): 235–240.

Fieschi, C. and Heywood, P. (2004). Trust, cynicism and populist anti-politics. *Journal of Political Ideologies*, 9(3): 289–309.

Finkelstein, N. G. (2003). The *Holocaust Industry. Reflections on the Exploitation of Jewish Suffering*. New York: Verso (2nd edition).

Fiske, J. (1987). *Television Culture*. London: Methuen.

Flam, E. (2013). The transnational movement for Truth, Justice and Reconciliation as an emotional (rule) regime? *Journal of Political Power*, 6(3): 363–384.

Flam, H. (2005). Emotions' map. A research agenda. In Flam, H. and King, D. (Eds) *Emotions and Social Movements*. (19–40). London: Routledge.

Flam, H. and Kleres, J. (Eds) (2015). *Methods of Exploring Emotions*. London: Routledge.

Flam, H. and King, D. (Eds) (2005). *Emotions and Social Movements*. London: Routledge.

Fleischacker, S. (1991). Philosophy in Moral Practice: Kant and Adam Smith. *Kant-Studien*, 82(3). https://doi.org/10.1515/KANT.1991.82.3.249.

Fleischer, H. (2008). *Wars of Memory. World War II in Public History.* Athens: Nefeli Publications.

Fleischer, H. (1986). *Im Kreuzschatten der Maechte: Griechenland 1941–1944. (Okkupation—Kollaboration—Resistance).* Frankfurt/Bern/New York: Peter Lang.

Freud, S. (2003). *An Outline of Psychoanalysis.* London: Penguin Books. [First published in 1940].

Freud, S. (1986). Moses and Monotheism: Three Essays. In *The Origins of Religion, The Pelican Freud Library,* vol. 13. London: Penguin Books. [First published in 1939].

Freud, S. (1949). *Group Psychology and the Analysis of the Ego* (translated by J. Strachey). London: The Hogarth Press (5th Impression).

Freud, S. (1930/2001). Civilization and its Discontents. In *The Standard Edition of the Complete Psychological Works,* Vol. XXI. London: Vintage, The Hogarth Press.

Freud, S. and Breuer, J. (1895/1956). *Studies on Hysteria.* Pelican Freud Library, Standard Edition. London: Penguin Books.

Friedman, J. R. and Strauss, C. (2018). Introduction: The Person in Politics and Culture. In Strauss, C. and Friedman, J. R. (Eds) *Political Sentiments and Social Movements. The Person in Politics and Culture.* (1–29). London: Palgrave/Macmillan.

Frijda, N. H. (2004a). The Psychologists' Point of View. In Lewis, M. and Haviland-Jones, J. M. (Eds) *Handbook of Emotions.* (59–74). New York, London: The Guilford Press (2nd edition).

Frijda, N. H. (2004b). Emotion and action. In Manstead, A., Fridja, N., and Fischer, A. (Eds) *Feelings and Emotions: The Amsterdam Symposium.* (158–173). Cambridge: Cambridge University Press.

Frings, M. (2005). Max Scheler and the Psychopathology of the Terrorist. *Modern Age,* 47(3): 210–219.

Fritsche, J. (1999). *Historical Destiny and National Socialism in Heidegger's Being and Time.* Berkeley: University of California Press.

Fukuyama, F. (2018). *Identity. Contemporary Identity Politics and the Struggle for Recognition.* New York: Farrar, Straus and Giroux.

Furedi, F. (2004). *Therapy Culture. Cultivating Vulnerability in an Uncertain Age.* London: Routledge.

Gadamer, H.-G. (1979). *Truth and Method.* London: Sheed and Ward.

Gamson, W. A. (1992). *Talking Politics.* Cambridge: Cambridge University Press.

Gao, R. (2013). Revolutionary Trauma and Representation of the War: The Case of China in Mao's Era. In Eyerman, R., Alexander, J. C., and Breese, E. (Eds) *Narrating Trauma. On the Impact of Collective Suffering.* (81–105). Boulder & London: Paradigm Publications.

Gaulejac, V. De (1987). *La Névrose de Classe.* Paris: Hommes et groups.

Gavrilidis, A. (2008). Two brotherless peoples: On the Constitutive Traumas of Class Struggle. *Psychoanalysis, Culture & Society,* 13: 143–162. DOI: 10.1057/pcs.2008.2.

Geertz, C. (1973). *The Interpretation of Cultures.* New York: Basic Books.

Gellner, E. (1994). *Encounters with Nationalism.* Oxford: Blackwell.

Gellner E. (1983). *Nations and Nationalism.* Oxford: Basil Blackwell.

Gellner, E. (1965). *Thought and Change.* London: Weidenfeld and Nicolson.

Gibbins, J. R. (1990). Contemporary Political Culture: An Introduction. In Gibbins, J. R. (Ed.) *Contemporary Political Culture. Politics in a Postmodern Age.* (1–30). London: Sage Publications.

Giddens, A. (1994). *Beyond Left and Right. The Future of Radical Politics.* Oxford: Polity Press.

Giddens, A. (1992). *The Transformation of Intimacy. Sexuality, Love & Eroticism in Modern Societies*. Stanford: Stanford University Press.

Giddens, A. (1990). *The Consequences of Modernity*. London: Polity Press.

Giddens A. (1985). *The Nation-State and Violence*. Cambridge: Polity Press.

Giddens, A. (1984). *The Constitution of Society*. Cambridge: Polity Press.

Giddens, A. (1981a). *The Class Structure of the Advanced Societies*. London: Hutchinson (2nd edition).

Giddens A. (1981b). *A Contemporary Critique of Historical Materialism.* London: The MacMillan Press.

Giesen, B. (2004a). *Triumph and Trauma*. Boulder: Paradigm Press.

Giesen, B. (2004b). The Trauma of Perpetrators. The Holocaust as the Traumatic Reference of German National Identity. In Alexander, J. C. *et al.* (Eds) *Cultural Trauma and Collective Identity*. (112–154). Berkeley: University of California Press.

Girard, R. (2005). *Violence and the Sacred*. London and New York: Continuum.

Girardet, R. (1997). *Myths and Political Mythologies*. Iaşi: European Institute.

Girvin, B. (1990). Change and Continuity in Liberal Democratic Political Culture. In Gibbins, J. R. (Ed.) *Contemporary Political Culture. Politics in a Postmodern Age.* (31–51). London: Sage Publications.

Glassner, B. (1999). *The Culture of Fear. Why Americans Are Afraid of the Wrong Things*. New York: Basic Books.

Glynos, J. and Mondon, A. (2019). The Political Logic of Populist Hype: The Case of Right-Wing Populism's 'Meteoric Rise' and Its Relation to the Status Quo. In Cossarini, P. and Vallespín, F. (Eds) *Populism and Passions. Democratic Legitimacy after Austerity.* (82–101). New York: Routledge.

Goldfarb, J. (1991). *The Cynical Society. The Culture of Politics and the Politics of Culture in American Life*. Chicago: The University of Chicago Press.

Goodwin, J. (2005). Revolutions and Revolutionary Movements. In Janoski, T., Alford, R. R., Hicks, A. M., and Schwartz, A. M. (Eds) *The Handbook of Political Sociology. States, Civil Societies, and Globalization*. (404–422). Cambridge: Cambridge University Press.

Goodwin, J. and Jasper, J. M. (2006). Emotions and Social Movements. In Stets, J. E. and Turner, J. H. (Eds) *Handbook of the Sociology of Emotions*. (611–635). New York: Springer.

Goodwin, J., Jasper, J. M., and Polletta, F. (2004). Emotional Dimensions of Social Movements. In Snow, D., Soule, S. A., and Kriesi, H. (Eds) *The Blackwell Companion to Social Movements*. (413–432). Malden, MA: Blackwell Publishing Ltd.

Goodwin, J., Jasper, J. M. and Polletta, F. (Eds) (2001a). *Passionate Politics. Emotions and Social Movements*. Chicago: The University of Chicago Press.

Goodwin, J., Jasper, J. M. and Polletta, F. (2001b). Why Emotions Matter. In Goodwin, J., Jasper, J. M. and Polletta, F. (Eds) *Passionate Politics. Emotions and Social Movements.* (1–24). Chicago and London: The University of Chicago Press.

Gordon, S. (1990). Social Structural Affects on Emotions. In Kemper, T. (Ed.) *Research Agendas in the Sociology of Emotions*. (145–179). New York: State University of New York Press.

Gordon, S. (1981). The sociology of sentiments and emotion. In Rosenberg, M. and Turner, R. H. (Eds) *Social Psychology: Sociological Perspectives*. (551–575). New York: Basic Books.

Gould, D. B. (2010). On Affect and Protest. In Staiger, J., Cvetkovitch, A. and Reynolds, A. (Eds) *Political Emotions. New Agendas in Communication*. (18–44). New York, London: Routledge.

Gouldner, A. (1976). *The Dialectic of Ideology and Technology*. New York: Seabury.

Gouldner, A. W. (1970). *The Coming Crisis of Western Sociology*. London: Heineman.

Gourgouris, S. (1996). *Dream Nation: Enlightenment, Colonization and the Institution of Modern Greece*. Stanford, CA: Stanford University Press.

Gramsci, A. (1971 mimeo). *On Machiavelli*. Athens.

Gransow, V. and Offe, C. (1982). Political Culture and Politics in the Social Democratic Government, *Telos* 53: 67–80.

Greco, M. and Stenner, P. (Eds) (2008). *Emotions: A Social Science Reader*. London, New York: Routledge.

Greene, J. D. (2013). *Moral Tribes: Emotion, Reason, and the Gap between Us and Them*. New York: Pelican Books.

Greenfeld, L. (2018). Revolutions. In Outhwaite, W. and Turner, S. P. (Eds) *The Sage Handbook of Political Sociology*. (vol. 2) (685–698). London: Sage Publications.

Greenfeld, L. (2001). *The Spirit of Capitalism. Nationalism and Economic Growth*. Cambridge: Harvard University Press.

Greenfeld, L. (1992). *Nationalism. Five Roads to Modernity*. Massachusetts: Harvard University Press.

Greenfeld, L. and Eastwood, L. (2005). Nationalism in Comparative Perspective. In Janoski, T., Alford, R. R., Hicks, A. M., and Schwartz, A. M. (Eds) *The Handbook of Political Sociology. States, Civil Societies, and Globalization*. (247–264). Cambridge: Cambridge University Press.

Griswold, C. (2007). *Forgiveness: A Philosophical Exploration*. New York: Cambridge University Press.

Guiso, L., Herrera, H., Morelli, M., and Sonno, T. (2017). *The Spread of Populism in Western Countries*. http://voxeu.org.

Guzzini, S. (2016). International political sociology, or: The social ontology and power politics of process. *Danish Institute for International Studies*, #6.

Hackett, J. (2016). *Ressentiment, Trump and His Supporters*. www.philpercs.com/2016/03/ressentiment-trump-and-his-supporters.html.

Haddad, W. (1977). Nationalism in the Ottoman Empire. In Haddad, W. W. and Ochesenwald, W. (Eds) *Nationalism in a Non-National State. The Dissolution of the Ottoman Empire*. (3–24). Columbus: Ohio State University Press.

Haidia, E. (2000). The Punishment of Collaborators in Northern Greece, 1945–1946. In Mazower, M. (Ed.) *After the War Was Over. Reconstructing the Family, Nation, and State in Greece, 1943–1960*. (42–61). Princeton and Oxford: Princeton University Press.

Haidt, J. (2013). *The Righteous Mind. Why Good People are Divided by Politics and Religion*. London: Penguin Books.

Halbwachs, M. (1992). *On Collective Memory*. Chicago and London: The University of Chicago Press.

Hall, S. (1980). Encoding/decoding. In Hall S., Hobson D., Lowe A., and Willis P. (Eds) *Culture, Media, Language*. (128–138). London: Hutchinson.

Halperin, E., Russell, A., Dweck, C., and Gross, J. (2011). Anger, Hatred, and the Quest for Peace: Anger Can Be Constructive in the Absence of Hatred. *The Journal of Conflict Resolution*, 55(2): 274–291. Retrieved from www.jstor.org/stable/23049702.

Hardin, R. (Ed.) (2004). *Distrust*. New York: Russell Sage Foundation.

Hardin, R. (1999). Do we want trust in government? In Warren, M. E. (Ed.) *Democracy & Trust*. (22–41). Cambridge: Cambridge University Press.

Hartman, G. H. (2003). Trauma within the limits of literature. *European Journal of English Studies*, 7(3): 257–274.

Harvey, D. (1989). *The Condition of Postmodernity*. Oxford: Blackwell.

Hawkins, K. A., Riding, S., and Mudde, C. (2012). Measuring populist attitudes. *Political Concepts Committee on Concepts and Methods Working Paper Series*, 55: 1–35.

Hayes, C. (1933). *Essays on Nationalism*. New York: Macmillan.

Heaney, G. J. (2019). Emotion as power: capital and strategy in the field of politics. *Journal of Political Power*. DOI: 10.1080/2158379X.2019.1618485.

Heaney, G. J. (2013a). Emotions and power: a bifocal prescription to cure theoretical myopia. *Journal of Political Power*, 6(3): 355–362.

Heaney, G. J. (2013b). Emotions and Nationalism: A Reappraisal. In Demertzis, N. (Ed.) *Emotions in Politics. The Affect Dimension in Political Tension*. (243–263). London, England: Palgrave Macmillan.

Heaney, G. J. and Flam, H. (Eds) (2015). *Power and Emotion*. New York: Routledge.

Hegel, G. W. F. (1975). *Hegel's Logic* (translated by William Wallace). Oxford: Oxford University Press.

Hegel, G. W. F. (1967). *The Phenomenology of Mind*. New York: Harper Torchbooks.

Heller, A. (1982). *A Theory of History*. London: Routledge Kegan Paul.

Heller A. and Fehér F. (1988). *The Postmodern Political Condition*. Oxford: Polity Press.

Hennessy, A. (1969). Latin America. In Ionescu, G. and Gellner, E. (Eds). *Populism. Its Meanings and National Characteristics*. (28–61). Hertfordshire: The Garden City Press.

Herman, J. (1992). *Trauma and Recovery: The Aftermath of Violence from Domestic Abuse to Political Terror*. New York: Basic Books.

Hertz, F. (1966). *Nationality in History and Politics. A Psychology and Sociology of National Sentiment and Nationalism*. London: Routledge & Kegan Paul (5th edition). (1st edition 1944).

Herzfeld, M. (2005). *Cultural Intimacy. Social Poetics of the Nation-State*. New York: Routledge (2nd edition).

Herzfeld, M. (1995). Hellenism and Occidentalism: The permutations of performance in Greek bourgeois identity. In Carrier, J. G. (Ed.) *Occidentalism: Images of the West*. (218–233). Oxford: Berg.

Herzfeld, M. (1989). *Anthropology Through the Looking-glass: Critical Ethnography in the Margins of Europe*. Cambridge: Cambridge University Press.

Herzfeld, M. (1986). *Ours Once More: Folklore, Ideology, and the Making of Modern Greece*. New York: Pella Publishing Company.

Hicks, A. M., Janoski, T. and Schwartz, M. A. (2005). Introduction. Political Sociology in the New Millennium. In Janoski, T., Alford, R. R., Hicks, A. M. and Schwartz, A. M. (Eds) *The Handbook of Political Sociology. States, Civil Societies, and Globalization*. (1–30). Cambridge: Cambridge University Press.

Hobsbawm, E. (1992). *Nations and Nationalism since 1780. Programme, Myth, Reality*. Cambridge: Cambridge Univ. Press.

Hochschild, A. R. (2016). *Strangers in their own Land. Anger and Mourning on the American Right*. New York: The New Press.

Hochschild, A. R. (2009). Introduction: An Emotions Lens on the World. In Hopkins, D. et al. (Eds) *Theorizing Emotions: Sociological Explorations and Applications*. (29–37). Berlin: Campus.

Hochschild, A. R. (1979). Emotion work, feeling rules, and social structure. *American Journal of Sociology*, 85(3): 551–575.

Hochschild, A. R. (1975). The sociology of feeling and emotion: Selected possibilities. In Millman, M. and Kanter, R. (Eds) *Another Voice*. (280–307). New York: Anchor.

Hoggett, P. (2018). *Resentment as a Political Force: From Nietzsche to Trump*. www. climatepsychologyalliance.org/explorations/papers/264-resentment-as-a-political-force-from-nietzche-to-trump.

Hoggett, P. (2006). Pity, Compassion, Solidarity. In Clarke, S. Hoggett, P., and Thompson, S. (Eds) *Emotion, Politics and Society*. (145–161). New York: Palgrave Macmillan.

Hoggett, P. and Thompson, S. (2012). Introduction. In Thompson, S. and Hoggett, P. (Eds) *Politics and the Emotions. The Affective Turn in Contemporary Political Studies*. (1–19). New York: Continuum.

Hoggett, P., Wilkinson, H., and Beedell, P. (2013). Fairness and the Politics of Resentment. *Journal of Social Policy*, 42(3): 567–585. DOI: 10.1017/S0047279413000056.

Höijer, B. (2003). The Discourse of Global Compassion and the Media. *Nordicom Review*, 24(2): 19–29.

Hokka, J. and Nelimarkka, M. (2019). Affective economy of national-populist images: Investigating national and transnational online networks through visual big data. *New Media & Society*, 1–23. https://doi.org/10.1177/1461444819868686.

Holcombe, R. G. (2012). Crony Capitalism: By-Product of Big Government. Working Paper, *Mercatus Center of George Mason University*, # 12–32.

Holmes, M. (2010). The Emotionalization of Reflexivity. *Sociology*, 44(1): 139–154.

Homer, S. (2019). *Greek Cinema in Times of Crisis: A Cultural Trauma in Becoming*. (Unpublished paper).

Hondros, J. (1983). *Occupation and Resistance: The Greek Agony 1941–44*. New York: Pella Publishing Company.

Hopkins, D., Kleres, J., Flam, H., and Kuzmics, H. (Eds) (2009). *Theorizing Emotions: Sociological Explorations and Applications*. Frankfurt: Campus Verlag.

Horton, R. (1982). Tradition and Modernity Revisited. In Hollis, M. and Lukes, S. (Eds) *Rationality and Relativism*. (201–260). Oxford: Basil Blackwell.

Horkheimer, M. and Adorno, T. (2002). *Dialectic of Enlightenment*. Stanford, CA: Stanford University Press.

Hughes, P. M. (2016). Forgiveness. *The Stanford Encyclopedia of Philosophy* (Winter 2016 Edition), Edward N. Zalta (Ed.). https://plato.stanford.edu/archives/win2016/entries/forgiveness/.

Hume, D. (1748/2018). *An Enquiry Concerning Human Understanding*. Global Grey globalgreyebooks.com.

Hume, D. (1739/1969). *A Treatise of Human Nature*. Middlesex: Penguin Books.

Huntington, S. and Dominguez, J. (1975). Political Development. In Greenstein, F. and Polsby, N. (Eds) *Handbook of Political Science* (Vol. III) (1–114). Massachusetts: Addison-Wesley Pub. Co.

Husserl, E. (1978). The Origin of Geometry. In Luckmann, T. (Ed.) *Phenomenology and Sociology. Selected Readings*. (42–70). Middlesex: Penguin Books.

Hutchinson, E. (2016). *Affective Communities in World Politics: Collective Emotions after Trauma*. Cambridge, U.K.: Cambridge University Press.

Hutchinson, E. and Bleiker, R. (2008). Emotional Reconciliation: Reconstituting Identity and Community after Trauma. *European Journal of Social Theory*, 11(3): 385–403.

Iatrides, J. (2002). The International Context of the Greek Civil War. In Nikolakopoulos, E., Rigos, A., and Psallidas, G. (Eds) *The Civil War. From Varkiza to Grammos. February 1945–August 1949*. (31–50). Athens: Themelio.

Illouz, E. (2007). *Cold Intimacies: The Making of Emotional Capitalism*. Cambridge: Polity Press.

Inglehart, F. R. (2018). *Cultural Evolution, People's Motivations Are Changing, and Reshaping the World*. Cambridge: Cambridge University Press.

Ioannou, G. (1984). *The Metropolis of Refugees*. Athens: Kedros Publications.

Israel, J. (1979). *The Language of Dialectics and the Dialectics of Language*. New York: Humanities Press.

James, W. (1931/1890). *The Principles of Psychology*, vol. 2. New York: Henry Holt.

Jankélévitch, V. (2005). *Forgiveness*. Chicago: University of Chicago Press.

Jankélévitch, V. (1996). Should We Pardon Them? *Critical Inquiry*, 22(3): 552–572.

Jankélévitch, V. (1987). *L'Ironie*. Paris: Flammarion.

Janoski, T., Alford, R. R., Hicks, A. M., and Schwartz, M. A. (Eds) (2005). The Handbook of Political Sociology: States, Civil Societies, and Globalization. Cambridge: Cambridge University Press.

Jasper, M. J. (2005). Culture, Knowledge, Politics. In Janoski, T., Alford, R. R., Hicks, A. M. and Schwartz, A. M. (Eds) *The Handbook of Political Sociology. States, Civil Societies, and Globalization*. (115–134). Cambridge: Cambridge University Press.

Johnson, W. M., and Davey, G. C. L. (1997). The Psychological Impact of Negative TV News Bulletins: The Catastrophizing of Person Worries. *British Journal of Psychology*, 88: 85–91.

Jonas, H. (1984). *The Imperative of Responsibility: In Search of an Ethics for the Technological Age*. Chicago: University of Chicago Press.

Kaindaneh, S. and Rigby, A. (2012). Peace-building in Sierra Leone: The emotional dimension. In Thompson, S. and Hoggett, P. (Eds), *Politics and the Emotions: The Affective Turn in Contemporary Political Studies*. (157–179). New York: Continuum International Publishing Group.

Kalyvas, A. and Katznelson, I. (2008). *Liberal Beginnings. Making a Republic for the Moderns*. Cambridge: Cambridge University Press.

Kalyvas, S. (2002). Forms, Dimensions and Practices of Violence in Civil War (1943–1949): An initial approach. In Nikolakopoulos, E., Rigos, A., and Psallidas, G. (Eds) *The Civil War. From Varkiza to Grammos. February 1945–August 1949*. (188–207). Athens: Themelio.

Kalyvas, S. (2000). Red Terror: Leftist Violence during the Occupation. In Mazower, M. (Ed.) *After the War Was Over. Reconstructing the Family, Nation, and State in Greece, 1943–1960*. (142–183). Princeton and Oxford: Princeton University Press.

Kalyvas, S. and Marantzidis, N. (2015). *Civil War Passions*. Athens: Metechmio.

Kansteiner, W. (2004). Genealogy of a category mistake: A critical intellectual history of the cultural trauma metaphor. *Rethinking History*, 8(2): 193–221.

Kaplan, A. (2005). *Trauma Culture. The Politics of Terror and Loss in Media and Literature*. New Brunswick: Rutgers UP.

Karapostolis, V. (1985). *The Impenetrable Society*. Athens: Polytropon Publications.

Karyotis, G. and Gerodimos, R. (Eds) (2015). *The Politics of Extreme Austerity: Greece in the Eurozone Crisis*. Basingstoke, New York: Palgrave Macmillan.

Keane, J. (2009). *The Life and Death of Democracy*. London: Simon and Schuster.

Kedourie, E. (1993). *Nationalism*. Oxford: Blackwell (4th edition).

Kemper, T. D. (2004). Social Models in the Explanation of Emotions. In Lewis, M. and Haviland-Jones, J. M. (Eds) *Handbook of Emotions*. (45–58). New York, London: The Guilford Press (2nd edition).

Kemper, T. (1991). An Introduction to the Sociology of Emotions. In Strongman, K. (Ed.) *International Review of Studies on Emotion.* Vol. 1 (301–349). New York: John Wiley.

Kemper, T. (1990) (Ed.) *Research Agendas in the Sociology of Emotions.* New York: State University of New York Press.

Kenny, M. (2017). Back to the populist future? Understanding nostalgia in contemporary ideological discourse, *Journal of Political Ideologies,* 22(3): 256–273. DOI: 10.1080/1 3569317.2017.1346773.

Kierkegaard, S. (1940). *The Present Age and of the Difference Between a Genius and an Apostle* (translated by Alexander Dru). New York: Harper Torchbooks.

Kinder, D. R. (2007). Curmudgeonly Advice. *Journal of Communication,* 57: 155–162.

King, R. R. (1973). *Minorities Under Communism. Nationalities as a Source of Tension among Balkan Communist States.* Cambridge, MA: Harvard University Press.

Kinnvall, C. and Nesbitt-Larking, P. (2011). *The Political Psychology of Globalization: Muslims in the West.* Oxford and New York: Oxford University Press.

Kitromilides, P. (1990). 'Imagined Communities' and the Origins of the National Question in the Balkans". In Blinkhorn, M. and Veremis T. (Eds) *Modern Greece: Nationalism and Nationality.* (23–66). Athens: Sage & ELIAMEP.

Kleres, J. (2009). Preface: Notes on the Sociology of Emotions in Europe. In Hopkins, D., Kleres, J., Flam, H., and Kuzmics, H. (Eds) *Theorizing Emotions: Sociological Explorations and Applications.* (7–27). Frankfurt: Campus Verlag.

Knauft, B. M. (2018). On the Political Genealogy of Trump after Foucault. *Genealogy,* 2(4): 2–18.

Koh, E. (2019). Cultural Work in Addressing Conflicts and Violence in Traumatized Communities. *New England Journal of Public Policy,* 31(1): 1–14.

Kohn, H. (1961). *The Idea of Nationalism. A Study in its Origins and Background.* (2nd edition). New York: The MacMillan Company.

Kolko, G. (1994). *Century of War. Politics, Conflict, and Society since 1914.* New York: The New Press.

Konstan, D. (2011). Before Forgiveness: Classical Antiquity, early Christianity and Beyond. *New England Classical Journal,* 38(2): 91–109.

Konstan, D. (2010). *Before Forgiveness. The Origins of a Moral Idea.* New York and Cambridge: Cambridge University Press.

Koulouri, C. (Ed.) (2002). *Clio in the Balkans. The Politics of History Education.* Thessaloniki: CDRSEE.

Koumandaraki, A. (2002). The Evolution of Greek National Identity. *Studies in Ethnicity and Nationalism,* 2(2): 39–53.

Kövecses, Z. (2000). The Concept of Anger: Universal or Culture Specific? *Psychopathology,* 33(4): 159–170.

Koziak, B. (2000). *Retrieving Political Emotion: Thumos, Aristotle, and Gender.* University Park: The Pennsylvania University Press.

Kyriakidou, M. (2017). Remembering Global Disasters and the Construction of Cosmopolitan Memory. *Communication, Culture & Critique,* 10: 93–111.

Kyriakidou, M. (2015). Media witnessing: exploring the audience of distant suffering. *Media, Culture & Society,* 37(2): 215–231.

Kyriakidou, M. (2014). Distant Suffering in Audience Memory: The Moral Hierarchy of Remembering. *International Journal of Communication,* 8: 1474–1494.

Lacan, J. (1977a). *Ecrits. A Selection.* (translated by A. Sheridan). London: Tavistock/ Routledge.

Lacan, J. (1977b). *The Four Fundamental Concepts of Psycho-Analysis.* (translated by A. Sheridan). London: Penguin Books.

Laclau, E. and Mouffe, C. (1985). *Hegemony and Socialist Strategy.* London: Verso.

Lafleur, M. (2007). Life and Death in the Shadow of the A-Bomb: Sovereignty and Memory on the 60th Anniversary of Hiroshima and Nagasaki. In Carpentier, N. (Ed.) *Culture, Trauma, and Conflict. Cultural Studies Perspectives on War.* (209–228). Newcastle: Cambridge Scholars Publishing.

Lakoff, G. and Johnson, M. (1980). *Metaphors We Live By.* Chicago: University of Chicago Press.

Lamb, R. (1987). Objectless Emotions. *Philosophy and Phenomenological Research,* 48(1): 107–117.

Laplanche, J. and Pontalis, J.-B. (1988). *The Language of Psychoanalysis* (translated by Donald Nicholson-Smith). London: Karnak Books.

Laplanche J., and Pontalis, J.-B. (1986). *Vocabulaire de la Psychanalyse.* Athens: Kedros Publications.

Larocco, S. (2011). Forgiveness: A quiet assault on the malicious. In Karabin, G. and Wigura, K. (Eds) *Forgiveness: Promise, Possibility & Failure.* (3–12). Oxford: Inter-Disciplinary Press.

Lašas, A. (2016). Shadow of Guilt: U.S.-Rwandese Relations after the 1994 Genocide. In Ariffin, Y., Coicaud, J.-M., and Popovski, V. (Eds) *Emotions in International Politics.* (254–276). New York: Cambridge University Press.

Lash, C. (1978). *The Culture of Narcissism. American Life in An Age of Diminishing Expectations.* New York: W.W. Norton & Company Inc.

Lash, S. (1994). Reflexivity and its Doubles: Structure, Aesthetics, Community. In Beck, U., Giddens, A., and Lash, S. *Reflexive Modernization. Politics, Tradition and Aesthetics in the Modern Social Order.* (110–173). Cambridge: Polity Press.

Latif, M., Blee, K., DeMichele, M., and Simi, P. (2018). How Emotional Dynamics Maintain and Destroy White Supremacist Groups. *Humanity & Society,* 42(4): 480–501.

Le Bon, G. (1918). *The Psychology of Revolution* (translated by Bernard Miall). Global Grey (globalgreyebooks.com).

Le Bon, G. (1912). *La Révolution Française et la Psychologie des Révolutions.* (http:// classiques.uqac.ca/classiques/le_bon_gustave/revolution_francaise/revolution_franc. html).

Leigh-Fermor, P. (1966). *Roumeli: Travels in Northern Greece.* London: John Murray.

Leiss, W., Kline, S., and Jhally, S. (1986). *Social Communication in Advertising. Persons, Products, and Images of Well Being.* Toronto: Methuen.

Lekas, P. (2005). The Greek War of Independence from the Perspective of Historical Sociology. *The Historical Review,* 2: 161–183.

León, R., Romer, C., Novara, J., and Quesada, E. (1988). "Una Scala Para Medir del Resentimiento". *Revista Latinoamericana Psicologia,* 20(3): 331–354.

Lev-Wiesel, R. (2007). Intergenerational Transmission of Trauma Across Three Generations. A Preliminary Study. *Qualitative Social Work,* 6(1): 75–94.

Levine, P. D. (2018). *Dark Fantasy. Regressive Movements and the Search for Meaning in Politics.* London: Karnac.

Levy, D. and Sznaider, N. (2002). Memory Unbound. The Holocaust and the Formation of Cosmopolitan Memory. *European Journal of Social Theory*, 5(1): 87–106.

Lévy, P. (1995). *Qu' est-ce que le virtuel?* Paris: Éditions La Découverte.

Leys, R. (2000). *Trauma: A Genealogy*. Baltimore, MD: Johns Hopkins University Press.

Lilla, M. (2016). *The Shipwrecked Mind. On Political Reaction.* New York: New York Review of Books.

Lincoln, A. (1995). First Inaugural Address, March 1861. In Dahbour, O. and Ishay, M. R. (Eds) *The Nationalism Reader.* (286–291). New Jersey: Humanities Press.

Lipovetsky, G. (1992). *Le Grepuscule du Devoir.* Paris: Gallimard.

Lipowatz, Th. (2014). *Die Trügerische Verführung und die Unheimliche Enthüllung des Bösen.* Berlin: Weidler Buchverlag.

Lipset, S. M. (1960). *Political Man: The Social Bases of Politics.* London: Heinemann.

Lipset, S. M. and Rokkan, S. (Eds) (1967). *Party Systems and Voter Alignments: Cross-National Perspectives.* New York: The Free Press.

Listhaug, O. (1995). *The Dynamics of Trust in Politicians.* In Klingemann, H.-D. and Fuchs, D. (Eds) *Citizens and the State.* (261–297). Oxford: Oxford University Press.

Luckhurst, R. (2008). *The Trauma Question.* London: Routledge.

Luostarinen, H. (2002). Journalism and the Cultural Preconditions of War. In Kempf, W. and Luostarinen, H. (Eds) *Journalism and the New World Order. Studying War and the Media*, Vol. 2. (273–83). Gothenburg: NORDICOM, Gothenburg University.

Lyon, D. (1994). *Postmodernity.* London: Open University Press.

Lyotard, J.-F. (1984). *The Postmodern Condition: A Report on Knowledge.* Manchester: Manchester University Press.

Lyrintzis, C. (1993). PASOK in Power: From 'Change' to Disenchantment. In Clogg, R. (Ed.) (1993). *Greece, 1981–89. The Populist Decade.* (26–46). New York: St. Martin's Press.

Lyrintzis, C. (1984). Political Parties in Post-Junta Greece: A case of 'bureaucratic clientelism? In Pridham, G. (Ed.) *The New Mediterranean Democracies. Regime Transition in Spain, Greece and Portugal.* (99–118). London: Frank Cass.

Magni, G. (2017). It's the emotions, Stupid! Anger about the economic crisis, low political efficacy, and support for populist parties. *Electoral Studies*, 50: 91–102.

Makrides, V. (1995). The Orthodox Church and the Post-War Religious Situation in Greece. In Roof, W., Carroll, J., and Roozen, D. (Eds) *The Post-War Generation and Establishment Religion. Cross-Cultural Perspectives.* (225–245). Oxford: Westview Press.

Maldonado, M. A. (2019). Political Affects in the Neuroscientific Age. In Cossarini, P. and Vallespín, F. (Eds) *Populism and Passions. Democratic Legitimacy after Austerity.* (15–30). New York: Routledge.

Mann, R. and Fenton, S. (2017). *Nation, Class and Resentment. The Politics of National Identity in England, Scotland and Wales.* London: Palgrave/MacMillan.

Mazzini, G. (1995). The Duties of Man. In Dahbour, O. and Ishay, M. R. (Eds) *The Nationalism Reader.* (87–97). New Jersey: Humanities Press.

Marantzides, N. (2002). Ethnic Dimensions of the Civil War. In Nikolakopoulos, E., Rigos, A., and Psallidas, G. (Eds) *The Civil War. From Varkiza to Grammos. February 1945–August 1949.* (208–221). Athens: Themelio.

Marantzides, N. and Antoniou, G. (2004). The Axis Occupation and Civil War: Changing Trends in Greek Historiography, 1941–2002. *Journal of Peace Research*, 41(2): 223–231.

Marcus, G. E. (2002). *The Sentimental Citizen. Emotion in Democratic Politics.* University Park, PA: Pennsylvania State University Press.

Marcus, G. E., Neuman, R. W. and MacKuen, M. B. (2000). *Affective Intelligence and Political Judgment.* Chicago: University of Chicago Press.

Margarites, G. (1989). Internecine conflicts in Occupation (1941–1944): Analogies and differences. In Fleischer, H. and Svoronos, N. (Eds) *Greece 1936–1944. Dictatorship, Occupation, Resistance.* (505–515). Athens: Cultural Institute of the Agricultural Bank of Greece.

Martin, E. (2013). The Potentiality of Ethnography and the Limits of Affect Theory. *Current Anthropology*, 54: S149–S158.

Marx, K. (1844/1978). *Critique to the Hegelian Philosophy of the State and Right.* Athens: Papazissis.

Massumi, B. (2015). *Politics of Affect.* Cambridge: Polity Press.

Massumi, B. (2002). *Parables for the Virtual; Movement, Affect, Sensation.* Durham, NC: Duke University Press.

Massumi, B. (1995). The Autonomy of Affect. *Cultural Critique*, 31: 83–109.

Mateos, S. and Laiz, Á.-M. (2018). International Relations and Political Sociology. In Outhwaite, W. and Turner, S. P. (Eds) *The Sage Handbook of Political Sociology.* (Vol. 1) (172–188). London: Sage Publications.

Maurras, C. (1995). The Future of French Nationalism. In Dahbour, O. and Ishay, M. R. (Eds) *The Nationalism Reader.* (216–221). New Jersey: Humanities Press.

Mavrogordatos, G. (1983). *Stillborn Republic. Social Coalitions and Party Strategies in Greece, 1922–1936.* Berkeley: University of California Press.

Mazower, M. (2018). *What You Did Not Tell. A Father's Past and a Journey Home.* London: Penguin Books.

Mazower, M. (2000). Three Forms of Political Justice: Greece, 1944–1945. In Mazower, M. (Ed.) *After the War Was Over. Reconstructing the Family, Nation, and State in Greece, 1943–1960.* (24–41). Princeton and Oxford: Princeton University Press.

Mazower, M. (1993). *Inside Hitler's Greece.* New Haven: Yale University Press.

McGuigan, J. (1992). *Cultural Populism.* London: Routledge.

Mead, G. H. (1934). *Mind, Self, and Society. From the Standpoint of a Social Behaviorist.* Chicago, London: The University of Chicago Press.

Meek, A. (2010). *Trauma and Media. Theories, Histories, and Images.* New York: Routledge.

Meinecke, F. (1908). *Weltbuergertum und Nationalstaat.* Leipzing: Koehler & Amelang.

Meltzer, B. and Musolf, G.-R. (2002). Resentment and Ressentiment. *Sociological Inquiry*. 72(2): 240–255.

Merton, R. (1957/1968). *Social Theory and Social Structure.* New York: The Free Press.

Meštrović, S. G. (1997). *Postemotional Society.* London: Sage Publications.

Milbrath, L. and Goel, M. (1977). *Political Participation.* New York: Univ. Press of America.

Milka, A. and Warfield, A. (2017). *News Reporting* and *Emotions Part 2: Reporting Disaster: Emotions, Trauma* and *Media Ethics.* https://historiesofemotion.com.

Miller, N. and Tougaw, J. (Eds) (2002). *Extremities: Trauma, Testimony and Community.* Urbana: University of Illinois Press.

Mills, C. W. (1959/2000). *The Sociological Imagination* (with a new Afterword by Todd Gitlin). New York: Oxford University Press.

Minkenberg, M. (2000). The Renewal of the Radical Right: Between Modernity and Antimodernity. *Government and Opposition*, 35: 170–188. DOI: 10.1111/1477-7053.00022.

Minogue, K. (1969). Populism as a Political Movement. In Ionescu, G. and Gellner, E. (Eds) *Populism. Its Meanings and National Characteristics*. (197–211). Hertfordshire: The Garden City Press.

Minow, M. (1998). *Between Vengeance and Forgiveness. Facing History after Genocide and Mass Violence*. Boston: Beacon Press.

Mishra, P. (2017). *Age of Anger: A History of the Present*. New York: Farrar, Straus and Giroux.

Mitter, R. (2017). *80 years later: can China, Japan overcome Nanking Massacre's legacy?* www.scmp.com/week-asia/opinion/article/2123261/80-years-later-can-china-japan-overcome-nanking-massacres-legacy.

Moeller, S. D. (1999). *Compassion Fatigue: How the Media Sell Disease, Famine, War and Death*. New York and London: Routledge.

Moffitt, B. (2016). *The Global Rise of Populism: Performance, Political Style, and Representation*. Stanford, California: Stanford University Press.

Moore, B. (1984). *Privacy. Studies in Social and Cultural History*. New York: M.E. Sharpe, Inc.

Moore, B. (1978). *Injustice. The Social Bases of Obedience and Revolt*. New York: Macmillan.

Moruno, D. M. (2013). Introduction. On Resentment: Past and Present of an Emotion. In Fantini, B., Moruno, D. M. and Moscoso, J. (Eds) *On Resentment: Past and Present*. (1–18). Newcastle: Cambridge Scholars Publishing.

Moulier-Boutang, Y. (2012). *Cognitive Capitalism*. Cambridge: Polity Press.

Mounk, Y. (2018). *The People vs. Democracy: Why our Freedom is in Danger and how to Save it*. Cambridge, Massachusetts: Harvard University Press.

Mourselas, K. (1996). *Red Dyed Hair* (translated by Fred A. Reed). Athens: Kedros Publications.

Mouzelis, N. (2008). *Modern and Postmodern Social Theorizing. Bridging the Divide*. Cambridge: Cambridge University Press.

Mouzelis, N. (1995). Greece in the Twenty-first Century: Institutions and Political Culture. In Constas, D. and Stavrou, G. T. (Eds) *Greece Prepares for the Twenty-first Century*. (17–34). Baltimore and London: The John Hopkins University Press.

Mouzelis, N. (1986). *Politics in the Semi-Periphery. Early Parliamentarism and Late Industrialism in the Balkans and Latin America*. London: Macmillan.

Mouzelis, N. (1985). On the Concept of Populism: Populist and Clientelist Models of Incorporation in Semiperipheral Polities. *Politics and Society*, 14(3): 329–348.

Mudde, C. (2017). Populism. An Ideational Approach. In Kaltwasser, C. R., Taggart, P., Espejo, P. O., and Ostiguy, P. (Eds) *The Oxford Handbook of Populism*. (27–47). Oxford: Oxford University Press.

Mudde, C. (2015). *SYRIZA. The Refutation of the Populist Pledge*. Athens: Epikentro Publications.

Mudde, C. (2004). The populist Zeitgeist. *Government and Opposition*, 39(4): 542–563.

Mudde, C. and Rovira-Kaltwasser, R. C. (2017). *Populism: A Very Short Introduction*. Oxford: Oxford University Press.

Müller, J.-W. (2016). *What is Populism?* Philadelphia: University of Pennsylvania Press.

Murphy, G. J. (2005). Forgiveness, Self-Respect, and the Value of Resentment. In Worthington, L. E. (Ed.) *Handbook of Forgiveness*. (33–40). New York: Routledge.

Murphy, G. J. (1998). Forgiveness and Resentment. In Murphy, G. J. and Hampton, J. (Eds) *Forgiveness and Mercy*. (14–34). Cambridge: Cambridge University Press.

Mylonas, H. (2003). The Comparative Method and the Study of Civil Wars. *Science and Society*, 11: 1–35. DOI: http://dx.doi.org/10.12681/sas.915.

Nabi, R. L. (2003). Exploring the Framing Effects of Emotion. *Communication Research*, 30(2): 224–247.

Nabi, R. L. (2002). Discrete emotions and persuasion. In Dillard, J. and Pfau, M. (Eds) *Handbook of Persuasion*. (289–308). Thousand Oaks, CA: Sage.

Nabi, R. L. (1999). A Cognitive-Functional Model for the Effects of Discrete Negative Emotions on Information Processing, Attitude Change, and Recall. *Communication Theory*, 9: 292–320.

Nabi, R. L., Jiyeon So, J., and Prestin, A. (2011). Media-based emotional coping. Examining the emotional benefits and pitfalls of media consumption. In Döveling, K., von Scheve, C. and Konijn, E. A. (Eds) *The Routledge Handbook of Emotions and Mass Media*. (116–132). London: Routledge.

Nadal, M. M. and Calvo, M. (Eds) (2014). *Trauma in Contemporary Literature. Narrative and Representation*. New York: Routledge.

Nash, K. (Ed.) (2000). *Readings in Contemporary Political Sociology*. Oxford: Blackwell Publishers.

Nash, K. and Scott, A. (Eds) (2001). *The Blackwell Companion to Political Sociology*. Oxford: Blackwell Publishing.

Natanson, M. (1978). Phenomenology as a Rigorous Science. In Luckmann, T. (Ed.) *Phenomenology and Sociology. Selected Readings*. (181–199). Middlesex: Penguin Books.

Neil, A. G. (1998). *National Trauma and Collective Memory: Major Events in the American Century*. New York: M. E. Sharpe.

Nesbitt-Larking, P., Kinnvall, C., Capelos, T., and Dekker, H. (Eds) (2014). *The Palgrave Handbook of Global Political Psychology*. London: Palgrave/Macmillan.

Nietzsche, F. (1895/2014). *The Antichrist*. Athens: Gutenberg.

Nietzsche, F. (1994/1887). *On the Genealogy of Morality* (translated by C. Deithe). Cambridge: Cambridge University Press.

Nobles, M. (2008). *The Politics of Official Apologies*. Cambridge, U.K.: Cambridge University Press.

Norris, P. and Inglehart, F. R. (2019). *Cultural Backlash: Trump, Brexit, and the rise of Authoritarian-populism*. New York: Cambridge University Press.

Novello, S. (2015). Max Scheler's *ordo amoris* in Albert Camus's philosophical work. *Thaumazein*, 3: 199–216.

Nussbaum, M. (2016). *Anger and Forgiveness. Resentment, Generosity, Justice*. Oxford: Oxford University Press.

Nussbaum, M. C. (2013). *Political Emotions. Why Love Matters for Justice*. Cambridge MA: The Belknap Press of Harvard University Press.

Nussbaum, M. C. (2006). Radical evil in the Lockean state: The neglect of the political emotions. *Journal of Moral Philosophy*, 3(2): 159–178.

Nussbaum, M. C. (2001). *Upheavals of Thought. The Intelligence of Emotion*. Cambridge: Cambridge University Press.

Offe. C. (1999). How can we trust our fellow citizens? In Warren, M. E. (Ed.) *Democracy & Trust*. (42–87). Cambridge: Cambridge University Press.

Offe, C. (1985). New Social Movements: Challenging the Boundaries of Institutional Politics. *Social Research*, 52(4): 817–868.

O'Keefe, J. D. (2000). Guilt and Social Influence. In Roloff, M. E. (Ed.) *Communication Yearbook 23*. (67–101). London: Sage Publications.

Olick, J. (2007). *The Politics of Regret: On Collective Memory and Historical Responsibility*. New York: Routledge.

Orum, A. M. and Dale, J. G. (2009). *Introduction to Political Sociology. Power and Participation in the Modern World* (5th Edition), New York and Oxford: Oxford University Press.

Ost, D. (2004). 'Politics as the Mobilization of Anger'. *European Journal of Social Theory*, 7(2): 229–244.

Ostiguy, P. (2017). Populism. A Socio-Cultural Approach. In Kaltwasser, C. R., Taggart, P., Espejo, P. O., and Ostiguy, P. (Eds) *The Oxford Handbook of Populism*. (73–97). Oxford: Oxford University Press.

Otto, L. P., Fabian, T., Maier, M., and Ottenstein, C. (2019). Only One Moment in Time? Investigating the Dynamic Relationship of Emotions and Attention Toward Political Information With Mobile Experience Sampling. *Communication Research*, 1–24. https://doi.org/10.1177/0093650219872392.

Outhwaite, W. and Turner, S. P. (Eds) (2018). *The SAGE Handbook of Political Sociology*. London: Sage Publications.

Oxford Advanced Learner's Dictionary of Current English (4th edition by A. P. Cowie, Oxford University Press, 1990).

Oxford Advanced Learner's Dictionary of Current English (A. S. Hornby, Oxford University Press, 1974).

Özkirimli, U. and Sofos, S. (2008). *Tormented by History. Nationalism in Greece and Turkey*. London: Hurst & Company.

Panagiotopoulos, P. (1994). The representation of suffering in the narration of two extreme experiences: Giorgis Pikros, *Makronissos Chronicle*, Primo Levi, *Si c'est un home*. *Dokimes*, 2: 13–32.

Panagiotopoulos, P. and Vamvakas, V. (2014). Acrobats on a Rope. Greek Society between European Demands and Archaic Cultural Reflexes. In Temel, B. (Ed.) *The Great Catalyst. European Union Project and Lessons from Greece and Turkey*. (113–134). Plymouth: Lexington Books.

Pantazopoulos, A. (2001). *For the People and the Nation. The Andreas Papandreou Moment 1965–1989*. Athens: Polis editions.

Papacharissi, Z. (2015). *Affective Publics: Sentiment, Technology, and Politics*. New York: Oxford University Press.

Papacosma, V. S. (1988). *Politics and Culture in Greece*. Ann Arbor: University of Michigan Press.

Papandreou, A. (1969). *Democracy at Gunpoint: The Greek Front*, New York: Doubleday.

Pappas T. S. (2016). Modern Populism: Research Advances, Conceptual and Methodological Pitfalls, and the Minimal Definition. *Oxford Research Encyclopedia of Politics*. DOI: 10.1093/acrefore/9780190228637.013.17.

Petersen, R. D. (2002). *Understanding Ethnic Violence: Fear, Hatred, and Resentment in Twentieth-Century Eastern Europe*. Cambridge: Cambridge University Press.

Peterson, J. B. (2018). *12 Rules for Life. An Antidote to Chaos*. London: Penguin Random House.

Petropoulos, E. (2000). *Songs of The Greek Underworld: The Rebetika Tradition*. London: Saqe Books.

Petropoulos, J. A. (1968). *Politics and Statecraft in the Kingdom of Greece, 1833–1843*. Princeton: Princeton University Press.

Pirc, T. (2018). The collective of *Ressentiment*: Psychopolitics, Mediocrity, and Morality. In Pirc, T. (Ed.) *Participation, Culture and Democracy. Perspective on Public Engagement and Social Communication*. (111–132). Newcastle: Cambridge Scholars Publishing.

Planalp, S., Hafen, S., and Adkins, A. D. (2000). Messages of Shame and Guilt. In Roloff, M. E. (Ed.) *Communication Yearbook 23*. (1–65). London: Sage Publications.

Portman, J. (2000). *When Bad Things Happen to Other People*. London: Routledge.

Posner, R. A. (2004). *Catastrophe: Risk and Response*. Oxford: Oxford University Press.

Postman, N. (1985). *Amusing Ourselves to Death. Public Discourse in the Age of Show Business*. London: Penguin Books.

Posłuszna, E. and Posłuszny, J. (2015). The Trouble with *Ressentiment*. *Ruch Filozoficzny*, 71(4): 87–104. DOI: http://dx.doi.org/10.12775/RF.2015.006.

Potamitis, N. Y. (2008). Antagonism and Genre. *Resistance, the Costume Romance and the Ghost of Greek Communism*. In Carpentier, N. and Spinoy E. (Eds) *Discourse Theory and Cultural Analysis Media, Arts and Literature*. (119–139). NJ: Hampton Press.

Protevi, J. (2014). Political Emotion. In von Scheve, C. and Salmela, M. (Eds) *Collective Emotions. Perspectives from Psychology, Philosophy, and Sociology*. (326–340). Oxford: Oxford University Press.

Protevi, J. (2009). *Political Affect. Connecting the Social and the Somatic*. Minneapolis: University of Minnesota Press.

Pupavac, V. (2007). *Therapeutic Governance: The Politics of Psychosocial Intervention and Trauma Risk Management* www.odi.org.uk/hpg/confpapers/pupavac.

Rawls, J. (1971/1991). *A Theory of Justice*. Oxford: Oxford University Press.

Rayburn J. D. II. (1996). Uses and Gratifications. In Salwen, M. and Stacks, D. (Eds) *An Integrated Approach to Communication Theory and Research*. (145–163). New Jersey: Lawrence Erlbaum Associates.

Reimer, D. J. (2007). Stories of Forgiveness: Narrative Ethics and the Old Testament. In Rezetko, R., Lim, T. H., and Auker, W. B. (Eds) *Reflection and Refraction. Studies in Biblical Historiography in Honour of A. Graeme Auld*. (Vol. 113) (359–378). Leiden and Boston: Brill. Supplements to VT.

Reinert, H. and Reinert, E. S. (2006). Creative Destruction in Economics: Nietzsche, Sombart, Schumpeter. In Backhaus, J. R. and Drechsler, W. (Eds) *Friedrich Nietzsche 1844–2000: Economy and Society*. (55–85). Boston: Springer/Kluwer.

Renan, E. (1995). What is a nation? In Dahbour, O. and Ishay, M. R. (Eds) *The Nationalism Reader*. (143–155). New Jersey: Humanities Press.

Revault d' Allonnnes, M. (2008). *L' Homme Compassionnel*. Paris: Éditions du Seuil.

Richards, B. (2019). *The Psychology of Politics*. New York: Routledge.

Richards, B. (2018). *What Holds Us Together. Popular Culture and Social Cohesion*. London: Karnac.

Richards, B. (2010). News and the emotional public sphere. In Allan, S. (Ed.) *The Routledge Companion to News and Journalism*. (301–311). London, UK: Routledge.

Richards, B. (2007). *Emotional Governance: Politics, Media and Terror*. Basingstoke: Palgrave Macmillan.

Richter, H. (1997). The *British Intervention in Greece. From Varkiza to the Civil War. February 1945-August 1946*. Athens: Estia.

Robinson, E. (2016). Radical Nostalgia, Progressive Patriotism and Labour's 'English Problem'. *Political Studies Review*, 14(3): 378–387.

Rooduijn, M. (2014). The Nucleus of Populism: In Search of the Lowest Common Denominator. *Government and Opposition*, 49(4): 572–598. DOI: 10.1017/gov.2013.30.

Rosas, O. V. and Serrano-Puche, J. (2018). News Media and the Emotional Public Sphere. *International Journal of Communication*, 12: 2031–2039.

Rico, G., Guinjoan, M., and Anduiza, E. (2017). The Emotional Underpinnings of Populism: How Anger and Fear Affect Populist Attitudes. *Swiss Political Science Review*. DOI: 10.1111/spsr.12261.

Ricoeur, P. (2004a). *Le Mal: Undéfit à la Théologie* (3rd edition). Geneva: Laboret Fides.

Ricoeur, P. (2004b). *Memory, History, Forgetting*. Chicago: University of Chicago Press.

Riggs, F. (1964). *Administration in Developing Countries: The Theory of Prismatic Society*. Boston: Houghton Mifflin Company.

Robin, C. (2004). *Fear. The History of a Political Idea*. Oxford: Oxford University Press.

Robins, K. (2001). Seeing the world from a safe distance. *Science as Culture*, 10(4): 531–539.

Rorty, R. (1989). *Contingency, Irony, and Solidarity*. Cambridge: Cambridge University Press.

Rosen, J. (2002). September 11 in the Mind of American Journalism. In Zelizer, B. and Allan, S. (Eds) *Journalism after September 11*. (27–35). London: Routledge.

Rosenau, P. M. (1992). *Post-Modernism and the Social Sciences. Insights, Inroads, and Intrusions*. Princeton/New Jersey: Princeton University Press.

Rosenwein, B. (2001). Writing Without Fear About Early Medieval Emotions. *Early Medieval Europe*, 10(2): 229–234.

Roth, M. S. (2012). *Memory, Trauma, and History. Essays on Living with the Past*. New York: Columbia University Press.

Roudometof, V. and Christou, M. (2013). 1974 and Greek Cypriot Identity: The Division of Cyprus as Cultural Trauma. In Eyerman, R., Alexander, J. C., and Breese, E. (Eds) *Narrating Trauma. On the Impact of Collective Suffering*. (163–187). Boulder & London: Paradigm Publications.

Rousso, H. (1991). *The Vichy Syndrome. History and Memory in France since 1944*. Cambridge: Harvard University Press.

Rovira-Kaltwasser, C., Taggart, P., Espejo, P. O., and Ostiguy, P. (2017). Populism. An Over view of the Concept and the State of the Art. In Rovira-Kaltwasser, C., Taggart, P., Espejo, P. O., and Ostiguy, P. (Eds) *The Oxford Handbook of Populism*. (1–24). Oxford: Oxford University Press.

Sadler, G. (2008). Forgiveness, anger, and virtue in an Aristotelean perspective. *Proceedings of the American Catholic Philosophical Association*, 82: 229–247.

Salmela, M. and von Scheve, C. (2018). Emotional Dynamics of Right- and Left-Wing Political Populism. *Humanity & Society*, 42(4): 434–454.

Salmela, M. and von Scheve C. (2017). Emotional roots of right-wing political populism. *Social Science Information*, 1–29. DOI: https://doi.org/10.1177/0539018417734419.

Sambanis, N. (2002). A Review of Recent Advances and Future Directions in the Literature on Civil War. *Defense and Peace Economics*, 13(2): 215–243.

Sarat, I. A. and Hussain, N. (2007). *Forgiveness, Mercy, and Clemency*. Stanford: Stanford University Press.

Sayer, A. (2005). *The Moral Significance of Class*. Cambridge: Cambridge University Press.

Scheff, T. J. (2000). *Bloody Revenge: Emotions, Nationalism and War*. Bloomington, IN: iUniverse.com, Inc.

Scheff, T. J. (1997a). *Emotions, the Social Bond, and Human Reality: Part/Whole Analysis*. Cambridge: Cambridge University Press.

Scheff, T. J. (1997b). *Alienation, Nationalism, and Inter-ethnic Conflict*. www.soc.ucsb.edu/faculty/scheff/main.php?id=5.html.

Scheff, T. J. (1994a). Emotions and Identity: A Theory of Ethnic Nationalism. In Calhoun, C. (Ed.) *Social Theory and the Politics of Identity*. (277–303). Oxford: Blackwell Publishers

Scheff, T. J. (1994b). *Bloody Revenge. Emotions, Nationalism and War*. Boulder: Westview Press.

Scheler, M. (1980). *Problems of a Sociology of Knowledge*. London: Routledge and Kegan Paul.

Scheler, M. (1961). *Ressentiment*. Glencoe: Free Press.

Scheler, M. (1954). *The Nature of Sympathy*. London: Routledge & Kegan Paul Ltd.

Scherer, K. R. (2009). The dynamic architecture of emotion: Evidence for the component process model. *Cognition & Emotion*, 23(7): 1307–1351.

Scherer, K. R. (2001). Appraisal considered as a process of multi-level sequential checking. In Scherer, K. R., Schorr, A., and Johnstone, T. (Eds) *Appraisal Processes in Emotion: Theory, Methods, Research*. (92–120). New York: Oxford University Press.

Scherke, K. (2018). Nostalgie und Politik. Eine emotionssoziologische Perspektive. *Zeitschrift für Politik*, 65(1): 81–96.

Scherke, K. (2015). Emotion: History of the Concept. In Wright, J. D. (Ed.) *International Encyclopedia of the Social & Behavioral Sciences*. (472–476). Vol. 7. Oxford: Elsevier (2nd edition).

Schieman, S. (2006). Anger. In Stets, J. and Turner, J. (Eds) *Handbook of the Sociology of Emotions*. (493–515). New York: Springer.

Schmitt, C. and Clark, C. (2006). Sympathy. In Stets, I. E. and Turner, J. H. (Eds), *Handbook of the Sociology of Emotions*. (467–492). New York: Springer.

Schoeck, H. (1969). *Envy: A Theory of Social Behavior*, (translated by Glenny and Ross). New York: Harcourt, Brace.

Schramm, H. and Wirth, W. (2008). A case for an integrative view on affect regulation through media usage. *Communications*, 33: 27–46. DOI: 10.1515/COMMUN.2008.002.

Schreiber, D. (2007). Political Cognition: Are We All Political Sophisticates? In Neuman, R. et al. (Eds) *The Affect Effect. Dynamics of Emotion in Political Thinking and Behavior*. (46–70). Chicago: The University of Chicago Press.

Schudson, M. (1997). The Sociology of News Production. In Berkowitz, D. (Ed.) *Social Meanings of New; A Text-Reader*. (7–22). London: Sage Publications.

Schutz, A. (1978). Phenomenology and the Social Sciences. In Luckmann, T. (Ed.) *Phenomenology and Sociology. Selected Readings*. (119–141). Middlesex: Penguin Books.

Schwab, F. and Schwender, C. (2011). The descent of emotions in media. Darwinian perspectives. In Döveling, K., von Scheve, C., and Konijn, E. A. (Eds) *The Routledge Handbook of Emotions and Mass Media*. (16–36). London: Routledge.

Schwartz, B. (1978). Vengeance and forgiveness: The uses of beneficence in social control. *School Review*, 86: 655–668.

Scott, A. (2000). Risk Society or Angst Society? Two Views of Risk, Consciousness and Community. In Adam, B. Beck, U., and Loon, J. V. (Eds) *The Risk Society and Beyond. Critical Issues for Social Theory*. (33–46). London: Sage Publications.

Scott, M. (2014). The mediation of distant suffering: An empirical contribution beyond television news texts. *Media, Culture & Society*, 36(1): 3–19.

Scouras, F., Chatjidemos, A., Kaloutsis, A., and Papademetriou, G. (1947). *The Psychopathology of Hunger, Fear and Anxiety. A Medical Chronicle of Occupation*. Athens: Exandas.

Scruton, R. (2006). *A Political Philosophy*. London/New York: Continuum.

Searle J. R. (1995). *The Construction of Social Reality*. London: Allen Lane, The Penguin Press.

Sears, D. (2001). The Role of Affect in Symbolic Politics. In Kuklinski, J. H. (Ed.) *Citizens and Politics*. (14–40). Cambridge: Cambridge University Press.

Sears, D. O., Huddy, L., and Jervis, R. (Eds) (2003). *Oxford Handbook of Political Psychology*. Oxford: Oxford University Press.

Sennett, R. (1977). *The Fall of Public Man*. London: Faber and Faber.

Sennett, R. and Cobb, J. (1972). *The Hidden Injuries of Class*. New York: W.W. Norton & Company.

Sewell, W. H. Jr. (1996). Historical events as transformations of structures: Inventing revolution at the Bastille. *Theory and Society* 25: 841–881.

Shils, E. (1956). *The Torment of Secrecy: the Background and Consequences of American Security Policies*. London: Heinemann.

Shoah Resource Center (1998). An Interview with Professor Jacques Derrida. January 8, 1998, Jerusalem, Interviewer: Dr. Michal Ben-Naftali (translated from French by Dr. Moshe Ron). The International School for Holocaust Studies. (www.yadvashem.org).

Shriver, D. W. (1998). *An Ethic for Enemies: Forgiveness in Politics*. Oxford: Oxford University Press.

Silverstone, R. (2007). *Media and Morality. On the Rise of Mediapolis*. London: Polity Press.

Skillington, T. (2015). Violence, memory, time: Towards a cosmopolitan model of learning from atrocity. In Marinopoulou, A. (Ed.) *Cosmopolitan Modernity*. (177–201). Oxford: Peter Lang.

Slaby, J. and Bens, J. (2019). Political Affect. In Slaby, J. and von Scheve, C. (Eds) *Affective Societies. Key Concepts*. (340–351). London and New York: Routledge.

Slaby, J. and Mühlhoff, R. (2019). Affect. In Slaby, J. and von Scheve, C. (Eds) *Affective Societies. Key Concepts*. (27–41). London and New York: Routledge.

Slaby, J. and von Scheve, C. (Eds) (2019a). *Affective Societies. Key Concepts*. London and New York: Routledge.

Slaby, J. and von Scheve, C. (2019b). Introduction. Affective Societies – Key Concepts. In Slaby, J. and von Scheve, C. (Eds) *Affective Societies. Key Concepts*. (1–24). London and New York: Routledge.

Sloterdijk, P. (2010). *Rage and Time. A Psychopolitical Investigation* (translated by Mario Wenning). New York: Columbia University Press.

Sloterdijk, P. (2006). *Zorn und Zeit: Politisch-psychologischer Versuch*. Frankfurt am Main: Suhrkamp Verlag.

Sloterdijk, P. (1988). *Critique of Cynical Reason*. London: Verso.

Smelser, N. (2004). Psychological trauma and cultural trauma. In Alexander J. C. *et al.* (Eds) *Cultural Trauma and Collective Identity*. (31–59). Berkeley: University of California Press.

Smith, A. (1976). *The Theory of Moral Sentiments*. Indianapolis: Liberty Press. [First published in 1759].

Smith, A. (1986). *The Ethnic Origins of Nations*. London: Blackwell.

Smith, A. (1971). *Theories of Nationalism*. Duckworth: London.

Smith, D. (2006). *Globalization. The Hidden Agenda*. Cambridge: Polity.

Snyder L. (1954). *The Meaning of Nationalism*. Connecticut: Greenwood Press (2nd reprint 1972).

Sokolon, M. (2006). *Political Emotions. Aristotle and the Symphony of Reason and Emotion*. Illinois: Northern Illinois University Press.

Soler, C. (2016). *Lacanian Affects: The Function of Affect in Lacan's Work* (translated by B. Fink). New York: Routledge.

Solomon, R. (1994). One Hundred Years of *Ressentiment*. Nietzsche's *Genealogy of Morals*. In Schacht, R. (Ed.) *Nietzsche, Genealogy, Morality*. (95–126). Berkeley: University of California Press.

Sombart, W. (1998). *The Quintessence of Capitalism: A Study of the History and Psychology of the Modern Business Man.* London: Routledge.

Sombart, W. (1998). *Ο Αστός. Πνευματικές Προϋποθέσεις και Ιστορική Πορεία του Δυτικού Καπιταλισμού* (μετ. Κ. Κουτσουρέλης). Αθήνα: εκδ. Νεφέλη.

Sontag, S. (2003). *Regarding the Pain of Others.* New York: Farrar, Strauss and Giroux.

Sotiropoulos, P. D. (2019). Historical Patterns of Greek Exoticism (Nineteenth-Twentieth Century). In Panagiotopoulos, P. and Sotiropoulos, P. D. (Eds) *Political and Cultural Aspects of Greek Exoticism.* (11–25). Cham: Palgrave/Macmillan.

Spasić I. (2013). The Trauma of Kosovo in Serbian National Narratives. In Eyerman, R. Alexander, J. C. and Breese, E. (Eds) *Narrating Trauma. On the Impact of Collective Suffering.* (81–105). Boulder & London: Paradigm Publications.

Stavrakakis, Y. (Ed.) (2019). *Routledge Handbook of Psychoanalytic Political Theory.* London: Routledge.

Stavrakakis, Y. (2017a). Populism and Hegemony. In Kaltwasser, C. R., Taggart, P., Espejo, P. O., and Ostiguy, P. (Eds) *The Oxford Handbook of Populism.* (535–553). Oxford: Oxford University Press.

Stavrakakis, Y. (2017b). Discourse theory in populism research. Three challenges and a dilemma. *Journal of Language and Politics* 1–12. DOI: 10.1075/jlp.17025.sta.

Stavrakakis, Y. (1999). *Lacan and the Political.* London: Routledge.

Stavrakakis, Y. Andreadis, I., and Katsambekis, G. (2016). A new populism index at work: Identifying populist candidates and parties in the contemporary Greek context. *European Politics and Society.* DOI: 10.1080.23745118.2016.1261434.

Stavrou, G. T. (1995). The Orthodox Church and Political Culture in Modern Greece. In Constas, D. and Stavrou, G. T. (Eds) *Greece Prepares for the Twenty-first Century.* (35–56). Baltimore and London: The John Hopkins University Press.

Stern, B. B. (1992). Historical and personal nostalgia in advertising text: The fin de siècle effect, *Journal of Advertising*, 21: 11–22.

Stewart, A. (1969). The Social Roots. In Ionescu, G. and Gellner, E. (Eds) *Populism. Its Meanings and National Characteristics.* (180–196). Hertfordshire: The Garden City Press.

Stivers, R. (1994). *The Culture of Cynicism. American Morality in Decline.* Oxford: Blackwell.

Stevenson, N. (1997). Media, Ethics and Morality. In: McGuigan, J. (Ed.) *Cultural Methodologies.* (62–86). London: Sage Publications Ltd.

Strawson, P. F. (1974). *Freedom and Resentment and Other Essays.* London: Methuen & Co.

Sturzo, D. L. (1946). *Nationalism and Internationalism.* New York: Roy.

Sugar, P. F. (1969). External and Domestic Roots of Eastern European Nationalism. In Sugar, P. F. and Lederer, I. (Eds) *Nationalism in Eastern Europe.* (3–54). Seattle and London: University of Washington Press.

Sullivan, B. G. (2014a). Collective pride, happiness, and celebratory emotions: aggregative, network, and cultural models. In von Scheve, Ch. and Salmela, M. (Eds) *Collective Emotions. Perspectives from Psychology, Philosophy, and Sociology.* (266–280). Oxford: Oxford University Press.

Sullivan, B. G. (Ed.) (2014b). *Understanding Collective Pride and Group Identity. New Directions in Emotion Theory, Research and Practice.* London and New York: Routledge.

Sznaider, N. (1998). The Sociology of Compassion: A Study in the Sociology of Morals. *Cultural Values*, 2(1): 117–39.

Sztompka, P. (2004). The Trauma of Social Change. A Case of Postcommunist Societies. In Alexander J. C., Eyerman, R., Giesen, B., Smelser, J. N., and Sztompka, P. (Eds) *Cultural Trauma and Collective Identity*. (155–195). Berkeley: University of California Press.

Taggart, P. (2000). *Populism*. Buckingham. Open University Press.

Taguieff, P.-A. (2007). *L' Illusion Populiste. De l' Archaïque au Mediatique*. Paris: Flammarion.

Tarchi, M. (2016). Populism: ideology, political style, mentality? *Czech Journal of Political Science*, 23: 95–109.

Tavuchis, N. (1991). *Mea Culpa: A Sociology of Apology and Reconciliation*. Stanford, CA: Stanford University Press.

Taylor, C. (1992). *Multiculturalism and 'the Politics of Recognition'. An Essay*. Princeton, NJ: Princeton University Press.

Taylor, G. (2010). *The New Political Sociology. Power, Ideology and Identity in an Age of Complexity*. Basingstoke: Palgrave Macmillan.

Tenenboim-Weinblatt, K. (2008). 'We will get through this together': Journalism, trauma and the Israeli disengagement from the Gaza Strip. *Media, Culture and Society*, 30(4): 495–513.

TenHouten, W. D. (2018). From *Ressentiment* to Resentment as a Tertiary Emotion. *Review of European Studies*, 10(4): 1–16.

TenHouten, W. D. (2007). *A General Theory of Emotions and Social Life*. London & New York: Routledge.

Tester, K. (2001). *Compassion, Morality and the Media*. Buckingham: Open University Press.

Tester, K. (1997). *Moral Culture*. London, Thousand Oaks, California: Sage.

The Concise Oxford Dictionary of Current English (8th edition by R. E. Allen, Clarendon Press, Oxford, 1990).

Thoits, P. A. (1989). The Sociology of Emotions. *Annual Review of Sociology*, 15: 317–342.

Thompson, B. J. (1995). *The Media and Modernity. A Social Theory of the Media*. Cambridge: Polity Press.

Thompson, S. and Hoggett, P. (Eds) (2012). *Politics and the Emotions. The Affective Turn in Contemporary Political Studies*. New York and London: Continuum International Publishing Group.

Ting-Toomey, S. (1994). Managing intercultural conflicts effectively. In Samovar, L. and Porter, R. (Eds) *Intercultural Communication: A Reader*. (360–372). Wadsworth, Belmont, CA.

Tivey, L. (1981) (Ed.) *The Nation-State. The Formation of Modern Politics*. Oxford: Martin Robertson.

Tocqueville, A. de (1969). *Democracy in America*. New York: Anchor Books.

Todorov, Tzv. (2010). *The Fear of Barbarians. Beyond the Clash of Civilizations*. Cambridge: Polity Press.

Todorov, Tzv. (1995). *Les Abus de la Mémoire*. Paris: Arléa.

Tollefsen, D. (2006). The rationality of collective guilt. *Midwest Studies in Philosophy*. 30: 222–239.

Tomelleri, S. (2013). The sociology of resentment. In Fantini, B., Moruno, D. M. and Moscoso, J. (Eds) *On Resentment: Past and Present*. (259–276). Newcastle: Cambridge Scholars Publishing.

Trevor-Roper, H. (1962). *Jewish and other Nationalisms Fifth Herbert Samuel Lecture, 2 Oct. 1961*. London: Weidenfeld and Nicolson.

Triandafyllidou, A., Gropas, R., and Kouki, H. (2013). Introduction: Is Greece a Modern European Country? In Triandafyllidou, A., Gropas, R., and Kouki, H. (Eds) *The Greek Crisis and European Modernity*. (1–24). Basingstoke: Palgrave Macmillan.

Tsoukalas, C. (1969). *The Greek Tragedy*. London: Harmondsworth, Penguin.

Turner, J. H. (1975). Marx and Simmel Revisited: Reassessing the Foundations of Conflict. *Social Forces*, 53(4): 618–627.

Turner, J. H. and Stets, J. E. (2005). *The Sociology of Emotions*. Cambridge: Cambridge University Press.

Ure, M. (2015). Resentment/*Ressentiment*. *Constellations: An International Journal of Critical and Democratic Theory*, 22(4): 599–613.

Van Stekelenburg, J. and Klandermans, B. (2011). *The Social Psychology of Protest*. Sociopedia.isa. Sage Publications.

Van Troost, D., Van Stekelenburg, J., and Klandermans, B. (2013). Emotions of Protest. In N. Demertzis (Ed.) *Emotions in Politics: The Affect Dimension in Political Tension*. (186–203). London: Palgrave Macmillan.

van Tuinen, S. (2020). The resentment-ressentiment complex: a critique of liberal discourse. *Global Discourse: An Interdisciplinary Journal of Current Affairs*. DOI: https://doi.org/10.1332/204378920X15828100918561.

Vaneigem, R. (2012). *The Revolution of Everyday Life*. Oakland: PM Press.

Vattimo, G. (1992). *The Transparent Society*. Cambridge: Polity Press.

Velleman, J. D. (2003). Don't Worry, Feel Guilty. In A. Hatzimoysis (Ed.) *Philosophy and the Emotions*. (235–248). Cambridge: Cambridge University Press.

Vogler, C. (2000). Social identity and emotion: the meeting of psychoanalysis and sociology. *The Sociological Review*, 48(1): 19–42. https://doi.org/10.1111/1467-954X.00201.

Voglis, P. (2008). Memories of the 1940s as a topic of historical analysis: methodological proposals. In Bouschoten P. B., Vervenioti, T., Voutira, E., Dalkavoukis, V., and Bada, K. (Eds) *Memories and Forgetting of the Greek Civil* War. (61–80). Salonica: Epikentron.

Voglis, P. (2002). *Becoming a Subject. Political Prisoners during the Greek Civil War*. PhD. European University Institute, Florence.

Voglis, P. (2000). Between Negation and Self-Negation: Political Prisoners in Greece, 1945–1950. In Mazower, M. (Ed.) *After the War Was Over. Reconstructing the Family, Nation, and State in Greece, 1943–1960*. (73–90). Princeton and Oxford: Princeton University Press.

Volkan, V. (2005). Large-group identity and chosen trauma. *Psychoanalysis Downunder*, 6. www.psychoanalysisdownunder.com.au/.

Volkan, V. (2004). *Blind Trust. Large Groups and Their Leaders in Times of Crisis and Terror*. Charlottesville, Virginia: Pitchstone Publishing.

Volkan, V. (2001). Transgenerational transmissions and chosen traumas: An aspect of large-group identity. *Group Analysis*, 34(1): 79–97.

Von Martin, A. (2016). *Soziologie der Renaissance und Weitere Schriften*. Klassiker der Sozialwissenschaften. Wiesbaden: Springer Fachmedien. DOI: 10.1007/978-3-658-10449-8_3.

Von Martin, A. (1944). *Sociology of the Renaissance* (translated by W. L. Luetkens). London: Kegan Paul.

von Scheve, C. (2013). *Emotion and Social Structures. The Affective Foundations of Social Order*. London and New York: Routledge.

von Scheve, C. and Salmela, M. (Eds) (2014). *Collective Emotions. Perspectives from Psychology, Philosophy, and Sociology*. Oxford: Oxford University Press.

von Scheve, C. and Slaby, J. (2019). Emotion, emotion concept. In Slaby, J. and von Scheve, C. (Eds) *Affective Societies. Key Concepts*. (42–51). London and New York: Routledge.

von Scheve, C. and von Luede, R. (2005). Emotion and Social Structures: Towards an Interdisciplinary Approach. *Journal for the Theory of Social Behavior*, 35(3): 303–328.

Vorderer, P. A. and Hartmann, T. (2009). Entertainment and Enjoyment as Media Effects. In Bryant, J. and Oliver, M. B. (Eds) *Media Effects. Advances in Theory and Research*. (3rd edition) (pp. 532–550). New York: Taylor & Francis.

Wade, G. N., Worthington, L. E., and Meyer, E. J. (2005). But Do They Work? A Meta-Analysis of Group Interventions to Promote Forgiveness. In Worthington, L. E. (Ed.) *Handbook of Forgiveness*. (423–439). New York: Routledge.

Wahl-Jorgensen, K. (2019). *Emotions, Media and Politics*. Cambridge: Polity Press.

Walicki, A. (1969). Russia. In Ionescu, G. and Gellner, E. (Eds). *Populism. Its Meanings and National Characteristics*. (62–96). Hertfordshire: The Garden City Press.

Wallerstein, E. (2000). *The Essential Wallerstein*. New York: The New Press.

Walzer, M. (1994). *Thick and Thin. Moral Argument at Home and Abroad*. Notre Dame: University of Notre Dame Press.

Ward, L. (2012). A new century of trauma? *Alluvium*, 1(7). http://dx.doi.org/10.7766/alluvium.v1.7.03.

Warren, M. (Ed.) (1999). *Democracy & Trust*. Cambridge: Cambridge University Press.

Wastell, C. (2005). *Understanding Trauma and Emotion*, Berkshire: Open University Press.

Weber, A-K. (2018). The Pitfalls of 'Love and Kindness': On the Challenges to Compassion/Pity as a Political Emotion. *Politics and Governance*, 6(4): 53–61.

Weber, M. (1978a), *Economy and Society* (Vol. I). Berkeley: University of California Press.

Weber, M. (1978b). *Economy and Society* (Vol. II). Berkeley: University of California Press.

Weil, S. (1962). *Selected Essays, 1934–1943: Historical, Political, and Moral Writings* (chosen and translated by Richard Rees). Oregon: Wipf and Stock Publishers.

Wenning, M. (2009). The Return of Rage. *Parrhesia*, 8: 89–99.

Wetherell, M. (2012). *Affect and Emotion. A New Social Science Understanding*. Los Angeles: Sage.

Williams, J. S. (2009). A 'Neurosociology' of Emotion? Progress, Problems and Prospects. In Hopkins, D., Kleres, J., Flam, H., and Kuzmics, H. (Eds) *Theorizing Emotions: Sociological Explorations and Applications*. (315–337). Berlin: Campus.

Williams, R. (1987). *Culture and Society. Coleridge to Orwell*. London: The Hogarth Press.

Williams, R. (1980). *Problems in Materialism and Culture*. London: Verso.

Williams, R. (1978). *Marxism and Literature*. Oxford: Oxford University Press.

Williams, R. (1965). *The Long Revolution*. Middlesex: Pelican Books.

Williams, W. D. (1952). *Nietzsche and the French. A Study of the Influence of Nietzsche's French Reading on his Thought and Writing*. Oxford: Basil Blackwell.

Wilson, T. (1993). *Watching Television. Hermeneutics, Reception and Popular Culture*. Cambridge: Polity Press.

Wimberly, C. (2018). Trump, Propaganda, and the Politics of Ressentiment. *The Journal of Speculative Philosophy*, 32(1): 179–195. DOI: 10.5325/jspecphil.32.1.0179.

Wirth, W. and Schramm, H. (2005). Media and Emotions. *Communication Research Trends*, 24: 3–39.

Worthington, L. E. and Cowden R.G. (2017). The psychology of forgiveness and its importance in South Africa. *South African Journal of Psychology*, 47: 292–304. DOI: 10.1177/0081246316685074.

Woshinsky, O. H. (1995). *Culture and Politics. An Introduction in Mass and Elite Political Behavior*. New Jersey: Prentice Hall.

Wouters, C. (2007). *Informalization: Manners and Emotions Since 1890*. London: Sage.

Wyschogrod, E. (1998). *An Ethics of Remembering: History, Heterology, and the Nameless Others*. Chicago & London: University of Chicago Press.

Xydis, S. (1980). Modern Greek Nationalism. In Sugar, P. and Lederer, I. (Ed.) *Nationalism in Eastern Europe*. (207–258). Washington D.C.: University of Washington Press.

Yankelovich, D. (2009). The New Pragmatism: Coping with America's Overwhelming Problems. *Bulletin of the American Academy* (Spring): 25–30.

Yankelovich, D. (1975). The Status of Ressentiment in America. *Social Research*, 42(4): 760–777.

Yates, C. (2019). Affect and Emotion. In Stavrakakis, Y. (Ed.) *Routledge Handbook of Psychoanalytic Political Theory*. (162–173). London: Routledge.

Yates, C. (2015). *The Play of Political Culture, Emotion and Identity*. Basingstoke: Palgrave Macmillan.

Zelizer, B. (2002). Finding aids to the past: bearing personal witness to traumatic public events. *Media, Culture and Society*, 24(5): 697–714.

Zelizer, B. and Allan S. (Eds) (2002). *Journalism after September 11*. London: Routledge.

Zillmann, D. (2011). Mechanisms of emotional reactivity to media entertainments. In Döveling, K., von Scheve, C., and Konijn, E. A. (Eds) *The Routledge Handbook of Emotions and Mass Media*. (101–115). London: Routledge.

Žižek, S. (2008). *Violence. Six Sideways Reflections*. New York: Picador.

Žižek, S. (1997). *The Plague of Fantasies*. London: Verso.

Žižek, S. (1996). *The Indivisible Remainder*. London: Verso.

Žižek, S. (1989). *The Sublime Object of Ideology*. London: Verso.

Index

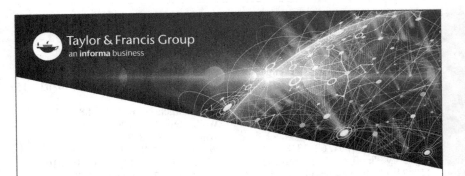

Printed in the United States
by Baker & Taylor Publisher Services

Printed in the United States
by Baker & Taylor Publisher Services